POWER BOILERS

A GUIDE TO SECTION I OF THE ASME BOILER AND PRESSURE VESSEL CODE

by

Martin D. Bernstein

Foster Wheeler Energy Corporation
Clinton, New Jersey

and

Lloyd W. Yoder

Wadsworth, Ohio

ASME Press   •   •   New York   •   •   1998

Copyright © 1998 by The American Society of Mechanical Engineers
Three Park Avenue, New York, NY 10016

Library of Congress Cataloging-in-Publication Data

Bernstein, Martin D., 1928–
    Power Boilers: A Guide to Section I of the ASME Boiler and Pressure
Vessel Code by Martin D. Bernstein and Lloyd W. Yoder.
      p.   cm.
    Includes index.
    ISBN 0-7918-0056-3
    1.   Steam-boilers—Standards—United States.
    2.   Pressure vessels—Standards—United States.
    I.   Yoder, Lloyd W.,   1930—
    II.   American Society of Mechanical Engineers. Boiler and Pressure
        Vessel Committee. ASME Boiler and Pressure Vessel Code.
    III. Title.
    TJ288.B47   1998   98-43261
    621.1'83'021873—dc21   CIP

Cover figure courtesy of Foster Wheeler Energy Corporation.

DEDICATION

The authors dedicate this book to their distinguished colleague on Subcommittee I, the late Walter Harding of Combustion Engineering. From the 1950s until his retirement in the 1980s, Walter Harding served with distinction on the ASME Boiler and Pressure Vessel Committee in many capacities, including Chairman of the Main Committee. An engineer of the greatest integrity, Harding was an inspiring role model for younger committee members, extremely knowledgeable, ever helpful, cheerful, and the possessor of a fine sense of humor. We were fortunate to have had the privilege of serving with him.

TABLE OF CONTENTS

LIST OF FIGURES

LIST OF TABLES

FOREWORD

In 1986 the ASME Professional Development Department approached author Martin Bernstein and asked him to develop and teach a two-day professional development course on Section I of the ASME Boiler and Pressure Vessel Code, whose title is now Rules for Construction of Power Boilers. The task of developing course notes and the prospect of teaching all day for two days with no relief was somewhat daunting, so Bernstein enlisted the help of Lloyd Yoder, his long-time friend and colleague on Subcommittee I, the ASME committee that governs Section I. The two then collaborated on the preparation of appropriate notes and since 1986 have been teaching a two-day course for the ASME at various locations within the United States and overseas. During that time they expanded and refined the course notes, covering material of interest to those who have been taking the course—design engineers; project managers; architect engineers; engineers from boiler manufacturers, boiler operators, and insurance and inspection agencies; and others involved with power boilers and the ASME Code governing their construction.

The introduction in the brochure for the two-day professional development course on Section I explains that Section I contains the rules for the construction of power boilers, but that those rules are accompanied by very little explanation. Thus the course is intended to provide a basic understanding of those rules, their intent, and how they are applied, interpreted, and revised. This book is a greatly expanded version of the course notes, with broader and more detailed coverage on various topics of interest to those involved with power boilers. Both authors have spent their careers working for major boiler manufacturers, and decades serving in various capacities on the ASME Boiler and Pressure Vessel Committee. This book is a compendium of the authors' knowledge and experience.

The design and construction of power boilers involves the use of other sections of the ASME Code besides Section I, and the use of those other sections is explained in this book when appropriate. Section II, Materials, provides detailed specifications for materials and welding consumables, as well as tabulations of design stresses and material properties, such as yield strength and tensile strength as a function of temperature. Section V, Nondestructive Examination, contains a series of standards that provide the methodology for conducting the various nondestructive examinations used in Section I construction. Section IX, Welding and Brazing Qualifications, provides the information necessary to qualify the weld procedures and the welders required for Section I construction. In a rather unusual arrangement, the construction rules for boiler piping are found partly in Section I and partly in the B31.1 Power Piping Code. This has led to considerable misunderstanding and confusion, and the authors have endeavored to provide a clear explanation of these rules and the potential pitfalls to be avoided.

Both authors are long-time members of the ASME committee that governs Section I, the Subcommittee on Power Boilers, Subcommittee I. In the course of its duties, that committee answers a steady flow of questions from all over the world on the application and interpretation of Section I. The authors have thus participated in or are familiar with Section I interpretations and Code changes going back more than a quarter of a century. Since 1978 the ASME has been publishing those interpretations, but they are not widely or readily available, nor is their background known by most people. A useful feature of this book is that as a particular subject is covered, many important interpretations dealing with that subject are cited and discussed. These interpretations form a very useful body of reference information, since they explain or clarify rules that have led to inquiries from the public and provide insight into the views of Subcommittee I.

Although Section I nominally covers only the new construction of power boilers, this book goes further and contains a chapter on rules covering boilers in service, such as those found in the National Board

Inspection Code. The authors have also found that participants in their course on Section I were interested in the subject of creep and fatigue damage during boiler life and the related subject of boiler life extension. Accordingly, a chapter providing an overview of those subjects has been included.

When the ASME in 1911 began the development of a boiler code, it addressed a serious safety problem of that time by formulating a uniform set of rules for the construction of steam boilers, rules that could be adopted by all the states. That set of rules eventually became Section I. The interesting history of steam power in the nineteenth century provides the background for the development of Section I and is described in Appendix I.

The organization and operation of the ASME Boiler and Pressure Vessel Committee is a subject not likely to be found on a best-seller list. Nevertheless, some knowledge of the workings of the committee may prove useful for those submitting inquiries or requests for Code changes, or for Code Cases. Appendix II provides a description of how the committee operates.

Those unfamiliar with the ASME Code may at first be confused by a number of terms commonly used in it. Examples include third party inspection, Authorized Inspector, Authorized Inspection Agency, jurisdiction, Maximum Allowable Working Pressure (MAWP), boiler proper, boiler proper piping, boiler external piping, interpretation, Code Case, accreditation, and Certificate of Authorization (to use a Code symbol stamp). The authors explain these terms in the text wherever appropriate.

Although the ASME Boiler and Pressure Vessel Code changes very slowly, it does change continuously. The rate of change in recent years seems to have increased, perhaps due to technological innovation and international competition. Thus although this book provides a substantial body of information and an explanation of the rules as they now exist, it can never provide the last word. The edition covered is the 1998 edition of Section I, which comprises the 1995 edition with Addenda through 1997. (In a departure from previous practice, the 1998 edition included the first Addenda to that edition, the 1998 Addenda, as explained further in Chapter 2.) The authors are confident that the book will provide the industry with a very useful reference and guide to Section I.

INTRODUCTION

It is helpful to begin the study of Power Boilers, Section I of the ASME Boiler & Pressure Vessel Code, with some discussion of its character and philosophy. According to the dictionary, the term **code** has several meanings: a system of principles or rules; a body of laws arranged systematically for easy reference. Section I is primarily a system of rules. When the ASME in 1911 decided that the country needed a boiler code, it assigned a committee and gave it a mandate to formulate standard rules for the construction of steam boilers and other pressure vessels. The first edition of what is now known as Section I was finally approved by the ASME in 1915 and incorporated what was considered the best current practice in boiler construction. However, the guiding principle then, as now, was that these be safety rules.

Part of the foreword of Section I explains the guiding principles and philosophy of Section I, and also of the Boiler and Pressure Vessel Committee, which continues to administer the Code:

> *The American Society of Mechanical Engineers set up a committee in 1911 for the purpose of formulating standard rules for the construction of steam boilers and other pressure vessels. This committee is now called the Boiler and Pressure Vessel Committee.*
>
> *The Committee's function is to establish rules of safety governing the design, fabrication, and inspection during construction of boilers and pressure vessels, and to interpret these rules when questions arise regarding their intent. In formulating the rules, the Committee considers the needs of users, manufacturers, and inspectors of pressure vessels. The objective of the rules is to afford reasonably certain protection of life and property and to provide a margin for deterioration in service so as to give a reasonably long, safe period of usefulness. Advancements in design and material and the evidence of experience have been recognized.*
>
> *This Code contains mandatory requirements, specific prohibitions, and nonmandatory guidance for construction activities. The Code does not address all aspects of these activities, and those aspects which are not specifically addressed should not be considered prohibited. The Code is not a handbook and cannot replace education, experience, and the use of engineering judgment. The phrase **engineering judgment** refers to technical judgments made by knowledgeable designers experienced in the application of the Code. Engineering judgments must be consistent with Code philosophy and such judgments must never be used to overrule mandatory requirements or specific prohibitions of the Code.*
>
> *The Boiler and Pressure Vessel Committee deals with the care and inspection of boilers and pressure vessels in service only to the extent of providing suggested rules of good practice as an aid to owners and their inspectors.*
>
> *The rules established by the Committee are not to be interpreted as approving, recommending, or endorsing any proprietary or specific design or as limiting in any way the manufacturer's freedom to choose any method of design or any form of construction that conforms to the Code rules.*

Certain points in these paragraphs should be stressed. Section I covers the design, fabrication, and inspection of boilers *during construction,* i.e., it covers new construction only. (Rules covering boilers in service will be discussed later, in Chapter 16.) Although there is general agreement that Section I should apply to new replacement parts, and such parts are usually specified that way, until the appearance of the

1996 Addenda, Section I had no clear provisions dealing with replacement parts other than how they should be documented. Those addenda included changes to PG-106.8 and PG-112.2.4, which require the manufacturer of replacement parts to state on the data report form (the documentation that accompanies the part; see Chapter 9) whether or not the company is assuming design responsibility for a replacement pressure part.

Also mentioned in the Foreword is the objective of the rules: reasonably certain protection of life and property, with a margin for deterioration in service, so as to provide a reasonably long, safe period of usefulness. This is an acknowledgment of the fact that no equipment lasts forever and that boilers do have a finite life. The mechanisms by which a boiler wears out and the means by which remaining life may be estimated are discussed in Chapter 14.

The Section I rules were based on the best design practice known when they were written and have continued to evolve on the same basis. They have worked well over many years. Rule changes have been made that recognize advances in design and materials, as well as evidence of satisfactory experience. The needs of the users, manufacturers, and inspectors are considered, but safety is always the primary concern.

Another basis for the success of Section I is its committee's insistence that the rules be general, and not be interpreted as approving, recommending, or endorsing any proprietary or specific design, or limiting a manufacturer's freedom to choose any design or construction that conforms to the Code rules. The committee considers that a manufacturer is ultimately responsible for the boiler design and leaves certain aspects not explicitly covered by Section I to the manufacturer. Traditionally, the manufacturer has recognized and borne this responsibility for such things as functional performance, providing for thermal expansion and support of the boiler and its associated piping, and for consideration of thermal stress, wind, and seismic loading.

The useful advice that Section I is not a handbook and cannot replace education, experience, and the use of engineering judgment was added to the Foreword of all sections of the Code in the 1992 Addenda. This overdue advice was added in an attempt to limit the ASME's entanglement in certain disputes among manufacturers, owners, and regulatory authorities regarding the application of the Code.

Further evidence of the flexibility and reasonableness of Section I and another key to its success as a living document is found in the second paragraph of the Preamble:

> *The Code does not contain rules to cover all details of design and construction. Where complete details are not given, it is intended that the manufacturer, subject to the acceptance of the Authorized Inspector, shall provide details of design and construction which will be as safe as otherwise provided by the rules in the Code.*

This important paragraph provides a way to accept new or special designs for which no rules are given, by allowing the designer to prove to the satisfaction of an Authorized Inspector that the safety of the new design is equivalent to that usually supplied.

According to one of the definitions cited above, a code is a body of laws arranged systematically for easy reference. Although this may be true of Section I, it is not at first easy to use or understand, because it is neither a textbook nor an engineering design manual. It is rather a collection of rules that have been revised and expanded over the years with very little accompanying explanation. These rules mandate the fundamental construction features considered necessary for a safe boiler (one that is a safe pressure container), but typically do not provide any advice on how to design a boiler from the standpoint of what size or arrangement of components should be used. There are no provisions dealing with the thermal performance and efficiency of the boiler or how much steam it will produce (other than for some approximate guidelines for judging the adequacy of the safety valve discharge capacity). It is assumed that the boiler manufacturer or designer already has this knowledge, presumably from experience or available technical literature. Many rules seem, and indeed are, arbitrary; but as explained before, they were originally written to incorporate what was considered good practice in the industry.

Section I's method of achieving safe boiler design is a relatively simple one. It requires all those features considered necessary for safety (e.g., water gage glass, safety valve, pressure gage, check valve, drain) and then provides detailed rules governing the construction of the various components comprising the boiler, such as tubes, piping, headers, shells, and heads. This approach is analogous to the old saying that a chain is no stronger than its weakest link. For a boiler, the links of the figurative chain can be considered to be the material, the design (formulas, loads, allowable stresses, construction details), fabrication techniques including welding, inspection, testing, and certification by stamping and data reports. If each of these elements meets the appropriate Section I rules, a safe boiler results. The boiler can then be described as a Section I boiler, meaning one constructed to meet all the requirements of Section I of the ASME Boiler & Pressure Vessel Code. An important element of this construction process is a quality control program, intended to assure that the Code has been followed. Each aspect of the process is discussed in this book.

Code construction under the rules of Section I takes place as follows: The ASME accredits a manufacturer (i.e., after appropriate review and acceptance of the manufacturer's facilities, organization, and quality control system, authorizes the manufacturer to engage in Code construction). The manufacturer then constructs, documents, certifies, and stamps the boiler in compliance with the rules of Section I. The manufacturer's activities are monitored and inspected by a third party (the Authorized Inspector, see Chapter 8). The boiler is then acceptable to jurisdictions with laws stipulating ASME construction of boilers. Section VIII construction (of pressure vessels) and Section IV construction (of heating boilers) is carried out in similar fashion.

ACKNOWLEDGMENT

We wish to acknowledge first the generous support of our companies, Foster Wheeler Energy Corporation and the Babcock and Wilcox Company, support which for so many years has provided us the opportunity and the honor of serving on the ASME Boiler and Pressure Vessel Committee.

The material presented in this book is based on knowledge we have gained during our careers in the industry and from our long participation on the Code committees. Much of what we have learned has come from present and former colleagues at our companies and on those committees. We have sought the assistance of many of them in reviewing this book. Although it is not possible to name them all, we do acknowledge with thanks the particular assistance and helpful suggestions of the following: Chuck Becht IV, Jeff Blough, Joe Brzuszkiewicz, D'or Doty, Jim Farr, Joel Feldstein, John Fishburn, Ron Friend, Ron Haupt, Ed Kistner, Tom McGough, Bob McLaughlin, Peter Molvie, Kim Morrison, T.V. Narayanan, Everett Rodabaugh, Blaine Roberts, Bob Schueler, Mark Sheehan, Will Somers, and Mitch Stanko.

SCOPE OF SECTION I, ORGANIZATION, AND SERVICE LIMITS

SCOPE

Section I applies to several types of boilers and components of boilers, such as economizers, superheaters, reheaters, and in some circumstances feedwater heaters. Although its title is Power Boilers, its scope is somewhat broader. The Preamble to Section I explains that it covers electric boilers, miniature boilers, high-temperature water boilers, and organic fluid vaporizers. Since the precise definitions of these various types of boilers are not generally known, the following definitions, found in footnotes to the Preamble, are helpful:

Power boiler—a boiler in which steam or other vapor is generated at a pressure of more than 15 psi for use external to itself.

Electric boiler—a power boiler or a high-temperature water boiler in which the source of heat is electricity.

Miniature boiler—a power boiler or high-temperature water boiler in which the limits in PMB-2 are not exceeded. (PMB-2 establishes limits on shell diameter, heating surface, gross volume, and maximum allowable working pressure. See further discussion under Part PMB later in this chapter.)

High-temperature water boiler—a water boiler intended for operation at pressures in excess of 160 psi and/or temperatures in excess of 250°F.

It happens that Section I doesn't provide an explicit definition of an organic fluid vaporizer. An **organic fluid vaporizer** is a boilerlike device that uses an organic fluid instead of steam as the working fluid. The last paragraph of the Preamble states that a pressure vessel in which an organic fluid is vaporized by the application of heat resulting from the combustion of fuel shall be constructed under the provisions of Section I. (Those provisions are found in Part PVG.) Thus, so far as Section I is concerned, an organic fluid vaporizer is a boilerlike device in which an organic fluid is vaporized as just described. (Note that a key factor is the vaporization of the organic fluid. If the organic fluid is merely heated without vaporizing, the device does not fall within the scope of Section I; it might fall instead under the scope of Section VIII as a pressure vessel.) The Preamble then provides a notable exception to the Section I definition of an organic fluid vaporizer: ''Vessels in which vapor is generated incidental to the operation of a processing system, containing a number of pressure vessels such as are used in chemical and petroleum manufacture, are not covered by the rules of Section I.'' Again, if Section I rules don't cover these vessels, what rules do? The answer is, the rules of Section VIII, Pressure Vessels. Those rules cover all kinds of pressure vessels, including in some cases fired pressure vessels.

Although the origin of the above exception to Section I coverage is uncertain, a possible explanation can be offered. Note that the vessel in question would probably be used in a chemical plant or petroleum refinery. Such plants are normally owned and operated by large companies with capable engineering staffs and well-trained operators. Those companies can usually demonstrate a good record of maintenance and safety. Furthermore, vaporizing organic liquids inside pressure vessels is a routine matter for them. They might also argue that it really doesn't make much difference whether a properly designed vessel is built to the rules of Section I or Section VIII, nor does it matter whether the source of heat is from direct firing, from hot gases that may have given up some of their heat by having passed over heat-transfer surface upstream, or from a hot liquid that is being processed. There are also economic reasons why an owner might prefer a Section VIII vessel over a Section I vessel, as explained later in the discussion of fired versus unfired boilers.

The Preamble doesn't explain the precise meaning of ''vapor generation incidental to the operation of a processing system.'' It apparently means the generation of vapor in a vessel or heat exchanger that is part of a processing system in a chemical plant or petroleum refinery where this vapor generation is only a minor or secondary aspect of the principal business of the plant, such as refining oil. Thus, certain equipment normally constructed to Section I rules could, under this exception, be constructed instead to the rules of Section VIII, provided the appropriate authorities in the jurisdiction where the equipment is to be installed have no objection. As will be explained later, these so-called jurisdictional authorities have the last word in deciding which Code section applies (see How and Where Section I Is Enforced in Chapter 2).

There is also some imprecision in the use and meaning of the term **power boiler**. That term is sometimes understood to mean a boiler whose steam is used for the generation of power, as opposed, for example, to a boiler whose steam is used for chemical processing or high-pressure steam heating. However, according to the definition in the Preamble, a boiler that generates steam or other vapor at a pressure greater than 15 psi for external use is considered by Section I to be a power boiler, irrespective of how the steam might be used. Although exceptions exist, most jurisdictional authorities follow the ASME Code in defining and categorizing boilers and pressure vessels.

Note from the definition of a power boiler that the steam or other vapor generated is for use external to the boiler. This is supposed to distinguish a power boiler from certain other pressure vessels such as autoclaves, which may similarly generate steam or vapor at a pressure greater than 15 psi, but not generally for external use. These pressure vessels, often used as process equipment in the chemical and petroleum industries, and for cooking or sterilization in other industries, are designed to meet the rules of Section VIII.

From the Preamble definitions, it is apparent that a high-temperature water boiler, which generally produces pressurized hot water for heating or process use, is not considered a power boiler. However, as a practical matter, the particular characterization of a device by Section I as a power boiler or something else is less important than the fact that it is indeed covered by Section I rules.

The Preamble explains that the scope of Section I covers the **complete boiler unit**, which is defined as comprising the **boiler proper** and the **boiler external piping**. This very important distinction needs further explanation. The term **boiler proper** is an unusual one, chosen to distinguish the boiler itself from its external piping. The **boiler proper** consists of all the pressure parts comprising the boiler, such as the drum, the economizer, the superheater, the reheater, waterwalls, steam-generating tubes known as the boiler bank, various headers, downcomers, risers, and transfer piping connecting these components. Any such piping connecting parts of the boiler proper is called **boiler proper piping**. The **boiler external piping** is defined by its extent, as that piping that begins at the first joint where the boiler proper terminates and extends to and includes the valve or valves required by Section I.

The importance of all these definitions and distinctions is that the construction rules that apply to the boiler proper and boiler proper piping are somewhat different from those that apply to the boiler external piping. This is explained in Chapter 4, Distinction Between Boiler Proper Piping and Boiler External Piping.

The Preamble also explains that the piping beyond the valves required by Section I is not within the scope of Section I. Thus these valves define the boundary of the boiler, and Section I's jurisdiction stops there. Note that the upstream boundary of the scope of Section I varies slightly, depending on feedwater valve arrangements. Different valve arrangements are required for a single boiler fed from a single source, as opposed to two or more boilers fed from a common source, with and without bypass valves around the required regulating valve (see PG-58.3.3, PG-58.3.4, and the solid and dotted lines shown in Fig. PG-58.3.1). The required boiler feed check valve is typically placed upstream of the feed stop valve, except that on a single boiler-turbine unit installation, the stop valve may be located upstream of the check valve. Changes in the Section I boundary can have some significance in the choice of design pressure for the feedwater piping and valves, which is governed by the rules of B31.1, Power Piping (see paragraph 122.1.3). These several topics are discussed further in Chapter 5.

An exception to this coverage of boiler piping is found in the treatment of the hot and cold reheat piping between the boiler and a turbine, which is excluded from the scope of Section I. Occasionally, someone asks how and why the reheat piping was left outside the scope of Section I. The explanation offered some years ago by a senior member of the committee was as follows: Although some use of reheaters dates back to the earliest days of Section I, the rising steam pressures used in large utility boilers in the 1940s led to their increased use. The reheat piping became larger, heavier, and more complex. In those days the General Electric Company and Westinghouse made virtually all the turbines used in the United States, and it was customary for those turbine manufacturers to take responsibility for the design of the reheat piping. When some committee members suggested that it might be time to consider bringing the reheat piping into the scope of Section I, those two companies objected. On the basis of their long, successful experience, they persuaded the committee that such a change was unnecessary and that they were perfectly capable of designing that piping. Thus the reheat piping remained outside the scope of Section I.

In the late 1980s, failures of hot reheat piping occurred at two major utilities, with injuries and loss of life. This reopened the question of Section I coverage for reheat piping and whether the failed piping, designed in the 1960s, might not have failed had it been within the scope of Section I. (Current Section I rules call for all boiler components to be manufactured, inspected, certified, and stamped under a quality control system, although the quality control system requirement wasn't imposed until 1973. Moreover, any piping within Section I's scope is generally inspected yearly, with the rest of the boiler, although finding potential leaks or failures under the insulation is not so readily accomplished.) However, investigators were not able to agree on the causes of the failures, and the idea of bringing the reheat piping within Section I's scope died for lack of support.

Code coverage of this reheat piping and also of piping beyond the boiler external piping varies. Normally in power plant design, the owner or the architect-engineer will select the B31.1, Power Piping Code, for the reheat piping and that code or the B31.3, Process Piping Code (formerly called the Chemical Plant and Petroleum Refinery Piping Code), for the piping beyond the boiler external piping. Or, the jurisdiction (state or province) may have laws mandating use of one or the other of these two codes to cover piping that isn't within Section I's scope.

Some unfired steam boilers are constructed to the requirements of Section VIII, as permitted by the Preamble. Section VIII rules deal only with the vessels themselves; they don't deal with any piping attached to the vessels. In such cases, the choice of an appropriate design code for the piping may be left to the plant designers, or the jurisdictional authorities may mandate a particular piping code.

THE ORGANIZATION OF SECTION I

The original edition of Section I was not well organized. The first half, some 36 pages, dealt with material specifications. This was followed by 25 pages of design formulas and construction details, 7 pages on safety valves, and 6 pages on piping, valves, fittings, hydrostatic testing, and stamping. Rules for heating boilers were contained in a second 7-page section. Appendix page A 20 contained rules for existing installations and further useful supplementary information on riveted joints and stayed surfaces. Over the years the contents changed as portions were moved into other sections of the Code, and more and more provisions were added. In 1965, following a precedent set by Subcommittee VIII, Subcommittee I completely reorganized Section I into its present, improved arrangement of parts, in which each part covers a major topic or particular type of boiler or other Section I device. That reorganization was merely a reshuffling of the existing requirements, with no technical change from the previous version. Here is that arrangement:

Foreword
Statements of Policy
Personnel
Preamble

Part PG	General Requirements for All Methods of Construction
Part PW	Requirements for Boilers Fabricated by Welding
Part PR	Requirements for Boilers Fabricated by Riveting
Part PB	Requirements for Boilers Fabricated by Brazing
Part PWT	Requirements for Watertube Boilers
Part PFT	Requirements for Firetube Boilers
Part PFH	Optional Requirements for Feedwater Heater (When Located Within Scope of Section I)
Part PMB	Requirements for Miniature Boilers
Part PEB	Requirements for Electric Boilers
Part PVG	Requirements for Organic Fluid Vaporizers
Appendix I	(Mandatory) Preparation of Technical Inquiries to the Boiler and Pressure Vessel Committee
Appendix	Explanation of the Code Containing Matter Not Mandatory Unless Specifically Referenced to in the Rules of the Code

Index

The front of the volume contains the Foreword and Preamble, which provide a fundamental description of Section I and its application. First-time users of Section I often skip over the Foreword and the Preamble and proceed directly to the body of the book. This is a mistake, because there is a good deal of useful information about Section I, its scope, how the Code works, and definitions of various types of power boilers in this introductory material. Also in the front of the book is a current listing of the personnel of the Boiler and Pressure Vessel Committee, which administers all sections of the Code.

THE PARTS OF SECTION I

Part PG

Part PG is the first major section of Section I and provides general requirements for all methods of construction. It covers such important topics as scope and service limitations; permitted materials; design; requirements for piping, valves, fittings, feedwater supply, and safety valves; permitted fabrication methods; inspection; hydrostatic testing; and certification by stamping and data reports. The general requirements of Part PG must be used in conjunction with specific requirements given in the remainder of the book for the particular type of construction or type of boiler used. Most Part PG topics are covered in later chapters of this book.

Part PW

The second major section is called Part PW, Requirements for Boilers Fabricated by Welding. Since almost all boilers are now welded, this is a broadly applicable part containing much important information. It covers such topics as responsibility for welding; acceptable weld joint designs; types of welding permitted; postweld heat treatment of welds; radiography, inspection, and repair of welds; and testing of welded test plates. Part PW is described and explained in detail in Chapter 6.

Part PR

With the gradual replacement of riveting by welding, Subcommittee I decided that no purpose was served by continuing to maintain and reprint in each edition the special rules applicable to riveted construction, Part PR. Accordingly, the 1974 and all subsequent editions of Section I mandate the use of Part PR rules as last published in the 1971 edition.

Part PB

The next section, Part PB, Requirements for Boilers Fabricated by Brazing, first appeared in the 1996 Addenda to the 1995 edition of Section I. Although brazing had long been used in the construction of certain low-pressure boilers, no brazing rules had ever been provided in Section I. Part PB brazing rules resemble Part PW welding rules, but are much less extensive. One notable difference is that the maximum design temperature depends on the brazing filler metal being used and the base metals being joined. Maximum design temperatures for the various brazing filler metals are given in Table PB-1. Brazing procedures and the performance of brazers must be qualified in accordance with Section IX, by methods similar to those described in Chapter 6 for qualifying weld procedures and welders. The design approach used for determining the strength of brazed joints is given in PB-9: the manufacturer must determine from suitable tests or from experience that the specific brazing filler metal selected can provide a joint of adequate strength at design temperature. The strength of the brazed joint may not be less than that of the base metals being joined. This strength is normally established by the qualification of the brazing procedure. If the manufacturer desires to extend the design temperature range normally permitted by Table PB-1 for the brazing filler metal selected, the manufacturer must conduct two tension tests of production joints, one at design temperature, T, and one at $1.05T$. The joints must not fail in the braze metal. Some acceptable types of brazed joints are illustrated in Fig. PB-15. Nondestructive examination of brazed construction relies primarily on visual examination supplemented by dye penetrant inspection if necessary. PB-49 provides guidance on inspection and any necessary repairs.

The remainder of Section I is composed of parts providing special rules applicable to particular types of boilers or other Section I components, such as feedwater heaters.

Part PWT

Part PWT, Requirements for Watertube Boilers, is a very brief collection of rules for this type of boiler, some of them pertaining to construction details rarely used today. All large high-pressure boilers are watertube boilers. Most construction rules for these boilers are found in Parts PG and PW, and Part PWT is merely a brief supplement. It is, however, the only place where rules for the attachment of tubes to shells and headers of watertube boilers can be found. Part PWT is an example of the retention by Section I of certain old rules and construction details because some manufacturers might still use them.

Part PFT

Part PFT, Requirements for Firetube Boilers, is a much more extensive set of rules, still very much in use for this popular type of boiler, which is quite economical for generating low-pressure steam. Except in special cases, the large-diameter shell places a practical limit on design pressure at about 400 psi, although most firetube boilers have lower pressures. Part PFT covers many variations within the type, in considerable detail. Requirements cover material, design, combustion chambers and furnaces, stayed surfaces, doors and openings, domes, setting, piping, and fittings. Part PFT is the only part of Section I where design for external pressure (on the tubes and on the furnace) is considered.

A good deal of Part PFT dates back to the original 1915 edition of Section I. However, extensive revisions were made in the 1988 Addenda. A much larger selection of materials was permitted, and allowable stresses for the first time became a function of temperature, as is the case elsewhere in the Code. In addition, the design rules for tubes and circular furnaces under external pressure were made consistent with the latest such rules in Section VIII, from which they were taken. In the mid-1990s Part PFT was further revised and updated, as more fully described in Chapter 4, under Stayed Surfaces.

Part PFH

Part PFH, Optional Requirements for Feedwater Heater, applies to feedwater heaters that fall within the scope of Section I by virtue of their location in the feedwater piping between the Code-required stop valve and the boiler. Under these circumstances, the heater may be constructed in compliance with the rules in Section VIII, Pressure Vessels, Division 1, for unfired steam boilers, which are stricter in a number of respects than those applicable to ordinary Section VIII vessels, as explained later in this chapter under Use of Section VIII Vessels in a Section I Boiler.

The primary side of the heater must be designed for a higher pressure than the Maximum Allowable Working Pressure (MAWP) of the boiler, per 122.1.3 of ASME B31.1. (Remember: this heater is in the feedwater piping, and rules for the design of feedwater piping are within the scope of B31.1.) Part PFH also stipulates how the heater is to be stamped and documented.

If a feedwater heater within the scope of Section I is equipped with isolation and bypass valves, there is a possibility that it could be exposed to the full shut-off head of the boiler feed pump. PG-58.3.3 cautions about this and notes that control and interlock systems are permitted in order to prevent excessive pressure. (It is impractical to provide sufficient safety valve capacity in these circumstances.) This is discussed further in Chapter 5: Piping Design.

Part PMB

Part PMB, Requirements for Miniature Boilers, contains special rules for the construction of small boilers that do not exceed certain limits (16-inch inside diameter of shell, 20 square feet of heating surface, 5 cubic feet gross volume, 100 psi MAWP). Because of this relatively small size and low pressure, many requirements normally applicable to power boilers are waived. These have to do with material, material marking, minimum plate thickness, postweld heat treatment and radiography of welds, and feedwater supply. To compensate somewhat for this relaxation of the normal rules, and to provide an extra margin of safety, the hydrostatic test must be conducted at a pressure equal to three times the MAWP.

Part PEB

This part, Requirements for Electric Boilers, was added to Section I in 1976. Up until that time, manufacturers had been building these boilers as Section VIII devices. However, Section VIII had no specific rules for electric boilers and the various openings, valves, and fittings normally mandated by Section I (blowoff, drain, water gage, pressure gage, check valve, etc.). Also, the manufacturers were not formally assuming design responsibility for the boiler.

Part PEB covers electric boilers of the electrode and immersion resistance type only and doesn't include boilers in which the heat is applied externally by electrical means. The boiler pressure vessel may be either a Section I vessel or may be constructed as an unfired steam boiler under the rules of Section VIII, Division 1. Note that those rules require radiography and postweld heat treatment, as explained later in this chapter under Use of a Section VIII Vessel in a Section I Boiler.

Under the provisions of Part PEB, a manufacturer can obtain an E stamp, which authorizes the manufacturer to assemble electric boilers (by installing electrodes and trim on a pressure vessel). The manufacturer must obtain the pressure vessel from an appropriate symbol stamp holder. The E stamp holder is limited to assembly methods that don't require welding or brazing. (Electric boilers may, of course, also be constructed by holders of an S or M stamp.) The E stamp holder becomes the manufacturer of record, who is responsible for the design of the electric boiler. This is stipulated in PEB-8.2.

The Data Report Form P-2A for electric boilers is in two parts, one for the vessel manufacturer and one for the manufacturer responsible for the completed boiler, who may be a holder of an E, S, or M stamp. If the vessel is constructed to the Section VIII, Division 1, rules for unfired steam boilers, it must be stamped with the Code symbol U and documented with a U-1 or U-1A Data Report. In such a case, the Section I master Data Report P-2A must indicate this fact, and the U-1A form must be attached to it.

Electric boilers of the resistance-element type must be equipped with an automatic low-water cutoff, which cuts the power before the surface of the water falls below the visible part of the gage glass. Such a cutoff is not required for electrode-type boilers (which use the water as a conductor), since these in effect cut themselves off when the water level falls too low.

Part PVG

This part provides rules for organic fluid vaporizers, which are boilerlike devices that use an organic fluid (such as Dowtherm™) instead of steam as the working fluid. The principal advantage of these organic fluids is that they have much lower vapor pressures than water at a given temperature. Thus they are particularly suitable for heating in industrial processes requiring high temperatures at low pressure. On the other hand, these liquids are flammable, and toxic, and Part PVG contains a number of special provisions because of these drawbacks.

In order to prevent the uncontrolled discharge of organic fluid or vapor to the atmosphere, the use of gage cocks is prohibited. Safety valves must be of a totally enclosed type that will discharge into a pipe designed to carry all vapors to a safe point of discharge. The safety-valve lifting lever normally required on Section I safety valves is prohibited, and valve body drains are not mandatory. Because the polymerization of organic fluids can lead to clogging or otherwise adversely affect the operation of safety valves (and because the valves have no lifting levers by which they can be periodically tested), PEB 12.2 requires the removal and inspection of these valves at least yearly. This is a rare instance of the Code reaching beyond its normal new-construction-only coverage.

As a further means of reducing unintentional discharge of organic fluid to the environment by leakage through safety valves, and to prevent the gumming up of these valves, Section I permits installation of rupture disks under the valves. This is the only use of such disks permitted by Section I. Special rules are also provided for calculating safety-valve capacity. The required capacity of these valves is based on the heat of combustion of the fuel and the latent heat of the organic fluid. This capacity is determined by the manufacturer.

The Appendix

The Appendix to Section I contains a great deal of miscellaneous information, some of it dating from the first edition. Much of this is nonmandatory explanatory material, unless specifically referred to in the main body of Section I. In recent years, Subcommittee I has been doing some long overdue housecleaning and has removed material in the Appendix considered obsolete or redundant. The diversity of the remaining subjects can be seen from this partial list of contents:

Braced and Stayed Surfaces
Method of Checking Safety-Valve Capacity by Measuring Maximum Amount of Fuel That Can Be Burned
Automatic Water Gages
Fusible Plugs

Proof Tests to Establish Maximum Allowable Working Pressure
Suggested Rules Covering Existing Installations:
    Safety Valves for Power Boilers
    Repairs to Existing Boilers
Examples of Methods of Computation of Openings in Vessel Shells
Examples of Computation of Allowable Loadings on Structural Attachments to Tubes
Preheating
Maximum Allowable Working Pressure—Thick Shells
Rounded Indication Charts
Quality Control System
Data Report Forms and Guides
Codes, Standards, and Specifications Referenced in Text (The dates of the currently approved editions
    are listed in this location only, rather than throughout Section I.)
Guide to Information Appearing on Certificate of Authorization
Sample Calculations for External Pressure Design

INTERPRETATIONS

Since 1977, the ASME has published all the replies from the ASME staff on behalf of the ASME Boiler and Pressure Vessel Committee to inquiries on the interpretation of Section I (and also the other book sections). These interpretations are issued twice a year with a cumulative index by subject and paragraph number. Many Section I users find it convenient to bind the interpretations in the back of the book, for easy reference. Despite the fact that interpretations are said not to be a part of the Code, they can be very useful in explaining its application and in resolving disputes as to what the Code intends, since Code users and Authorized Inspectors generally accept interpretations as the equivalent of Code rules. Many interpretations are quoted throughout this book to explain various provisions of Section I. Further discussion about interpretations, their official status, and how Subcommittee I formulates them is found in Appendix II, under Committee Operations.

An example of the designation of a Section I Interpretation is I-95-27. The I signifies that the interpretation pertains to the rules of Section I; the 95 indicates that it is an interpretation of the 1995 Edition of Section I; and the 27 means that it is the 27th published interpretation of that edition. Before 1983, the year of the interpretation referred to the year it was issued rather than the edition of Section I to which it pertained. For example, Interpretation I-81-26 was the 26th interpretation issued in 1981. Throughout this book, interpretations are designated in this same fashion. One means of finding an interpretation on a particular subject is to find that subject in the general index and then go to the chapter on that subject. Also, a numerical index of all interpretations mentioned in this book, with key words and abstracts, has been provided to help in finding any desired interpretation.

PRESSURE RANGE OF SECTION I BOILERS

As explained in the Preamble, Section I covers boilers in which steam or other vapor is generated at a pressure of more than 15 psi for use external to the boiler (all pressures used in Section I are gage pressure). For high-temperature water boilers, Section I applies if either the pressure exceeds 160 psi or the temperature exceeds 250°F. Since the vapor pressure of water at 250°F is approximately 15 psi, the two lower limits of coverage are related. Boilers designed for pressures below 15 psi are usually constructed to the rules of Section IV, Heating Boilers. However, such units could be built, certified, and stamped as Section I boilers if all the requirements of Section I were met. Section I has no upper limit on boiler design pressure.

FIRED VERSUS UNFIRED BOILERS

Section I rules are intended primarily for fired steam or hot water boilers and organic fluid vaporizers. There are however boilers called unfired steam boilers that don't derive their heat from direct firing. These unfired steam boilers may be constructed under the provisions of either Section I or Section VIII, Pressure Vessels. The definition of an unfired steam boiler is not as clear as it might be, which has led to some confusion and occasional disagreement between Subcommittees I and VIII. The Preamble to Section I states that:

> A pressure vessel in which steam is generated by the application of heat resulting from the combustion of fuel (solid, liquid, or gaseous) shall be classed as a fired steam boiler.
> Unfired pressure vessels in which steam is generated shall be classed as unfired steam boilers with the following exceptions:
> (a) vessels known as evaporators or heat exchangers;
> (b) vessels in which steam is generated by the use of heat resulting from operation of a processing system containing a number of pressure vessels such as used in the manufacture of chemical and petroleum products.

It is these exceptions that cause the confusion, for several reasons. It is sometimes not apparent whether a boiler using waste heat is fired or unfired. It may also be difficult to distinguish between an unfired steam boiler using waste heat and certain heat exchangers also using waste heat. Another difficulty arises from the source of the waste heat; is it from the combustion of fuel? If so, the device is generally considered fired. However, some jurisdictions have permitted boilers using waste heat from a combustion turbine to be considered unfired. The presence of auxiliary burners in a boiler using waste heat from another source would cause that boiler to be considered fired. This was affirmed in Interpretation I-92-20. Remember also that all ASME Code sections routinely caution users that laws or regulations at the point of installation may dictate which Code section applies to a particular device and that those regulations may be different from or more restrictive than Code rules. The jurisdictions may also have rules that define whether a device is considered a boiler or a heat exchanger and whether it is considered fired or unfired.

Electric boilers can be considered fired or unfired, depending on the circumstances. An inquiry on this subject was answered by Interpretation I-81-01, as follows:

> Question: The Preamble of Section I defines an electric boiler as "a power boiler or a high-temperature boiler in which the source of heat is electricity." Would you define whether or not an electric boiler is considered to be a fired or an unfired steam boiler?
> Reply: PEB-2 provides criteria for determining if an electric boiler is considered to be a fired or an unfired steam boiler.
> An electric boiler where heat is applied to the boiler pressure vessel externally by electric heating elements, induction coils, or other electrical means is considered to be a fired steam boiler.
> An electric boiler where the medium (water) is directly heated by the energy source (electrode type or immersion element type) is considered to be an unfired steam boiler.

An unfired steam boiler constructed to Section VIII must meet rules more stringent than run-of-the-mill Section VIII vessels. For example, these rules require additional radiographic examination, postweld heat treatment, and possibly impact testing of materials and welds, as more fully described under the next topic, The Use of Section VIII Vessels in a Section I Boiler.

Some owners of steam generating vessels would prefer, if possible, to avoid the classification of boiler, or fired boiler, because most states and provinces require an annual shutdown for inspection of boilers, although some states permit longer periods of operation. The interval between inspections mandated for Section VIII vessels is typically much longer, permitting longer uninterrupted use of the equipment in question.

One of the distinctions some committee members perceive between fired and unfired boilers is whether a flame from a poorly adjusted burner can impinge directly on pressure parts and overheat them. This is clearly a possibility in a fired boiler, and it is one reason for some of the conservatism of Section I rules, and for requiring annual outages for inspection. By contrast, hot gases from some upstream source are less likely to cause overheating of those same pressure parts, because they are less likely to be misdirected and, presumably, no actual flames can reach the downstream steam-generating pressure parts.

A question was once asked as to whether the determination of fired service (versus unfired) involves the transfer of heat by all three means—radiation, convection, and conduction—or whether fired service is defined simply as the transfer of heat resulting from combustion no matter where or how it takes place. Unfortunately, Section I provides no criteria by which such a distinction could be made, and it is unlikely that Subcommittee I could formulate such criteria. It is possible that the committee could be persuaded that a boiler heated by gases below some temperature could be considered an unfired boiler. Just what temperature that might be (850°F, 700°F?) is a matter of conjecture. Such an idea might be tried as a Code Case, but so far no one has requested such a case.

THE USE OF SECTION VIII VESSELS IN A SECTION I BOILER

Usually all parts of a Section I boiler are constructed to Section I rules. There are a few exceptions worth noting. Vessels used as the pressure vessels for electric boilers (see PEB-3), the drums or other parts of unfired steam boilers (see Code Case 1855), and feedwater heaters that fall under Section I jurisdiction by virtue of their location in the feedwater piping (see PFH-1.1) are among those that may be constructed to Section VIII under that section's special rules for unfired steam boilers in UW-2(c). One other type of Section VIII vessel that can be used in a Section I boiler is mentioned in the next-to-last paragraph of the Preamble. That vessel is an expansion tank used in connection with what is defined in footnote 4 to the Preamble as a high-temperature water boiler. This is the only mention of such tanks in Section I, and nothing is said about their construction having to meet the special Section VIII requirements for unfired steam boilers. Thus it seems that an ordinary Section VIII vessel would serve the purpose.

Although it might seem that components built to Section VIII are just as safe as those built to Section I and that therefore they should be virtually interchangeable, this is unfortunately not the case. When Section I does permit the use of Section VIII vessels, it insists on invoking certain special Section VIII rules for unfired steam boilers. These rules are a little more stringent than those for run-of-the-mill Section VIII vessels. The rules are found in U-1(g), UG-16(b)(3), UG-125(b), UW-2(c), and UCS-25 of Section VIII. The most important requirements called out in these several paragraphs are the following:

- Safety valves are to be furnished in accordance with the requirements of Section I insofar as they are applicable.
- For design pressures exceeding 50 psi, radiography of all butt welded joints is required, except for circumferential welds that meet the size and thickness exemptions of UW-11(a)(4). (Those exemptions cover welds in nozzles that are neither greater than NPS 10 nor thicker than 1⅛ inches.)
- Postweld heat treatment is required for vessels constructed of carbon and low alloy steel.
- A minimum thickness of ¼ inch is required for shells and heads, exclusive of any corrosion allowance.

Another extra requirement imposed on an unfired steam boiler constructed to Section VIII rules applies to all Section VIII vessels made of carbon and low alloy steels. The designer must establish a Minimum Design Metal Temperature (MDMT), the lowest temperature expected in service at which a specified design pressure may be applied. Unless the combination of material and thickness used is exempt by the curves of UCS-66 at the MDMT, impact testing of the material is required, and the Weld Procedure Qualification would also have to include impact testing of welds and heat-affected zones. This is not explicitly mentioned in the paragraphs dealing with unfired steam boilers because it is a requirement that was added to Section

VIII in the mid-1980s, long after those other paragraphs were written, and probably because Subcommittee VIII members assumed that everyone knows that impact testing is routinely required unless the vessel materials meet certain exemptions provided in UCS-66.

What is also notable about these rules for unfired steam boilers is the fact that the ordinary Section VIII exemptions for postweld heat treatment (PWHT) are not available, even for welds in relatively thin P-No. 1 materials. Moreover the Section I PW-41.1.1 exemption from radiography of circumferential butt welds in steam-containing parts up to NPS 16 or 1⅝ -inch wall thickness is also not available.

An example of the consequences of overlooking these somewhat obscure rules occurred recently when a boiler manufacturer placed an order for a sweetwater condenser with a manufacturer of Section VIII pressure vessels without specifying that the condenser, as a type of feedwater heater, had to meet the UW-2(c) rules for unfired steam boilers. (A sweetwater condenser is a shell and tube heat exchanger connected directly to the boiler drum that receives saturated steam from the drum on the shell side and has feedwater going to the drum inside the tubes. The steam condenses on the tubes, giving up its heat to the feedwater and providing a supply of water pure enough to use as spray water in the superheater. These condensers are used where the regular feedwater supply is not of sufficient quality to use as spray water.)

After the condenser was installed and the boiler was ready for the hydrostatic test, the Authorized Inspector (AI) inquired about the presence of a Section VIII vessel in a Section I boiler. He was told that it was a feedwater heater furnished to the rules of Part PFH of Section I. The AI reviewed those rules and asked about the required radiography and postweld heat treatment, which it turned out had not been done because the boiler manufacturer had ordered the condenser as an ordinary Section VIII vessel. The condenser had to be cut out of the installation, shipped back to the manufacturer, radiographed, and given a postweld heat treatment. This PWHT had to be conducted very slowly and carefully, using thermocouples during heating and cooling to avoid too great a temperature difference between shell and tubes, which could cause excessive loading on the tubes, tubewelds, or tubesheet. Fortunately the radiography and PWHT were quickly and successfully accomplished, and the condenser was reinstalled in the boiler with minimal delay.

Three circumstances have been mentioned in which Section VIII vessels can be used as part of a Section I boiler: feedwater heaters under Part PFH; pressure vessels for electric boilers covered under Part PEB; and as a ''Section VIII Unfired Steam Boiler in a Section I System,'' covered under the provisions of Code Case 1855. The title of Case 1855 is a little confusing, as is the case itself. Its ostensible purpose is to permit the inclusion in a Section I boiler of parts built to Section VIII and construed to be components of an unfired steam boiler. Case 1855 was originally developed for application to waste heat boilers in petroleum refineries, which use one or more drums and one or more arrays of heat-exchange surface to generate steam. Case 1855 permits what might be called a hybrid Section I boiler, with some portions built to Section VIII. The separate portions must be stamped and documented as called for by the respective sections. In addition, a P-5 Master Data Report for process steam generators must be completed in accordance with PG-112.2.6 of Section I.

STANDARDS REFERENCED BY SECTION I

Users of Section I soon realize that it is not a stand-alone volume; it depends to a considerable extent on other sections of the Code and on certain other standards. In the development of the Code, it was recognized that there was no need to duplicate already available and acceptable standards used in the industry. Section I, like most other product codes, reviews and adopts material, product, and testing standards developed by other standards groups. This process makes use of the expertise available from organizations that specialize in these other areas.

For example, standard pressure parts such as fittings, flanges, and valves are accepted by reference to the ANSI product standards listed in PG-42. (See also Standard Pressure Parts, in Chapter 11.) Another example is nondestructive examination, where PW-11 refers to Section V, Article 2 for radiographic examination and Article 5 for ultrasonic examination. To obtain quality castings, PG-25 invokes Section V, Article 7 for magnetic particle examination. Within Section V, mention is made also of ASTM E-125-63 for reference photographs to determine which defects must be removed. PW-51 requires personnel performing and evaluating radiographic examinations to be qualified and certified in accordance with a written practice developed using a document entitled *Recommended Practice for Nondestructive Testing Personnel Qualification and Certification*, SNT-TC-1A. That document is published by the Society for Nondestructive Testing. An alternative document published by the same society, with a similar purpose, was approved by the Main Committee in early 1997 and appeared in the 1997 Addenda (see further discussion in Chapter 7, under Qualification and Certification of NDE Personnel). It is entitled *Standard for Qualification and Certification of Nondestructive Testing Personnel*, CP-189.

An indispensable code for use in conjunction with Section I is ASME B31.1, Power Piping, which in 1972 was assigned the rules for material, design, fabrication, installation, and testing of that part of the boiler known as boiler external piping. (See discussion in Chapter 4.) This transfer came about when Subcommittee I decided that certain rules were needed for the flexibility analysis of this piping, which extends beyond the boiler setting. After recognizing the magnitude of the task of developing its own rules, Subcommittee I somewhat reluctantly agreed that this piping could henceforth be covered by B31.1, which already had the necessary rules. Accordingly, a number of Section I paragraphs dealing with what is now called boiler external piping were transferred to B31.1, by the Summer 1972 Addenda of Section I, and the B31.1.0.d Addenda to B31.1 were issued on the same date.

As explained in the discussion of fired versus unfired boilers earlier in this chapter, Section I and Section VIII both have rules for unfired steam boilers, and both sections refer to each other's rules. Section VIII vessels may also be used in electric boilers, under the provisions of Part PEB (see description of Part PEB in Organization of Section I, above).

It is a policy of the Committee to strive for consistency in the rules of the various book sections, while recognizing the wide differences in the construction rules suitable for such diverse components as heating boilers and nuclear vessels. Accordingly, the various sections typically use the same reference standards for such activities as nondestructive examination and welding.

Specific (dated) editions of standards referenced in Section I are grouped together in one table, in Appendix A-360. This avoids having to revise the date at every mention of the standard throughout the text each time a later edition of the standard is adopted.

HOW AND WHERE SECTION I IS ENFORCED AND EFFECTIVE DATES

UNITED STATES AND CANADA

Section I is a set of rules for the construction of boilers. These rules are not mandatory unless adopted into the laws of a government subdivision. These territorial and political units are usually referred to as jurisdictions, a term which reflects the range of their governmental authority. A jurisdiction can be a government subdivision, i.e., state, province, county, or city. If there happen to be no jurisdictional requirements regarding Code coverage, the purchaser of a boiler may specify to the manufacturer construction to Section I rules as a requirement.

Currently, Section I has been adopted by all but two states (South Carolina and Alabama are the exceptions) and by all of the provinces of Canada. Wyoming has a boiler law, but has not developed boiler rules and regulations, nor does it have a chief inspector with a staff to enforce the boiler law. Because of Wyoming's relatively sparse population and limited industrial facilities, the legislature apparently does not consider the expense of implementing the law to be justified, since as a practical matter, all boilers in the state probably meet the ASME Code. Thirty cities and counties also have laws requiring Section I construction (see Fig. 2.1, which is a map of the jurisdictions with boiler and pressure vessel laws).

Except as noted above, states and provinces with boiler laws typically enforce the Code by means of an appointed board or commission that writes and revises boiler rules and an appointed chief inspector with the necessary enforcement staff. The jurisdictional authorities also depend on the efforts of Authorized Inspectors working for Authorized Inspection Agencies, whose function is to assure that the boiler manufacturer has complied with Section I. (See Chapter 8: Third Party Inspection.) These same authorities also have the responsibility for overseeing the care and operation of boilers once they are completed.

In 1997, because of international trade agreements that require removal of non-technical barriers to trade, the situation just described, in which most states and provinces required boilers and pressure vessels installed within their jurisdictions to meet the ASME Code, began to change. At that time, the National Board of Boiler and Pressure Vessel Inspectors (a regulatory agency comprised of the chief inspectors of states in the U.S. and provinces of Canada; see description and discussion in Chapter 16) changed its policy regarding equipment that could be registered with the National Board. The National Board now permits registration of boilers and pressure vessels manufactured to codes of construction other than the ASME Code, provided that the code of construction has been accepted by the National Board, the manufacturing organization has implemented a quality system, and the manufacturing organization has provided for third party inspection.

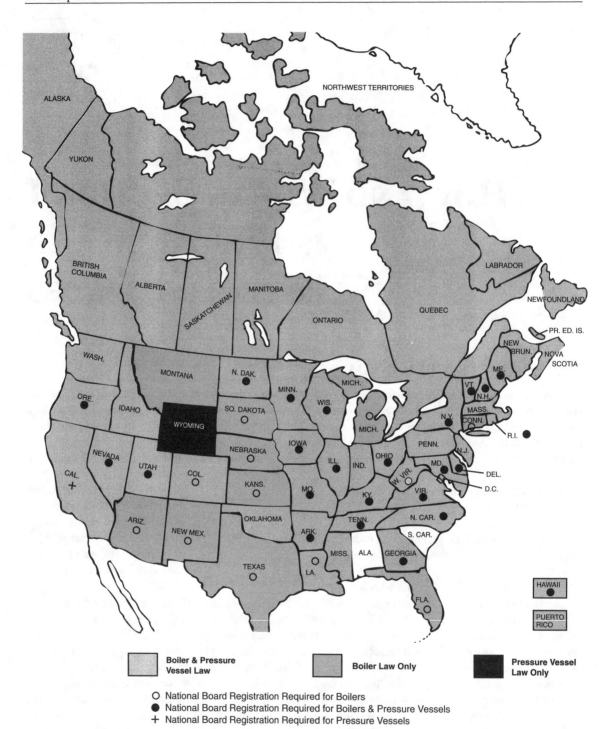

O National Board Registration Required for Boilers
● National Board Registration Required for Boilers & Pressure Vessels
+ National Board Registration Required for Pressure Vessels

FIG. 2.1
MAP OF STATES AND PROVINCES WITH BOILER AND PRESSURE VESSEL LAWS
(COURTESY OF THE UNIFORM BOILER AND PRESSURE VESSEL LAWS SOCIETY, INC.)

The National Board has established criteria for the acceptance of codes of construction, quality systems, and third party inspection in a document entitled *Criteria for Registration of Boilers, Pressure Vessels and Other Pressure Retaining Items*, NB264, Rev. 1, dated October 28, 1997. The introduction to this criteria document explains that the purpose of registration with the National Board is to provide owners, users, and jurisdictional authorities charged with public safety with certification by the manufacturers of boilers, pressure vessels, and other pressure-retaining items that those registered items have been manufactured in accordance with a nationally or internationally recognized code of construction accepted by the National Board.

Before a manufacturing organization may register its equipment with the National Board it must obtain a Certificate of Authorization to Register issued by the National Board. Such a certificate is issued only after the National Board determines that the requirements established in NB264 covering the code of construction, quality system, and third party inspection have been met. (ASME Code construction meets those requirements.) By mid 1998, the National Board had approved certain other construction codes, from Canada, Great Britain, and the European Community. The Japanese and the Korean codes were the next candidates being considered for approval, and many others were expected to follow. Since the National Board is governed by influential state officials, by July 1998 several states had already changed their laws to accept equipment built to codes other than the ASME Code if that equipment met the National Board criteria for registration and is so registered. It remains to be seen how soon the European Community and other foreign jurisdictions will reciprocate and accept equipment built to the ASME Code.

INTERNATIONAL ACCEPTANCE

Section I is recognized internationally as an acceptable design code, along with other national codes such as the German code TRD and the Japanese code JIS. At one time the Mexican mechanical engineering society AMIME translated Section I into Spanish for use in Mexico, but it was apparently little used, in part due to the difficulty of keeping up with the ongoing changes in Section I. Before the 1970s, the ASME allowed only U.S. and Canadian companies to engage in ASME Code construction. However, as a result of legal action at that time, the ASME has been under specific instructions from a consent decree requiring cooperation with non-U.S. manufacturers to allow their accreditation as ASME Code symbol stamp holders, provided those manufacturers' quality control systems pass the ASME review process. There are now many foreign manufacturers who are ASME Code symbol stamp holders and can engage in Code construction of boilers and pressure vessels. More recently, the provisions of NAFTA and GATT require a level playing field for domestic and foreign manufacturers, and the ASME has been active in seeking acceptance of its Code by the European Community.

For field-assembled boilers (those too large to be completed in the shop), full compliance with the ASME Code requires that field assembly be accomplished by appropriate Code symbol stamp holders and also that an Authorized Inspection Agency provide the necessary field inspection. Foreign jurisdictions other than Canada do not generally insist on full compliance with the ASME Code. Since foreign purchasers want to reduce costs, they usually accept so-called ASME boilers that have not been field assembled by appropriate ASME Code symbol stamp holders under the surveillance of an Authorized Inspector and thus cannot be furnished with full ASME Code stamping and certification. Consequently, although the components of such boilers may have been designed and manufactured in accordance with Section I, the completed boilers do not meet all the requirements that would permit full Code stamping and certification. See further discussion at the end of Chapter 9.

UNIFORM BOILER AND PRESSURE VESSEL LAWS SOCIETY

To provide the advantages of uniformity of construction and inspection, the Uniform Boiler Laws Society, Inc. was founded in 1915, just three months following the official adoption of the first ASME Boiler Code.

This society provides model legislation to jurisdictions and supports nationally accepted codes, such as Section I and Section VIII, as standards for the construction of boilers and pressure vessels. In general, the society encourages adoption of uniform requirements by the jurisdictional authorities. Manufacturers strongly support this effort because it promotes the manufacture of economical standard products and obviates custom designed features for different jurisdictions.

Now that most states have boiler laws and many have pressure vessel laws, the original mandate of the UBLS has been fulfilled. The Society is considering changing its name to the Pressure Equipment Laws Society, the better to address new challenges posed by international trade and competition from overseas.

NATIONAL BOARD INSPECTION CODE

Section I covers only new construction, and once all of its requirements for a new boiler have been met and the necessary data reports have been signed, Section I no longer applies. In jurisdictions that have adopted the National Board Inspection Code (NBIC), that code is applicable for inspection, repair, or alteration. See further discussion in Chapter 16: Rules Covering Boilers in Service.

EFFECTIVE DATES OF THE CODE AND CODE REVISIONS

Since the Code is in a constant state of evolution, to apply it properly, the user needs to understand the effective dates of the Code and addenda. (The addenda contain additions and revisions to the Code that have been made since the last new edition.)

A new edition of the Code is issued every three years. At this writing, 1997, the latest edition is the 1995 edition, issued July 1, 1995. That edition, printed on white pages, included all addenda to the previous edition, that is, the 1995 Code was the equivalent of the 1992 Code with all three of its annual addenda incorporated. That status continued until the first addenda to the 1995 Code were issued at the end of 1995. The ASME used to issue addenda every six months, as Summer and Winter Addenda, until 1986, when it was decided for reasons of economy that only one set of addenda per year would suffice, issued to be effective as of December 31. For example, the first addenda to the 1989 Code were simply called the 1989 Addenda, A89. All successive addenda are now identified in similar fashion by year only. Each Code edition eventually has three sets of addenda, issued on different colored pages to help keep track of which came first. Purchasers of any section of the Code receive the addenda for the next three years. Thus a purchaser of the 1995 edition of Section I would have eventually received the 1995, 1996, and 1997 addenda, nominally issued at the end of each of those years (although they are not usually mailed out until well into January).

In the normal course of events, the 1998 edition of Section I would have been the equivalent of the 1995 edition with the next three addenda incorporated. However at this writing, in 1997, the ASME intends to change this system. The new Code editions will continue to be issued on a three-year cycle (July 1 of 1998, 2001, 2004, etc.), but starting in 1998, the ASME intends to issue the addenda on July 1 of each year, rather than December 31. In years between editions, the addenda will be colored pages as before. But in years of the new editions, the new edition will contain white pages, which will serve two functions: they will represent the new edition, comprising the mandatory rules of the previous edition with its three sets of addenda incorporated, and in addition they will show the first addenda to the new edition. Revisions in the new 1998 edition will be identified and a description of each change will be provided. First-time users of the Code may want to know how it read before the revision, and they will be able to obtain the previous version from the ASME. Those who have been working with the 1995 Edition and all its addenda already have that information. Knowledge of the previous edition is important because revisions do not become mandatory immediately upon issuance.

Revisions (addenda) become mandatory six months after the date of issue, except for boilers contracted for prior to the end of the six-month period. In effect, this gives a manufacturer a six-month grace period during which any Code changes affecting new contracts are optional. Note that the key date is the contract date; this can be a formal contract, a letter of intent, or other legally accepted contractual date. Since boiler external piping may be contracted for long after the boiler, the parties involved may use either the Section I edition as of the boiler contract date or any mutually acceptable subsequent addenda for this piping. This was addressed in Interpretation I-78-13, Applicable Code Edition for Boiler External Piping:

> Question: *When boiler external piping is contracted for later than the boiler is, to what rules is the boiler external piping to be built?*
> Reply: *Boiler external piping is a part of the boiler and, as such, the applicable Code Editions are those which correspond to the contract date for the boiler. However, when contracts for boiler external piping are placed later than the boiler contract date, the parties have the option of using Section I as of the boiler contract date or any subsequent Addenda of Section I.*

The edition and addenda of B31.1 to which the boiler external piping is ordinarily to be constructed (unless a later version is selected) is found in the appendix of Section I in Table A-360: Codes, Standards, and Specifications Referenced in Text.

Note that once the edition of Section I to be used for the construction of the boiler is established, the manufacturer is not obliged to implement any changes made in subsequent addenda or editions. This is a practical approach which recognizes that any changes to Section I are now incremental and are very unlikely to have any influence on safety. One exception to this general principle is if the allowable stresses for a material were significantly reduced. In such cases it might be prudent for the manufacturer of as yet uncompleted components to use the new, lower allowable stresses. See further discussion under the related issue of ASME Code Edition Applicable to Repairs and Alterations in Chapter 16.

MATERIALS

HOW MATERIAL IS ORDERED

Material is a fundamental link in the chain of ASME Code construction, and great care is taken to assure its quality. The ASME accomplishes this by adopting material specifications that have been developed and adopted first by the American Society for Testing and Materials, the ASTM. The ASME material specifications are thus usually identical to those of the ASTM. ASTM issues specifications designated by letter and number, for example, A-106, Seamless Carbon Steel Pipe for High Temperature Service. When the ASME adopts ASTM specifications, it adds the letter S. The equivalent ASME pipe specification is thus SA-106.

The ASME or SA material specifications are used as purchase specifications. Each specification contains a variety of information appropriate to that product, dealing with how it is ordered, the manufacturing process, heat treatment, surface condition, chemical composition requirements, tensile requirements, hardness requirements, various test requirements, and how the material is to be marked. The purchaser can also specify nonmandatory supplementary requirements, dealing with such things as stress relieving, nondestructive examination, and additional testing. All of these requirements have evolved over the years in response to the needs of the users.

The purchaser orders by specification number, and the supplier certifies that the material complies with that specification. In most cases, furnished with the material are material test reports that provide the results of various tests required by the specification.

USING SECTION II

Section II of the ASME Code contains information on materials. It is a four-part compendium of materials data, almost a foot thick in its entirety. The four volumes are as follows:

Part A—Ferrous Material Specifications
Part B—Nonferrous Material Specifications
Part C—Specification for Welding Rods, Electrodes, and Filler Metals
Part D—Properties

Parts A, B, and C are, as their names indicate, a compendium of ASME material specifications for all the pressure vessel and welding materials permitted for use in Code construction by Sections I, III, IV, and VIII. It is important to remember that not every material listed in Section II can be used in Section I construction. In general, only materials listed in the material paragraphs of Section I, i.e., PG-5 through PG-13 and PW-5, may be used for Section I pressure retaining parts. PG-11 also permits the use of most materials listed in the ASME/ANSI standards recognized by Section I in PG-42, such as the ASME/ANSI standard Pipe Flanges and Flanged Fittings, B16.5. Materials not otherwise listed for Section I service may also be used if they are listed in a Section I Code Case (see Code Cases in Appendix II). Section I also sets temperature limits for the use of the various materials. They must be used within the temperature range for which allowable stress values are tabulated in Section II, Part D, Properties. The maximum temperature listed for Section I use is 1500°F, for high alloy steels used in superheaters.

With the publication of the 1992 edition of the Code, the allowable stresses that were in Table PG-23 of the earlier editions of Section I were transferred to Section II, Part D, a thick volume subtitled Properties. The allowable stresses from Section I and those from Section III and Section VIII were consolidated and reformatted into this single volume. This consolidation facilitates uniformity among those Code sections that use the same criteria for establishing allowable stress. (Note that some sections, e.g., Section VIII, Division 2, use different criteria, which generally result in higher allowable stresses than those of Section I and Section VIII, Division 1. Conversely, Section IV, Heating Boilers, uses lower stresses, in part to make up for the fact that less nondestructive examination is used in the construction of these relatively low-pressure boilers.) The allowable stresses for ferrous materials used in Section I construction are listed as a function of temperature in Table 1A of Section II, Part D. Allowable stresses for copper, copper alloys, nickel, and nickel alloys are listed in Table 1B. These Part D tables have many important footnotes covering conditions of use, footnotes that formerly appeared with the separate stress tables in the various book sections.

Section II, Part D is organized into subparts covering stress tables, physical-property tables, and external-pressure charts. Also included are several appendices. Two appendices describe the bases for establishing the allowable stress and allowable design stress intensity values (the latter are used in the design of so-called Class 1 nuclear power plant components). Another describes how charts for external-pressure design are established and what criteria are used to determine allowable compressive stresses when designing for external pressure. A last appendix explains the means by which a new material can be approved for Code use.

A useful feature of Section II, Part D is the physical-property tables. They provide such information as the coefficient of thermal expansion, thermal conductivity, thermal diffusivity, modulus of elasticity, Poisson's ratio, modulus of rigidity, density, melting-point range, and specific heat for a fairly large variety of materials. Much of this information is provided over a large temperature range. Also included are tables of yield and ultimate tensile strength as a function of temperature. Yield strength at design temperature is used in the design calculations of PFT-51 for cylindrical furnaces and tubes under external pressure.

Unfortunately, there are several disadvantages in having all these material properties in a single heavy volume. The convenience of having the allowable stresses included in Section I has been lost. At this writing, there are separate stress lines for the same material if it is used in more than one book section, although some thought is being given to consolidating many of these lines to facilitate the designer's choice of the correct line. The multiplicity of lines comes from the fact that the consolidation of stress lines into Section II, Part D was accomplished by gathering all the lines from Sections I, III, and VIII and listing them together. The stress information stretches out over four pages rather than just two, as it did formerly, when it was in Section I. Also, it is difficult to find a particular material specification because there are so many, and they are not listed in numerical order. Instead, they are listed by some not readily apparent rules having to do with increasing content of certain alloying elements. The listing is based on the sorting order the committee used. The sorting order for materials differs between the tables for ferrous materials (SA specifications) and nonferrous materials (SB specifications). The five-level sorting sequence for ferrous materials is as follows:

1. Nominal composition (more on this below)
2. Tensile strength
3. Yield strength
4. Specification number
5. Grade or type

While this seems simple enough, establishing a rule to sort materials in an ascending order of nominal composition is difficult; no single rule works universally, so several arbitrary choices were made. Ferrous materials were listed in the following order of nominal composition:

1. Carbon steels; C, C-Si, C-Mn, and C-Mn-Si steels.
2. Microalloy carbon steels; C-steels with Cb, V, and Ti added.

3. Low alloy steels (C-1/2 Mo, followed by Cr-xx—where xx represents a second element such as Mo, Ni, W, Ti, etc.—ordered on the basis of increasing Cr to 9Cr-1Mo-V).
4. High alloy ferritic, martensitic, and duplex stainless steel ordered by increasing Cr content, beginning at 11Cr-Ti and finishing at 29Cr-4Ni-Ti. Within the same Cr content, a second sort on the second element, e.g., Ni or Mo, was performed. Within this group, the second element never exceeds 4%.
5. A hard-to-categorize group of low alloy ferritics that have Mn and Si identified in their nominal compositions, beginning with Mn-1/4Mo and ending in 1 1/2Si-1/2Mo.
6. Nickel alloy steels ordered according to nickel compositions beginning with 1/2Ni-1/2Cr-1/4Mo-V and ending with 28Ni-19Cr-Cu-Mo.
7. The Cr-Mn and Cr-Ni austenitic (and perhaps some duplex) stainless steels beginning with 16Cr-9Mn-2Ni-N and ending with 25Cr-22Ni-2Mo-N, sorted by increasing Cr content and, within the same Cr content, increasing second element (e.g., Ni) content. In general, this sort on second-element content within the same major-element content applies to the previous categories, i.e., 2 through 6 above.

For the nonferrous materials, the order of the sort is as follows:

1. Alloy/UNS number (alpha-numeric)
2. Tensile strength
3. Yield strength
4. Class/condition/temper
5. Specification number

It is thus apparent that unless a designer knows the nominal composition of a material or its UNS number, he or she will have some difficulty in finding that material in the tables of Part D. However, there are a number of reference books that can help. One of these is the *Practical Guide to ASME Section II—1996 Materials Index*, by R.A. Moen, which lists and cross-indexes a great deal of useful information about Section II, such as nominal material compositions, material properties, and which materials are permitted for use in the various book sections. Another useful book is *Metals and Alloys in the Unified Numbering System,* which lists and cross-indexes materials of chemically similar specifications. The Unified Numbering System for Metals and Alloys (UNS) provides a means of correlating many nationally used metal and alloy numbering systems currently administered by societies, trade associations, and individual users and producers of metals and alloys. A UNS designation is not, in itself, a specification, since it is based on chemical composition and establishes no limits for form, condition, property, or quality. It is rather a unified identifier of a metal or alloy for which controlling limits have been established in specifications published elsewhere, by ASME, ASTM, SAE, and others.

FINDING AND USING DESIGN STRESSES IN SECTION II, PART D

The designer must determine which stress line for a particular material pertains to Section I (or one of the other sections) by referring to the second page of the four pages of information under three columns designated "Applicability and Maximum Temperature Limits." There is one column for Section I, one for Section III, and one for Section VIII, Division 1. Only if there is a maximum use temperature listed in one of these columns may the material be used for construction covered by that section of the Code. Otherwise, the letters NP appear, meaning not permitted. This has led to a number of problems. In Chapter 11, under Standard Pressure Parts, it is explained that such parts (covered in PG-11) which comply with any American National (ANSI) product standard accepted by reference in PG-42 may be made of any material listed in that standard, "but not of materials specifically prohibited, or beyond the use limitations of this Section."

Section I does have some constraints on the use of various materials, but it has no outright prohibitions against the use of any specific materials. For example, footnote 1 of PG-5 formerly cautioned that austenitic stainless steels are not to be used for boiler pressure parts that are water-wetted in normal service, although there are exceptions to this rule. (The advice in footnote 1 was moved to a new PG-5.5, in the 1998 Addenda.) The expression "beyond the use limitations of this Section" pertains to temperature limitations. One is the maximum design temperature for any given material, now established by the highest temperature at which a stress value for Section I use is listed in Section II, Part D. Materials such as cast iron and bronze have relatively low temperature limits, which are stipulated in PG-8 and PG-9.

The letters NP (not permitted) that appear in the Section II, Part D columns on applicability of material for a given book section were applied by default if that material had not been listed for use in that book section before the advent of Part D. Therein lies a problem, because there are many materials listed in the ASME/ANSI product standards accepted by Section I by reference in PG-42 that had never been listed in the Section I stress tables. No Authorized Inspector had previously challenged the use of those materials in these ANSI product standards because their use was (and is) explicitly sanctioned by PG-11. However, a problem arose once the designation NP appeared in Part D. One of the authors' companies found itself facing the following situation: It had designed and subcontracted to another stamp holder the fabrication of an austenitic stainless steel superheater whose return bends were purchased as standard elbows made in accordance with ANSI B16.9, Factory-Made Wrought Steel Buttwelding Fittings. The manufacture of B16.9 elbows starts with pipe or tube made from a material that is designated only by class and grade (e.g., WP 304) and that meets the chemical and tensile requirements of SA-403, Specification for Wrought Austenitic Stainless Steel Pipe Fittings. When the elbow is completed in accordance with the provisions of SA-403 and ANSI B16.9, it is given a new material designation, A-403 or SA-403, as a fitting. Unfortunately, Section I had never listed that material, but it had been listed in Section VIII. Accordingly, the Section II, Part D column for Section I applicability listed SA-403 as NP, not permitted; it was apparently permitted only for Section VIII construction.

The AI at the fabricator saw that listing and proclaimed that the elbows could not be used in Section I construction, stopping fabrication on a contract with heavy daily penalties for late completion. Fortunately, the Authorized Inspection Agency providing the AI at the job site had knowledgeable Code committee members who quickly agreed that the designation NP should not apply in these circumstances, or rather, shouldn't mean NP if PG-11 provides otherwise. Section I subsequently issued Interpretation I-92-97, which solved the problem for these particular fittings, but did nothing about the generic problem of the potentially misleading designation Not Permitted in Section II, Part D. An attempt to have Subcommittee II modify the NP designation with a note referring to PG-11 proved unavailing, so the unwary designer faces a potential trap in a situation similar to the one just described.

Interpretation I-92-97 is as follows:

> Question: *May SA-403 austenitic fittings made to ASME/ANSI standards accepted by reference in PG-42 be used for Section I steam service?*
> Reply: *Yes.*

An earlier interpretation that happened to apply to boiler external piping, and to only three of the many ANSI product standards listed in PG-42, supports the basic philosophy of PG-11 that materials listed in those standards are acceptable for Section I use. That Interpretation was I-79-01:

> Question: *May flanges and pipe fittings made of materials conforming to ASTM specifications, produced and identified in accordance with ANSI B16.5, B16.9, and B16.11 be used in boiler external piping?*
> Reply: *Flanges and pipe fittings made to ASTM specifications and conforming to the referenced ANSI standards may be used in Boiler External Piping provided the requirements of PG-11 are met.*
> Question: *If these components are used as provided for above, are certified test reports or additional inspection required?*
> Reply: *Certified test reports or additional inspection will not be required provided the requirements of PG-11 are met.*

The reply to the first question in the above interpretation illustrates that the material listings in the ASME/ANSI B16 standards for components such as valves and pipe flanges and fittings are given ASTM designations, not ASME designations. One reason for this is that these standards are used in many applications having nothing to do with the ASME Code. PG-11.1.1 explicitly permits the use of these ASTM materials when listed in ANSI product standards accepted by reference in PG-42.

MARKING OF MATERIALS

The ASME material specifications typically contain marking requirements intended to provide proper identification and assurance that the material is what it is supposed to be. These marking requirements are found both in a general specification that gives general requirements for the particular product form, such as piping or tubing, and in the specification covering the material in question. Thus, for example, pipe furnished to SA-106 must be marked as prescribed in the general specification covering carbon steel pipe (A 530) and with the additional marking called for in paragraph 23 of the SA-106 specification, which is quite detailed. That paragraph also stipulates that when pipe sections are cut into shorter lengths by a subsequent processor for resale as material, that processor must transfer complete identifying information to each unmarked cut length, or to metal tags securely attached to bundles of unmarked, small-diameter pipe. The material designation must be included with the information transferred, and also the name, trademark, or brand of the processor.

It is important that a boiler manufacturer have some provision in its quality control system to assure the proper identification of material during fabrication, so that the correct material is used. Section I has some very old and specific provisions in PG-77 regarding the identification of plate material used in the boiler. They call for the material identification information to remain visible on shell plates, furnace sheets, and heads after the boiler is completed. It is permissible to use coded markings instead of the full identification for this purpose. The system of transferring markings when plate is cut must be acceptable to the AI. Section I is silent on maintenance of markings on other product forms, such as pipe or tube. These are usually marked at regular intervals by the material manufacturer, in accordance with the requirements of the material specification, and it suffices if the fabricator has a system for assuring that the proper material has been used.

There have been a number of clarifying interpretations on marking requirements. Interpretation I-79-02 dealt with the subject:

> Question: *Does Section I require any additional information or marking (e.g., heat number marking, mill test reports, etc.) beyond that required by a Section I referenced specifiction in Section II?*
> Reply: *Section I has no supplementary requirements in this area.*

The usual marking requirements do not apply to piping smaller than NPS 2 or to tubing smaller than 1¼ inches in diameter. The generic product form specifications A-450 and A-530 permit the required information to be marked on a tag securely attached to the bundle or box in which such small-diameter piping or tube is shipped.

Another interpretation dealing with marking was I-92-83, dealing with nonpressure parts:

> Question: *If nonpressure parts, such as lugs, hangers, or brackets, are cut from weldable material (PG-55.2), is it a requirement of Section I that the product marking be transferred to each piece?*
> Reply: *No.*

A related topic is whether the markings have to remain visible during construction or upon completion of the boiler. This was addressed in a lengthy but informative interpretation entitled Identification of Materials and Miscellaneous Parts During and and After Construction, I-95-08:

Question: *Identification markings required for material by the material specification and for miscellaneous parts by PG-11 are verified by the Manufacturer during receipt inspection. During construction of the boiler, the material and miscellaneous pressure parts are controlled in accordance with the Manufacturer's approved material control system. The material control system provides methods acceptable to the Authorized Inspector for assuring that only the intended material is used in Code construction. Under these conditions and except for the special identification requirements of PG-77, must the identification markings required by the material specification and by PG-11 remain visible during construction of the boiler?*
Reply: *No.*

USE OF NON-ASME MATERIAL SPECIFICATIONS AND MATERIAL NOT FULLY IDENTIFIED

Both Section I and Section VIII, Division 1 have long had provisions (PG-10 and UG-10) dealing with the acceptance of material made to specifications not permitted by Section I or Section VIII (which could, for example, be an ASME specification or a foreign material specification) and with the acceptance of material of any kind that is not fully identified. In 1987 these provisions were expanded to provide more guidance on what must be done, and who may do it, when a material is requalified as the equivalent of an acceptable ASME material. The recertification or requalification process differs, depending on which of two categories of organization is doing the recertifying and whether the material is fully identified with a complete certification from the material manufacturer. The first category of organizations comprises only boiler or boiler-parts manufacturers. The second category comprises any other organizations. When PG-10 and UG-10 were revised in 1987, the burden of proof that a given material was the equivalent of a material acceptable to Section I (or Section VIII) was made greater for organizations that were not either boiler or boiler-parts manufacturers, for two reasons. These other organizations were often material warehouses that bought material from material manufacturers and resold it to users. Under the less stringent pre-1987 recertification rules, these organizations were in many cases found to be recertifying materials as the equivalent of other materials so that they could sell material they had in stock even if full equivalency of all details of manufacture, chemistry, and mechanical properties couldn't be shown. Accordingly, the rules for those organizations were made more stringent. Boiler and boiler-parts manufacturers were allowed a lower burden of proof because they had more to lose if the recertified material turned out to be less than satisfactory; it was they, after all, who were taking design responsibility for the boiler and/or the parts.

Thus to recertify a material that had complete certification from the material manufacturer as another material, an organization other than a boiler or boiler-parts manufacturer must meet the rules of PG-10.1.1, documenting complete equivalency of the respective specifications. Moreover, such an organization must certify that it has done this and must provide its certification, with copies of documentation proving conformance, to the boiler or boiler-parts manufacturer. This is a very heavy burden of proof. In contrast, the rules of PG-10.1.2 allow a boiler or boiler-parts manufacturer to meet a somewhat lower burden of proof of equivalency, simply by making available to the Authorized Inspector documentation showing that the chemical requirements of the desired specification have been met and that no conflict exists between the requirements of the two specifications. Documentation must also be made available to the AI demonstrating that any other requirements of the desired specification have been met.

When material cannot be requalified as just described, because, for example, complete certification from the material manufacturer is lacking, the rules of PG-10.2 and PG-10.3 may be used to demonstrate acceptability of a specification not permitted by Section I. These rules may be used only by a boiler or boiler-parts manufacturer, for the reasons just explained. In such cases, chemical analyses and mechanical-property tests are employed to demonstrate that the desired material is the equivalent of one that is normally allowed for Section I construction.

Once a material has been established as the the equivalent of an acceptable specification, it may be marked as required by that specification.

The provisions of PG-10 may also be used in the recertification of material provided to foreign specifications, such as DIN or JIS, as the equivalent of ASME specifications suitable for Section I or Section VIII construction.

With the increasing emphasis on global competition, the ASME Boiler and Pressure Vessel Committee has sought ways in which to gain wider international acceptance of ASME Code construction. This is a two-way street, and the ASME recognizes the need to remove unnecessary barriers to the use of ASME construction, in the hope that other countries will reciprocate. One of these barriers is the ASME's previous insistence that only ASME or ASTM materials could be used in ASME construction. Recognizing the many areas of the world where economic constraints or local rules may necessitate the use of non-ASME materials, the ASME relaxed its policy, and in 1997 Section II approved the first two foreign material specifications for inclusion in the 1998 Addenda. Those specifications cover a Canadian structural steel, CSA-G40.21, and a European carbon steel plate, EN 10028-2. (At this writing, requests for inclusion of Japanese and Chinese material specifications are pending.) It happens that these two materials are very similar to two existing ASME materials, SA-36, a structural steel, and SA-516 Grade 65, a common material used for pressure vessels. In this instance, the Canadian material was given its own stress line, with the same stresses as the equivalent ASME material, SA-36. The committee hasn't yet agreed on the stress values for the European material. Should a foreign material not be a close equivalent of an ASME material, Subcommittee II will either require all the usual information (yield, tensile, creep, and rupture strength at temperatures of interest, weldability, etc.) necessary to establish allowable stresses, or it will ask to review the data used by the other countries in setting their allowable stresses. There are also some copyright issues to be worked out, since the foreign materials organizations have so far refused ASME permission to reprint the foreign specifications in Section II.

SPECIAL CONCERNS

A number of Section I material applications deserve special mention. One of these, having to do with the use of austenitic stainless steels in water-wetted service, used to be mentioned in footnote 1 to PG-5. (In late 1997 the Main Committee approved Section I's action to move and clarify the advice in that footnote into a new paragraph PG-5.5. This new paragraph was published as part of the 1998 Addenda incorporated into the 1998 edition of Section I.) Austenitic steels are particularly susceptible to stress corrosion cracking. Consequently, Section I generally limits the use of these steels to what it calls steam-touched service and prohibits their use for water-wetted service, where chlorides or sulfides might be present. Those stainless steels listed in PG-9.1 for boiler parts (normally water-wetted) are ferritic stainless steels, which are not susceptible to stress corrosion cracking. There are, however, several exceptions to the rule.

An exception is made for surfaces that are water-wetted only during start-up of the boiler, since this is considered a short enough time that the risk of corrosion is quite limited. This exception was granted by means of Interpretation I-79-09, in response to a request from a manufacturer regarding a special type of boiler in which the steam headers contained water during start-up. Here is that interpretation:

> Question: *May austenitic stainless steel be used for portions of the boiler proper and the boiler external piping which are water-wetted only in start-up service?*
> Reply: *Austenitic stainless steel may be used for portions of the boiler proper and the boiler external piping which are water-wetted only for start-up service.*

Note the mention of boiler external piping in this reply. It happens that no such mention was needed, since the material requirements for boiler external piping are found in Power Piping, B31.1, and that code has no prohibition against austenitic stainless steel in water-wetted service. This probably comes about from the fact that usually piping is not transferring heat, thus no boiling takes place inside the pipe. Without boiling to concentrate any deleterious species such as chlorides in the water, the likelihood of chloride-induced stress corrosion cracking of austenitic steel is greatly reduced. A second reason for any lack of a prohibition in B31.1 against the use of austenitic steels for water-wetted service is that such a prohibition is hardly needed; it would be extremely unusual, unnecessary, and uneconomical for such material to be used for feedwater piping. Finally, Section I's prohibition against the use of austenitic stainless steel in water-wetted service was buried in a footnote to PG-5 and was none too clear. Consequently, it was probably overlooked when most of the rules covering boiler external piping were transferred to B31.1 in 1972. Occasionally, a designer wants to use austenitic stainless steel for a chemical feed line to the boiler and may feel constrained from doing so by Section I's apparent prohibition. However, the chemical feed line is a part of the boiler external piping, and the prohibition doesn't apply.

Another exception allowing the use of austenitic stainless steels in water-wetted service had been granted under the provisions of Code Case 1896, Miniature Electric Boilers Fabricated of Austenitic Stainless Steel (refer to Appendix II for an explanation of Code Cases). These boilers are used in special applications, such as in the pharmaceutical industry, where steam of great purity is required. One of the provisions of the case is a requirement that very pure water is to be used, thereby making it unlikely that chlorides will be present. Case 1896 was incorporated into a new PEB-5.3, in the 1998 Addenda. A number of other cases, such as Code Case 1325, also permit the use of very high alloy austenitic stainless steels, such as Incoloy 800 (a trade name that the ASME designates as Nickel-Iron-Chromium Alloy 800, UNS N08800) in water-wetted service. In the early 1990s, a warning about the potential susceptibility of these austenitic steels to various kinds of stress corrosion cracking in water-wetted service was added to all such Code Cases. However, the high content of chromium and nickel in these steels is sufficient to prevent most corrosion problems in water-wetted service; their application has generally been quite successful. A similar warning was added as a footnote to the new PG-5.5. It appeared in the 1998 Addenda.

The last exception in which austenitic steel is allowed in water-wetted service came in through the back door, as follows: Some time in the early 1980s, it came to the attention of Subcommittee I that gage glass bodies of austenitic stainless steel had been used for many years, despite the footnote in PG-5. The manufacturers of these devices stated that this water-wetted service presented no problem of chloride stress corrosion cracking because no boiling takes place at the gage glass body (and thus chlorides were not concentrating there the way they would in a tube subject to heating). They also claimed that the use of stainless steel improved the visibility of the water in the glass, because there was less iron oxide to cloud the glass. Moreover, they noted that to provide firm support for the relatively fragile gage glass, the body of the device is always overdesigned to achieve stiffness. As a result, the pressure stresses in the body were below the 7000-psi threshold normally considered necessary to initiate chloride stress corrosion (this threshold is actually a function of alloy and temperature). Finally, they cited quite a few years of successful experience. After considering all this evidence, Subcommittee I approved a new paragraph, PG-12, specifically permitting the use of austenitic stainless steel for gage glass bodies. A new PG-5.5 now identifies this and other exceptions to the general prohibition against the use of austenitic stainless steel in water-wetted service.

Another concern is the problem of **graphitization**, a type of microstructural deterioration that occurs in carbon and carbon-molybdenum steels from the decomposition of iron carbide into ferrite and graphite. This decomposition is a time- and temperature-dependent relationship that begins to become significant only after long service at temperatures above about 840°F (it occurs more quickly at higher temperatures). Graphitization can severely affect the strength and ductility of these steels, particularly the heat-affected zone of welds, where the graphite usually lines up as chain graphite. Note G10 in the allowable stress table (Table 1A of Section II, Part D) cautions: "Upon prolonged exposure to temperatures above 800°F, the carbide phase of carbon steel may be converted to graphite." Some members of Subcommittee I are in favor of prohibiting any use of carbon steels above 800°F, but others think this is not necessary, that the designers should be allowed to use their own judgment. They note also that even with a design temperature slightly above 800°F the components are likely to be operating at 800°F or lower and that graphitization is not a significant problem at these relatively low temperatures, especially if excursions over 800°F are infrequent and of short duration.

The use of cast iron and nonferrous metal castings is permitted by Section I only for those applications mentioned in PG-8, for relatively low temperature and pressure.

MATERIAL REQUIREMENTS FOR BOILER EXTERNAL PIPING

These are contained in 123.2.2 of B31.1, Power Piping. They are virtually the same as those of Section I in calling for the use of ASME SA, SB, or SFA specifications (Section IX doesn't actually require the use of SFA specifications for welding consumables). Material produced under an ASTM specification may be used, provided the requirements of that specification are identical or more stringent than the equivalent ASME specification for that particular category of material (grade, class, or type). For nonboiler external piping, which is entirely outside of Section I's jurisdiction, B31.1 allows material produced to ASTM A or B specifications to be used interchangeably with material specified to ASME SA or SB specifications of the same number. This more relaxed approach hasn't caused any problems of which the authors are aware, and it raises the question as to why the ASME Code bothers to insist on ASME or equivalent materials.

Miscellaneous pressure parts (see PG-11) such as valves, fittings, welding caps, etc., are often furnished as standard parts. As already explained, when those parts meet the requirements of the ANSI standards adopted by reference in PG-42, Section I accepts the materials listed in those standards, unless they are specifically prohibited or are beyond the use limits of Section I. However, standard parts made to a *parts manufacturer's standard* must be made only of materials explicitly permitted by Section I. This is discussed in Chapter 11, under Standard Pressure Parts.

ELECTRIC RESISTANCE WELDED (ERW) MATERIALS

When the Code first accepted tubes and pipes formed of electric resistance welded (ERW) materials, it imposed a 15% penalty on allowable stress, except for tubes within the boiler setting whose design temperature was 850°F or lower. There was apparently some doubt about the welding, and the use of the normal allowable stress was permitted only for tubes inside the protective barrier of the setting. Because of the good service record of these tubes in recent years, some thought is being given to removing the penalty on ERW tubes. As a first step in this direction, in 1988 several Code Cases were passed (CC 2041, 1942, 1996, 1984, and 2026) permitting the use of ERW tubes within the allowed design temperature range without the 15% reduction in allowable stress, if they meet three conditions: the tubes are given extra nondestructive examination (angle beam ultrasonic inspection and an electric test in accordance with ASME specification SA-450); the tubes are no larger than 3½ inches OD; and the tubes are enclosed within the setting.

Code users often wonder just what the term ''within the setting'' means since this term is not defined in Section I. Various books on boiler construction typically define the setting as the construction surrounding the boiler and/or the tubes: refractory, insulation, casing, lagging, or some combination of these. In former years, with heavy casing and refractory, it could be argued that a failed tube would not represent much of a danger to someone standing nearby. As simpler and lighter construction has evolved, there may be only insulation and relatively thin lagging outside the tubes or enclosing a header. Although such construction may not provide as strong a barrier as formerly, it is the industry's position, accepted by the Authorized Inspectors, that the tubes are still ''enclosed within the setting,'' and if they meet the other provisions of the Code cases cited, they would be entitled to the full allowable stress without the 15% penalty of former years. That penalty was, in effect, a vote of no confidence in the electric resistance weld and/or the inspection of the weld.

Any grease or other contamination of the surfaces to be welded has the potential for preventing a sound weld, and careful inspection has always been important in assuring that any bad welds will be found. However, the 15% penalty was always an arbitrary one. If the weld were truly bad, it would leak. Lowering the stress by 15% wouldn't really help. There has now been quite a lot of good experience with ERW tubes, and the additional NDE called for in the Code Cases should henceforth provide some further assurance that any bad welds will be detected. After further satisfactory use, the provisions of the Code Cases may be incorporated into Section I, although some think that the cost to a utility of even a single tube failure in service, requiring shutdown for repair, far outweighs any initial cost advantage of ERW tubing.

NONPRESSURE PART MATERIALS

Material for nonpressure parts, such as lugs, hangers, brackets, or fins, is not limited to material listed in the stress tables, but must be of weldable quality. "Weldable quality" is a rather vague term, and the choice of nonpressure material is thus left to the designer. The fact that a material is of weldable quality is established when the procedure used to weld it is qualified. For carbon and low alloy steels, PW-5.2 also stipulates another requirement for weldability: the carbon content may not exceed 0.35%.

MATERIAL TEST REPORTS

Although there is no specific reference within the text of Section I to a requirement that a material test report be furnished, PG-105.4 calls for the quality control system to be in accordance with the requirements of Appendix A-300. Thus that appendix becomes mandatory, and paragraph A-302.4 on Material Control stipulates that the boiler manufacturer (or any other stamp holder who is involved in fabrication or assembly) "shall have a system of receiving control that will insure that material received is properly identified and has documentation, including required material certifications or material test reports (MTR's), to satisfy Code requirements as ordered." Typically, ASME material specifications have a paragraph called Ordering Information, which usually says, "Orders for materials under this specification shall include the following, as required, to describe the material adequately:...." For pipe products for example, the ordering information stipulated includes a report described as a "Test Report Required (Section on Certification of Specification A-530/A-530M)." Specification A-530/A-530M is entitled "Specification for General Requirements for Specialized Carbon and Alloy Steel Pipe."

Paragraph 23 of that specification is entitled "Certified Test Report." The substance of this paragraph is that the producer or supplier of the pipe material must furnish a report certifying that the material was manufactured, sampled, and tested in accordance with the governing specification and any other requirements in the purchase order or contract and that those requirements were met. The paragraph advises also that a signature or notarization is not required on the Certified Test Report, but that the document must be dated and must clearly identify the organization submitting it. Also, the submitting organization is responsible for the contents of the report, even in the absence of a signature or notarization.

In the case of tube products, the comparable reference for the test report is to Specification A-450/A-450M, Specification for General Requirements for Carbon, Ferritic Alloy, and Austenitic Alloy Steel Tubes. Paragraph 26 of A-450 deals with the Certified Test Report and is identical to paragraph 23 of SA-530, just described. Plate materials and forgings also have specifications (A-20 and A-788) that give general requirements for those product forms. The test report paragraphs of those specifications are similar to those just described. The terminology used for these various test reports in the general requirements specifications and the individual material specifications varies somewhat, because it evolved over a long period of time: Material Test Reports (MTR's), Certificates of Compliance/Conformance, Certificates of Inspection, and Certified Material Test Reports. There is little actual difference among them, except that a Certificate of Conformance or Compliance is merely a statement of compliance or conformance with a particular specification, without test results. However even in specifications that call for such certificates (e.g., SA-234 and SA-403), there is a further statement that, if requested, the manufacturer shall provide reports of all tests called for in the specification. In most cases, an MTR serves to provide all the information needed, but the material specification and the general requirements specification should be reviewed to confirm this. Most boiler manufacturers routinely require the material manufacturer or supplier to furnish MTR's for all pipe, tube, and plate materials, since most A.I.'s and ASME survey team members want to see those as evidence of the certificate holder's compliance with the Code.

Thus, although the advice in A-302.4 on material control could be a little more definitive, the industry construes it to mean that manufacturers are expected to specify in their material purchase orders that either material test reports or material certifications are to be furnished with the material. These documents provide a description of the material sufficient to determine that it satisfies the specification as to chemical composition, mechanical properties, and any other information that the specification requires to be reported. The Authorized Inspector routinely expects and wants to review the MTR's to help assure that the right material has been received and used.

Sometimes one of the parties involved in constructing a boiler gets the mistaken notion that the boiler manufacturer, or a sub-tier stamp holder furnishing parts or piping for the boiler, is supposed to furnish material test reports as part of the documentation making up the Manufacturer's Data Report package. This is not the case. An Interpretation was issued on this subject in 1977 and then revised and reissued on December 20, 1982. It appears on page 5 of Interpretations Volume 12 as I-77-28R, Section I, Identification of Pipe Material:

> Question: *May A-53 and A-106 pipe be used to fabricate Section I boiler external piping without the need to provide certified mill test reports or manufacturer certification provided the pipe is marked in accordance with the ASTM "A" designation?*
> Reply: *The manufacturer who orders pipe material for a Code application is responsible for ordering it to Code requirements and shall have a material control system that ensures that the proper material is used in Code construction. The manufacturer's certification includes acknowledgment that the material requirements of the Code have been met. This certification by the manufacturer properly countersigned by the Authorized Inspector who provides inspection service for that manufacturer, may be accepted by you or others as sufficient evidence of Code compliance.*

The original version of this reply, I-77-28, had included the following additional advice:

> *The verification of the Authorized Inspector may be made by any of the following methods:*
> (1) *Review of the material test reports or Certificate of Compliance when available as provided for in the Code approved specifications.*
> (2) *Review of product markings furnished as provided for in the Code approved specification.*
> *In these circumstances there is no Code obligation on you as a subsequent user or installer of such subassemblies to examine mill test reports for material covered by ASME Data Forms.*

STRENGTH OF MATERIALS AFTER FABRICATION

The fabrication of material into boiler components may involve all sorts of hot or cold metal working, such as bending, rolling, forging, flaring, or swaging. Heat treatment such as solution annealing or postweld heat treatment may be employed following various forming and fabrication operations. As a consequence, the as-received mechanical properties of the material, such as yield strength, hardness, or ductility, may have changed. Typically, the ASME material specifications have been developed to provide material suitable for the various operations involved in boiler or pressure vessel manufacture, and the change in material properties presents no problem, especially for experienced manufacturers who know what, if any, intermediate or post-fabrication heat treatment may be appropriate. Occasionally, a dispute arises about the strength or other properties of a material after it has been fabricated into components of a boiler. A purchaser may have the mistaken idea that the properties of the material after fabrication must match its properties in the as-received condition, when it met the specification requirements. Section I has issued several interpretations to correct this misapprehension. The first was I-80-19, which harks back to the days when Subcommittee

I still gave what today might be called consulting advice, especially if the inquirer resided overseas and was thought to be having difficulty in understanding the Code.

> Question: *In construction of a power boiler in accordance with Section I, cracking has been observed in the ends of waterwall tubes that had been cold reduced from 3-in. diameter to 2-in. diameter; the material had not been heat treated following cold reduction. Does Section I require a heat treatment of cold reduced ends of waterwall tubes?*
>
> Reply: *Section I does not specifically address the heat treatment of cold reduced ends of waterwall tubes which are to be used in rolled tube joints. However, it is the responsibility of the Section I stamp holder to use only material that is free of rejectable imperfections (those that encroach on required minimum wall thickness). Further, it is the Section I stamp holder's responsibility to use only material that is suitable for the intended installations, such as tube expansion (see PG-10.2.2).* (Note by authors: At that time PG-10.2.2 contained the following sentence, subsequently deleted when rules for acceptance of unidentified material were tightened in 1987: ''Material specified as suitable for welding, cold bending, close coiling, and similar operations shall be given sufficient check tests to satisfy the inspector that it is suitable for the fabrication procedure to be used.'' Similar language is usually included in the material specifications.) *To provide tubes that can be expanded into drums without cracking, it is usually required to heat treat tube ends which have been cold reduced to the degree indicated in the inquiry.*

Another interpretion dealing with post-fabrication strength of materials is I-83-01, Hardness of Swaged Portion of Boiler Tubes:

> Question: *Is it required that carbon steel tubes such as SA-210 A-1 which have undergone fabrication operations by the boiler manufacturer meet the maximum hardness specified in the specifications?*
>
> Reply: *No. The maximum hardness specified in the Section II material specification is one of a number of requirements intended to ensure that the tubing furnished by the material manufacturer will withstand fabrication and installation requirements. Section I does not specify hardness limits for the tubing as fabricated and installed in a boiler.*

A last and more recent interpretation on this subject, I-95-03, was issued in response to an inquiry about chromized tubes:

> Question: *Do the materials listed in PG-9 for Section I construction have to comply with the individual specification requirements after the manufacturer has performed fabrication on the material?*
>
> Reply: *No.*

DOES NEW MATERIAL GET OLD?

The Foreword to all sections of the Code used to explain that revisions to the Code are issued as yearly Addenda on December 31. (However this changed to July 1, starting in 1998. See Effective Dates of the Code and Code Revisions in Chapter 2.) These revisions become mandatory six months later, except for boilers and pressure vessels contracted for prior to the end of the six-month period. However, the Foreword has never specifically addressed what must be done about revisions to the material specifications listed in Section II. These have traditionally been treated differently from revisions to the Code. As explained below, ASME now provides in Appendix A to Parts A and B of Section II a list of all the materials approved for Code construction and the book sections in which they are approved for use. The list shows the latest material specifications adopted by the ASME and the dates of earlier versions considered equivalent and acceptable, going back in some cases to 1983. Thus, except for unusual situations, manufacturers can readily determine the acceptability of any particular material on hand.

The story of the development of Appendix A is instructive. Until the mid-1980s, it had generally been understood that old (but previously unused) material could be used in boiler and pressure vessel construction. That is, if a manufacturer happened to have in inventory a lot of SA-213 tubes or SA-515 plate that had been manufactured years earlier, the manufacturer could use those materials for construction under the latest edition of the Code.

Confirmation of this practice is seen in the declarations the manufacturer signs on the Data Report Form, for example, line six on the P-3 form, which defines the year edition of Section I to which the boiler is constructed. That line declares that the chemical and physical properties of all the parts meet the requirements of the ASME Code, but of no specified date of issue. Thus the manufacturer is not required to list on the Data Report Form the date of the specifications covering the pressure part material. However, the manufacturer does have to declare that the design, construction, and workmanship conform to a particular edition and addenda of Section I. Further confirmation of the above practice is found in a Section VIII interpretation that was issued on March 27, 1980, under the subject Material Usage under Section VIII, Division 1, and UG-84(b)(2), Item BC-251. One of the inquirer's questions was as follows:

> Question: *May material which is produced to any version of the Code be used so long as there have not been any changes made to the specification except for editorial changes?*
>
> A key paragraph in the reply was this:
>
> *Specifications to editions or addenda earlier than those in effect at the time of vessel contract may be used provided the requirements are identical (excluding editorial differences) or more stringent for the grade, class, or type produced. Material produced to earlier specifications with requirements different from the specification in effect at the time of the contract may also be used provided the material manufacturer or vessel manufacturer certifies with evidence acceptable to the Authorized Inspector that the requirements of the version in effect at the time of contract will be met.*

This was a reasonable policy, since material specifications change very slowly and there is virtually no possibility that a material proven by years of satisfactory service would suddenly be found to be unsafe. At worst, allowable stresses at elevated temperature might be reduced, but the material itself would still be acceptable. Changes to material specifications are initiated by the American Society for Testing and Materials (ASTM) and are then usually adopted by ASME. Often the changes are little more than an updating of some of the many other specifications and documents typically referenced in the material specifications. Occasionally, however, a new grade is added to a particular specification, or the yield strength of a particular grade may be increased slightly to make its properties and allowable stresses consistent with other similar alloys.

In the 1980s a number of factors changed this situation. One of these was the development of improved technology, which allowed the major steel producers to produce cleaner steels with reduced maximum permissible phosphorous and sulfur content. In recognition of this advance, the ASTM reduced the maximum permissible limits on phosphorous and sulfur from 0.050% and 0.060%, respectively, to 0.035%. When the ASME adopted those changes, the Committee realized that a great deal of heretofore perfectly good material was still in inventory at material manufacturers, pressure vessel equipment manufacturers, and other users such as utilities. The prohibition of the further use of this material in inventory would have represented a huge economic penalty and could not be justified on the grounds of safety. Accordingly, the Committee decided that a gradual shift to the use of the new, cleaner steels would be appropriate, and issued Code Case 2053, For Material in Inventory, Sections I, IV, and VIII, Divisions 1 and 2. The case permits continued use of material in inventory, as explained below.

Another important change that took place in the 1980s was the increasing use of imported steel. The principal source of steel for the large domestic integrated steel producers was blast-furnace product derived from raw ore. This yielded steel with a fairly consistent content of residual elements. For various reasons, perhaps because of more variable ores or a greater use of scrap material, overseas steel producers sometimes produced steel with a much higher content of certain elements, called residual elements, not usually listed in the material specifications. These residual elements affect properties such as weldability, fracture toughness, and strength levels.

An example of the problems caused by residual elements occurred at one of the authors' companies during the welding of some pipe that was thought to be carbon steel. Unexpected cracks developed in the welds, despite the fact that all the normal welding procedures were used. After a careful check of the welding variables showed nothing amiss, a spectographic examination of the pipe material showed it to contain enough chromium to change its material category from P-1 (plain carbon steel, the easiest to weld) to a low-chromium category that required preheat to assure crack-free welds. Although the pipe had been bought in Texas, it turned out that the hollow from which it was drawn had been made in Europe. At that particular time, in the late 1980s, neither the ASTM nor the ASME material specifications for carbon steel pipe had any limit for residual elements such as chromium, copper, nickel, molybdenum, and vanadium, which were not normally expected in plain carbon steel.

Another problem that occurred occasionally with the use of steel manufactured in Europe was the cracking of cold-bent carbon steel pipe after a few years of service. Studies under the sponsorship of the Material Properties Council found some evidence that the problem may have been caused by the presence of small amounts of residual elements, such as copper and tin. These elements may have come from scrap used in the manufacture of the pipe.

Due to the welding and other problems with unexpected and unwanted residual elements in carbon steels, the ASTM (and the ASME) in the late 1980s began a program of revising the specifications for some of these materials, to restrict their chemistry. Limits were placed on the permitted percentages of certain residual elements. For example, the current specifications for SA-53 and SA-106 carbon steel pipe call for a maximum chromium, copper, or nickel content of 0.4%, and a maximum combined total of 1% for five named residual elements. The certified material test report routinely furnished with pressure-part material lists this material composition. As a consequence of the restricted chemistry, materials manufactured to the new requirements changed slightly. Again the problem of material in inventory was raised, since it would not be possible without expensive and tedious analysis to demonstrate that the old material met the requirements of the latest specification for residual elements. Fortunately, Code Case 2053, dealing with materials in inventory, again applied and permitted the continued use of those materials, as explained below.

USE OF MATERIALS IN INVENTORY

The inquiry in Code Case 2053 asks whether it is permissible to use material in inventory which meets the requirements of a Section II specification that was superseded, but which also meets the requirements of the current specification, except for the more restrictive requirements for chemical composition. In its reply, the Committee allowed the use of such material subject to a number of stipulations, the essence of which are these:

1. There must be no prohibition on the use of the superseded specification.
2. The specification has to have been accepted by the Code section for which it is to be used, and stress values for that section have to have been published in Section II.
3. The material has to have been melted (manufactured) when the less restrictive requirements were in effect. (This provision was added to prevent the manufacture of any more of the old material.)
4. The Code Case number has to be included on the Manufacturer's Data Report and also on the required specification marking of the material. (So far as the authors are aware, there have been no questions raised about this latter requirement. It is their opinion that a material supplier would have to mark the Code Case number on any material in inventory furnished to a boiler manufacturer, but a boiler manufacturer would not. Marking in accordance with the specification is a duty of material suppliers, not of boiler manufacturers. It is incumbent

upon the latter only to have a quality control system that maintains the proper identity of the material in question.)

CODE GUIDANCE ON THE USE OF OLD MATERIALS

In the early 1990s the Subcommittee on Materials, Subcommittee II, was asked to address and provide guidance on the problem of using materials made to previous editions of the specifications. Subcommittee II prepared an Appendix A to both Parts A and B of Section II, listing the latest material specifications adopted by the ASME and the dates of earlier versions considered equivalent and acceptable, going back in some cases to 1983. The task of comparing and accepting earlier specifications was lengthy and tedious, so Subcommittee II went back only to 1988 as a base year for comparison, but listed some earlier years when it was apparent that those earlier editions were identical to the 1988 editions.

The introduction to Appendix A is rather brief, but it explains several important policies that differ slightly from those explained in the 1980 Section VIII interpretation mentioned above. Appendix A states that an ASME or an ASTM specification with requirements different from the current or other acceptable specification (e.g., the one in effect at the time of the contract) may be used, provided the material manufacturer or vessel manufacturer certifies with evidence acceptable to the Authorized Inspector that the requirements of one of the acceptable versions of the specification have been met.

There is other useful information to be found in the introduction to Appendix A of Section II, Parts A and B. This appendix lists the book sections of the ASME Boiler Code that approve and permit the use of the various specifications. Stated another way, it is possible to find any Section II material in Appendix A in a numerical list and quickly determine which book sections permit its use. Appendix A also states that a material produced under an ASTM specification may be used in lieu of the corresponding ASME specification listed in Appendix A (if it is identical). If it is not identical, the manufacturer would have to demonstrate to the AI that the ASTM material met the requirements of the corresponding ASME specification. Again, Code Case 2053 could permit use of material in inventory that met all current specification requirements other than its more restrictive chemical requirements.

A situation illustrating many of the problems that can arise when using old material happened at one of the authors' companies. That company had in inventory a considerable quantity of SA-178 grade C tubing material from a job that had been cancelled in 1985 and decided to use it for a boiler being built to the 1995 edition of Section I. When the old tubes were shipped from one of the company's plants to another, they were challenged by the receiving plant's quality control inspectors because the list of acceptable specifications in Appendix A of Section II, Part A went back only as far as the 1989 SA-178 specification.

Upon careful review of the 1985 and 1995 specifications, it was found that a number of requirements had been deleted in the 1995 version, and the permitted chemistry ranges for sulfur and phosphorous had been reduced to 0.035% from the 1985 values of 0.060% and 0.050%, respectively. The certified material test reports for the old tubes showed that the sulfur content was just over the new 0.035% limit for at least some of the tubes. Thus it would not be possible to demonstrate that the old material met the requirements of one of the currently acceptable versions of the SA-178 specification listed in Appendix A. Moreover, if the old tubes could not be used, new ones would have to be purchased, and the job would suffer costly delay. Fortunately, Code Case 2053, dealing with materials in inventory, was written to provide relief for just this sort of situation. When Case 2053 was brought to the attention of the QC inspector, he dropped all further objections to the use of the old tubes.

A caveat should be offered here. Code Cases have a limited life. Code Case 2053 was intended to relieve economic hardship and allow old material in inventory to be used up. It is unlikely that the case will ever be incorporated into the text of the Code, and there is some sentiment among the Committee members to let it expire, on the grounds that sufficient time has been allowed to exhaust the old stocks of material. The authors, who share the viewpoint of the manufacturers, disagree, on the grounds that there are many boilers and pressure vessels that were manufactured using these now superseded material specifications. In the authors' opinion, the very fact that this equipment continues to give safe and satisfactory service shows that no significant safety issue is involved. At worst, some unanticipated residual elements in the old material might cause a few welding problems, but this disadvantage is outweighed by saving the cost of purchasing new material.

BOILER DESIGN

DESIGN LIFE

The Foreword to Section I states that the objective of the rules is to afford reasonable protection of life and property and to provide a margin for deterioration in service so as to give a reasonably long safe period of usefulness. Just how long is that period? There are many 40-year-old boilers that are still giving good service, and sometimes 40 years is cited as the expected life of a boiler, assuming of course that it is properly maintained. However, Section I gives no explicit indication of what constitutes a reasonably long design life. Sometimes the criteria used to establish allowable design stresses at elevated temperature are mistakenly assumed to imply a limited design life of only 100,000 hours (about 14 years if the boiler is in service 80% of the time). These criteria are found in paragraph 1–100(b) of Appendix 1 of Section II, Part D, from which the following excerpt is taken:

> At temperatures in the range where creep and stress rupture strength govern the selection of stresses, the maximum allowable stress value for all materials is established by the Committee not to exceed the lowest of the following:
> (1) 100% of the average stress to produce a creep rate of 0.01%/1000 hr;
> (2) 67% of the average stress to cause rupture at the end of 100,000 hr;
> (3) 80% of the minimum stress to cause rupture at the end of 100,000 hr.

The criterion that usually governs is the second, 67% of the average stress to cause creep rupture at the end of 100,000 hours of service. Note that this stress is *not* one that would cause creep rupture after 100,000 hours of service; it is only 67% of that stress. Creep life (i.e., the time it takes to cause creep rupture in elevated temperature service) is a complex nonlinear function of stress and temperature. By experiment and analysis, the relationship among temperature, stress, and time to creep rupture has been established and plotted as a series of creep rupture curves. These are log stress versus log time curves, with temperature as a parameter. When plotted this way, the "curves" are almost straight lines (see Fig. 14.2 in Chapter 14: Creep and Fatigue Damage During Boiler Life).

A review of creep rupture curves for typical steels used in boiler construction shows that for a given temperature, a one-third reduction in stress results in an increase in creep life (time to stress rupture) by a factor approaching 10. Thus if the allowable stress is chosen as two thirds (67%) of the average stress to cause rupture in 100,000 hours, the actual time to rupture will be far greater, perhaps approaching 1,000,000 hours. Such a time is considerably greater than the 40-year design life sometimes cited for boilers, which represents less than 300,000 hours for a boiler assumed to be in service 80% of the time. This estimate is confirmed by experience, which shows that boilers and pressure vessels designed using the ASME Code have very long safe lives.

Two inquiries on the design life of boilers have been answered in recent years. The first of these was Interpretation I-89-30, under the heading Section I, Basis for Establishing Allowable Stress Values:

Question: *Does the ASME Boiler and Pressure Vessel Committee establish a specific design life for components designed to Section I?*
Reply: *No.*

A similar but more elaborate version of the same question, posed by a foreign manufacturer, was answered in March 1996 by Subcommittee II, the Subcommittee on Materials. That Interpretation was published with the second group of 1997 Interpretations, as IID-95-01:

Question: *Do the criteria of Appendix 1 of Section II, Part D, for establishing stress values in Tables 1A and 1B, imply an explicit design life for Section I construction, using the allowable stresses in Tables 1A and 1B for materials permitted for Section I construction?*
Reply: *No. There is neither an explicit nor an implicit design life associated with the allowable stresses in Tables 1A and 1B for Section I construction. The criteria of Appendix 1 of Section II, Part D, have been established with the intention that sufficient margin is provided in the allowable stresses to preclude failure during normal operation for any reasonable life of boilers constructed according to Section I rules.*

It is probably appropriate to add a final comment that is based on committee discussion but was excluded from the above interpretation because it deals with matters that are clearly beyond new construction. The committee's response was worded with the understanding that boilers are subject to periodic inspections. These inspections provide reasonable assurance of finding and repairing areas that may have become damaged due to erosion, corrosion, or other wastage phenomena, or due to any unexpected operating conditions that may have reduced the wall thickness or load-carrying capability of particular components.

DESIGN METHODS

For the most part, Section I uses an experience-based design method known as design-by-rule. (Certain other sections of the Code, namely Section III and Section VIII, Division 2, use a newer method, known as design-by-analysis.) Design-by-rule is typically a process requiring the determination of loads, the choice of a design formula, and the selection of an appropriate design stress for the material to be used. Rules for this kind of design are found throughout Section I, with most being in Part PG (the general rules). Other design rules are found in those parts of Section I dealing with specific types of boilers or particular types of construction.

The principal design rules are found in PG-16 through PG-55. There are formulas for the design of cylindrical components under internal pressure (tube, pipe, headers, drums), heads (dished, flat, stayed and unstayed), stayed surfaces, and ligaments between holes. Rules are also provided for openings or penetrations in any of these components, based on a system of compensation in which the material removed for the opening is replaced by reinforcing the region immediately around the opening. All of these formulas involve internal pressure, except the rules for support and attachment lugs of PG-55, for which the designer chooses the design loads on the basis of the anticipated weight or other loads to be carried. These various design topics are discussed later in this chapter.

Another method of design permitted by Section I is a hydrostatic deformation or proof test (PG-18, Appendix A-22). An experience-based method, it is used to establish a safe design pressure for components for which no rules are given or when strength cannot be calculated with a satisfactory assurance of accuracy. In this type of proof test, a full-size prototype of the pressure part is carefully subjected to a slowly increasing hydrostatic pressure until yielding or bursting occurs (depending on which type of test it is). The maximum allowable working pressure (the design pressure; see definition under Design Loads, below) is then established by an appropriate formula, including the strength of the material and a suitable safety factor. Proof testing may not be used if Section I has design rules for the component, and in actual practice such testing is seldom employed. However, it can be a simple and effective way of establishing an acceptable design pressure for unusual designs, odd shapes, or special features that would be difficult and costly to analyze, even with the latest computer-based methods. The use of proof testing used to be common for the design of marine boiler headers of D-shaped or square cross section.

Tests which are used to establish the maximum allowable working pressure of pressure parts must be witnessed and approved by the Authorized Inspector, as required by Appendix A-22.10. The test report becomes a permanent reference to justify the design of such parts should the manufacturer want to use that design again for other boilers.

DESIGN LOADS

When Section I was written, its authors understood so well that the internal working pressure of the boiler was the design loading of overriding importance that virtually no other loading was mentioned. To this day there is very little discussion of any other kind of loading, except for PG-22, which merely cautions that other loads causing stresses greater than 10% of the allowable stress shall be taken into account. Because internal pressure is the principal loading of concern, it appears in all Section I design formulas. The pressure inside the pressure parts of the boiler is called the working pressure. For design purposes, Section I uses the term **maximum allowable working pressure**, or MAWP, to define what engineers usually understand to mean design pressure.

There are many other loads on a boiler in addition to pressure. Among these loads are the weight of the boiler and its contents, seismic loads, wind loads, and thermal loads. Section I has no provisions specifically dealing with the last three, but it is the responsibility of the designer to consider them. As boiler designs evolved over the years, the manufacturers have recognized and made provision for these loads.

Recently, as nuclear plants have assumed more of the utilities' base load, fossil-fired boilers are increasingly being used in cyclic service, with frequent start-ups and shutdowns. This type of operation has the potential for causing repeated, relatively high transient thermal stress and may require careful analysis for fatigue and creep damage. Although Section I has successfully considered creep and creep rupture in setting allowable stresses, it has no specific rules for creep–fatigue interaction, or even for simple fatigue. Nevertheless, consideration of this phenomenon remains the responsibility of the manufacturer. As discussed in Chapter 14, elevated-temperature design for fatigue and creep is a complex subject, for which no complete rules have yet been developed. Such Code rules as are available appear in Section III, Subsection NH, Class 1 Components in Elevated Temperature Service. Subsection NH is based on Code Case N-47, which dealt with nuclear power plant components in elevated-temperature service. The earliest version of Case N-47 dates back to the late 1960s. Elevated-temperature service means service above that temperature at which creep effects become significant. This threshold is usually taken as 700°F for many ferritic materials (carbon and low alloy materials) and 800°F or higher for austenitic materials (the high alloy materials such as stainless steel). With the exception of parts of superheaters, most boilers are made of ferritic materials.

Choice of Design Pressure and Temperature

Section I does not tell the designer what the design pressure (MAWP) and design temperature of a boiler should be. These design conditions must be chosen by the manufacturer or specified by an architect-engineer, user, or others. In most instances Section I requires only the MAWP, not the design temperature, to be stamped on the boiler. (When discussing design temperature, it is necessary to distinguish between metal temperatures and steam temperatures, either of which may vary within the boiler.) As a practical matter, the designer must know and choose an appropriate design (metal) temperature for each element of the boiler, because the allowable stress values are based on these temperatures. In many cases, the design temperature of an individual element such as a superheater tube may be much higher than the design steam temperature of the boiler. This stems from the fact that the metal temperature of a heated tube is always higher than that of the steam within and from the need to provide a design margin for potential upsets in gas temperature or steam flow.

In a natural circulation boiler, most of the pressure parts are designed for the MAWP. However, slightly different design pressures are used for some components, since PG-22 stipulates that hydrostatic head shall be taken into account in determining the minimum thickness required, "unless noted otherwise." The formulas for determining the thickness of cylindrical components under internal pressure found in PG-27 are of two types: one applicable to "tubing—up to and including 5-inch outside diameter," and the other applicable to "piping, drums, and headers." Having only these two categories of design formula sometimes puzzles inexperienced designers, who aren't sure which formula to use for tubing larger than 5-inch outside diameter. They soon learn the accepted practice for such tubing, which (by the process of elimination) is to use the formula for "piping, drums, and headers." Advice to this effect used to appear in P-22(a) of the 1962 edition of Section I. However, during the extensive rearrangement and renumbering of paragraphs carried out for the 1965 edition, this explicit guidance was somehow moved into PWT-10.2, where it is easily overlooked. The two types of wall thickness formulas have other differences worth noting, as they are a source of confusion to designers and have led to quite a number of inquiries to Subcommittee I.

Avoidance of Thermal Shock

Experience has shown that pressure parts may experience thermal fatigue damage if cooler feedwater is introduced in such a way that it impinges on the walls of a hot vessel. PG-59.2 addresses this problem by prohibiting the direct discharge of feedwater against vessel surfaces exposed to high-temperature gases or radiation from the fire. Moreover, for pressures over 400 psi, a device such as a thermal sleeve or shield is required on the feedwater inlet through the drum, to protect local areas of heads and shells from excessive temperature differentials; the latter can cause significant cyclic thermal stress and potentially lead to fatigue cracking.

Hydrostatic Head

The tubing design formula refers to Note 10, which states that the hydrostatic head need not be included when this formula is used. Thus, the design pressure P is taken as just the MAWP for all tubes 5-inch OD and under. The term $0.005D$ in the tube design formula was added by Subcommittee I when the current version of the tube design formula was adopted in the 1960s. The new formula initially proposed at that time gave somewhat thinner walls than the previous formula, leading to a concern on the part of some committee members that tubes for low-pressure boilers would lack adequate strength to sustain any occasional mechanical loading to which they might be subjected. The $0.005D$ term was a compromise value selected by the committee in response to this concern, to provide a little extra strength in the tubes. No similar note excluding consideration of hydrostatic head applies to the design formulas for piping, drums, and headers. Accordingly, the design pressure P used in those formulas must include the hydrostatic head, which on a large utility boiler could add as much as 50 psi to the MAWP for a lower waterwall header or a downcomer, since the height of such units may equal that of a 15-story building. Of course, the additional hydrostatic head would be much less for a header near the top of the boiler, such as a roof-outlet header. Incidentally, it is traditional to measure hydrostatic head from the elevation of the drum centerline, since that is the normal water line.

The exclusion of hydrostatic head in the tube design formula but not in the pipe design formula came about from the adoption of slightly different formulas. The formula for tubing had evolved for equipment of modest height, where hydrostatic head was considered negligible; in contrast, it was thought necessary to include hydrostatic head in the formula for piping. This arbitrarily different treatment of hydrostatic head has little significance in most practical instances. A much larger potential difference in wall thickness for pipe versus tube results from a number of factors prescribed by PG-27 to account for the effects of expanding tubes, threading pipe, or the need for structural stability.

Other Factors Applicable to Tube and Pipe Design

The first of these several factors is *e*, defined as a thickness factor for expanded tube ends, that is applied to tubes below certain minimum sizes. Note 4 of PG-27.4 lists *e* values required for various sizes of tubes. The thickness factor is an extra 0.04-inch thickness required over "the length of the seat plus 1 inch for tubes expanded into tube seats." This provides a slightly heavier wall for that portion of the tube that is rolled, or expanded, into the tube hole. In the rolling process, the tube undergoes considerable cold working and some thinning, and the extra 40 mils of wall thickness is intended to assure a sound joint no weaker than the unexpanded portion of the tube. This extra 40 mils applies only to very thin tubes, as stipulated in Note 4. Such tubes are usually used only for relatively low-pressure boilers.

Another of these factors is *C*, described as a "minimum allowance for threading and structural stability." It is intended for application to pipe, but as will be explained later, it can sometimes also apply to tubes. When used as a threading allowance, *C* provides extra wall thickness to make up for material lost in threading. For threaded pipe, Note 3 of PG-27.4 stipulates a *C* value of 0.065 inches for pipe NPS 3/4 and smaller and a *C* value equal to the depth of the thread for pipe NPS 1 and larger. For unthreaded pipe, an extra wall thickness of 0.065 inches, called an **allowance for structural stability**, must be provided for pipe smaller than NPS 4. This allowance is traditionally explained to new members of Subcommittee I as having been chosen to provide sufficient bending resistance in the smaller size pipes to withstand loads imposed by a 200-pound worker on a ladder leaning against the middle of a 10-foot span of pipe to do some maintenance on the boiler. The larger size pipes presumably are strong enough to take care of themselves under such occasional mechanical loads. While the worker on the ladder story may not be exactly true, the effect of these several factors is to strengthen the tubes and pipes of very low pressure boilers, whose walls might otherwise be so thin that the boiler would be too flimsy to resist any occasional mechanical loads. However, Note 3 of PG-27.4 makes clear that the allowances for threading and structural stability are not intended to provide for conditions of misapplied external loads or for mechanical abuse.

DISTINCTION BETWEEN BOILER PROPER PIPING AND BOILER EXTERNAL PIPING

For many years, Section I made no distinction regarding the piping of a boiler; all of it was just treated as piping. In the late 1960s some members of the committee decided that the piping rules of Section I were no longer adequate to cover the design of modern high-pressure, high-temperature, steam piping. After due consideration, it was decided to divide all piping within the scope of Section I into two categories. The piping that was actually part of the boiler (such as downcomers, risers, transfer piping, and piping between the drum and an attached superheater) was designated as **boiler proper piping**. Construction rules for this piping were retained in Section I. The piping that led to or from the boiler (such as feedwater, main steam, vent, drain, blowoff, chemical feed piping) was designated a new category, **boiler external piping**, an appropriate name since it is usually external to the boiler. This piping was defined by its extent: from the boiler to the valve or valves required by Section I (the boundary of the boiler). These valves also were defined as part of the boiler external piping. In 1972, responsibility for most aspects of the construction of this piping was transferred from Section I to the Power Piping Code, ASME B31.1. That is, rules for material, design, fabrication, installation, and testing of boiler external piping were now contained in B31.1, but Section I retained its usual certification requirements (with Authorized Inspection and stamping).

To clarify the distinction between the two categories of piping, Subcommittee I expanded Fig. PG-58.3.1 to show each category. A comparable figure appears in the front of B31.1. Within B31.1, still another new category of piping then had to be established, called **nonboiler external piping**, i.e., the piping beyond the scope of Section I that is connected to the boiler external piping. Figure PG-58.3.1 is reproduced in Chapter 5.

Transferring coverage of the boiler external piping to B31.1 made available, at a stroke, detailed and comprehensive rules not provided by Section I for such important aspects of piping design as flexibility analysis and hanger design. By retaining complete coverage of boiler proper piping, Section I continued its policy of providing minimal guidance in the design of this piping, which has traditionally been left to the boiler manufacturer.

There have been some negative consequences of the transfer of the design of boiler external piping to B31.1. For one thing, an entirely different committee governs the Power Piping Code. Inquiries about boiler external piping typically are answered by that committee, and occasional differences develop between the B31.1-section committee and Subcommittee I. In the early 1990s this problem was apparently resolved by an action of the Board of Pressure Technology Codes and Standards, which reaffirmed B31.1's primary responsibility for the rules governing the design of boiler external piping. Also, despite the best efforts of both committees, minor differences that exist between the respective rules result in different treatment of piping on the same boiler, depending on whether it is boiler external piping or boiler proper piping (e.g., requirements for postweld heat treatment and radiography differ for the two categories of piping). While both sets of rules are doubtless more than adequate, unwary designers who don't pay sufficient attention to what type of piping they are designing can run into expensive delays should an Authorized Inspector notice at the last minute that the piping doesn't meet the appropriate rules. Further discussion of piping design is found in Chapter 5.

SOME DESIGN DIFFERENCES: SECTION I VERSUS B31.1

As explained in the previous section, design rules for boiler external piping are found in ASME B31.1, Power Piping. The formula for the design of straight pipe under internal pressure in 104.1 of B31.1 looks the same as the design formula for piping, drums, and headers in PG-27.2.2 of Section I, except that the additive term C of Section I is called A in B31.1. In either case (A or C), the intention is to account for material removed in threading, provide some minimum level of mechanical strength, and, if needed, provide an allowance for corrosion or erosion. There are, however, some slight differences between Section I and B31.1 that should be mentioned, since occasionally designers forget that certain piping—e.g., piping from the drum to water-level indicators or vent and drain piping—is boiler external piping for which B31.1 rules apply. Unlike Section I, B31.1 does not mandate a specific 0.065-inch extra thickness for mechanical strength or structural stability of plain end pipes smaller than NPS 4. Rather, B31.1, 102.4.4 recommends an increase in wall thickness where necessary for mechanical strength or, alternatively, requires the use of design methods to reduce or eliminate superimposed mechanical loading. Thus, B31.1 gives the designer a choice: provide some (unspecified) increase in wall thickness or take measures to limit mechanical loads. In either case, the details are left to the judgment of the designer. Also, according to 104.1.2(C) of B31.1, pipe used for service above certain pressures or temperatures must be at least schedule 80 pipe. This paragraph is similar to Note 5 of Section I's PG-27.2.2.

BEND DESIGN

Another difference in the design of piping according to Section I versus B31.1 is in their respective treatment of bends. The B31.1 rule, found in 102.4.5, is that the minimum wall thickness at any point in a bend must not be less than that required for the equivalent straight pipe. Ordinary bending methods usually result in some thinning of the wall on the outside of bends and some thickening on the inside. Consequently, under the B31.1 rule, a bend thinning allowance must be added to the wall of a straight pipe so that the finished bend will meet the required minimum thickness. Table 102.4.5 of B31.1 provides recommendations for minimum thickness prior to bending, as a function of the radius of the pipe bend. If good shop practices are followed, these allowances should result in bends of the required minimum thickness. Section I has no similar rule, and over the years quite a few inquiries have been received asking for guidance on this subject. Before the advent of published Interpretations in January 1977, Subcommittee I responded to questions by a letter known as an "unofficial communication." Since 1977, committee responses have been published as Interpretations, as explained in the Introduction and Appendix II. Section I's design approach to bends, for both pipe and tube, is explained in Interpretation I-83-23, a composite of earlier informal replies. Since the acceptance of pipe or tube bends by the Authorized Inspector as meeting the rules of either Section I or B31.1 is crucial to the timely completion of the boiler, the designer should always be aware of which rules apply. Here is Interpretation I-83-23:

> Question: *What rules does Section I impose with respect to wall thinning and flattening of bends in boiler tubing?*
>
> Reply: *Section I does not provide specific rules directly applicable to wall thinning and flattening of bends in boiler tubing. However, the Subcommittee on Power Boilers intends that the manufacturer shall select a combination of thickness and configuration such that the strength of the bend under internal pressure will at least equal the minimum strength required for the straight tube connected to the bend.*
>
> *The Subcommittee on Power Boilers' position, based on many years of service experience, has been that thinning and flattening of tube bends are permissible within limits established by the manufacturer, based upon proof testing.*

To understand the reason for the difference in design approach used by B31.1 and Section I for bends, remember that Subcommittee I has always recognized satisfactory experience as an acceptable basis for design. When the issue of the wall thickness of tube and pipe bends was raised in the late 1960s, the committee had available a long record of satisfactory experience to justify the design approach traditionally used for bends by the major boiler manufacturers. There are, however, further arguments to support such an approach. First, bends are components with two directions of curvature, having inherently greater strength than straight pipe, which has only one direction of curvature. Second, a pipe bend resembles a portion of a torus. Membrane stresses in a torus happen to be lower on the outside diameter than on the inside, so the change in wall thickness from bending a straight pipe tends to put the material just where it is needed. Third, the bending process usually increases the yield strength of the metal by work hardening. Unless the bend is subsequently annealed, it retains this extra strength in service. When bends are proof tested to failure, they typically fail in the straight portion, not in the bend, showing that the bend is stronger. This is the basis on which Section I accepts them. It is also true that bends in Section I tubes and pipe are more likely to be within the setting of the boiler, where there is less chance that a bend failure would be dangerous to people.

Interchanging Pipe and Tube

As explained above, Section I's two different design formulas for tubes and pipe mandate the addition of various wall thickness allowances. Over the years, a number of designers have realized that using the tube formula to design pipe, or vice versa, or substituting one of these product forms for the other might provide more economical construction. For example, if pipe is available at a cost lower than the equivalent tube, why not use the pipe if it satisfies the tube formula? The next question then is whether it would still be necessary to add the 0.065-inch wall thickness allowance required by the pipe design formula for mechanical stability. Subcommittee I decided that in doing the work of tubes, pipe need only meet the thickness requirements of the tube formula, since tubes are generally within an enclosure and there is less likelihood that the 200-pound worker mentioned earlier would be leaning a ladder against the middle of a 10-foot tube span. Looked at another way, this approach appeals to common sense: if a heat-transfer tube designed by the tube formula requires a certain wall thickness and some pipe size provides the necessary thickness, the pipe has the required strength. The fact that it is a different product form should not matter. Conversely, if tubing were used as pipe, it would require the addition of any wall thickness allowances called for by the piping design formula. One note of caution, however: pipe and tube typically come in different sizes and have different manufacturing tolerances on the wall thickness; these differences must be recognized to assure that the required design thickness is obtained.

Over the years, Subcommittee I has received quite a few inquiries on the subject of pipe versus tube and on which formulas could be used. Having in mind the experience and rationale just described, the committee twice issued an interpretation in response to the following two, related, questions. The first, on the definition of pipe and tube, was issued in 1979, but by an oversight was not published until 1983 as Interpretation I-83-47. The second was issued in 1986 as I-86-25 on the subject of wall thickness of pipe and tube.

> Question: *Do the rules of PG-27.2.2 define tubing and piping, respectively, by size only, or are there other criteria?*
>
> Question: *Under what circumstances do the equations under PG-27.2.1 (for tubing) and PG-27.2.2 (for piping) apply?*

The two replies were essentially the same. This is the 1986 reply:

> Reply: *The formulas in ASME Section I, PG-27.2.1, for tubing, are intended primarily for applications such as boiler tubes, superheater and reheater tubes, or economizer tubes in which groups of such tubular elements are arranged within some enclosure for the purpose of transferring heat to or from the fluid within the tubes. For this heat-transfer function, the tubular elements ordinarily do not need fittings or valves.*
>
> *When the function to be performed is mainly the conveying of a pressurized fluid from one location to another, with little or no intentional heat transfer, the formulas in PG-27.2.2 should be used. The elements commonly used for this purpose are called piping. The standard piping sizes are available with matching sizes of fittings (tees, elbows, etc.) and valves to facilitate the installation of piping systems to direct and control the flow.*

While this reply doesn't explicitly say that tubes and pipe may be designed using either formula, that is its intent. A close reading shows that the choice of formula is not mandatory, that the formulas for tubing are intended primarily (but not exclusively) for such applications as boiler tubes, etc. Similarly, the second paragraph, in explaining what piping does, notes only that the piping formulas should (not shall) be used. In Code language **shall** is construed as mandatory, **should** is nonmandatory.

Another source of confusion that has elicited inquiries to Subcommittee I is the question of where should any thickness allowance be added to pipe (i.e., on the inside surface or the outside surface, or perhaps both). Unless the proper assumptions are made, the two pipe design formulas in PG-27.2.2, one in terms of inside radius, R, and the other in terms of outside diameter, D, will give different results. The advice in Note 3 of PG-27.4 was added in the early 1980s to address this problem, but the confusion persists. At the December 1996 Subcommittee I meeting, in response to still another inquiry on this subject, the committee approved the following question and reply, which were published as I-95-35 in the 1997 Interpretations:

> Question: *The equations of PG-27.2.2 are presented in terms of either outside diameter, D, or inside radius, R. There appears to be a discrepancy between the two formulas. If the equivalent of (2R + 2t) is substituted into the diameter-based formula, the radius-based formula cannot be derived from the results. Are the formulas of PG-27.2.2 based on radius and diameter correct and consistent with each other?*
>
> Reply: *Your attention is directed to the last sentence of the first paragraph of Note 3 in PG-27.2.4. Evaluating the formulas in PG-27.2.2, using either a diameter or radius which has been adjusted by an amount equal to that part of C which is being applied to the outside diameter or inside radius will result in correct and consistent values for maximum allowable working pressure.*

Corrosion/Erosion Allowance

Section I notes in PG-27.4, Note 3 that the design formulas do not include any allowance for corrosion and/or erosion and that additional thickness should be provided where they are expected. Thus, the designer is given the responsibility for deciding whether such allowances are needed and, if so, their amount and location.

U.S. boiler manufacturers do not normally add a corrosion or erosion allowance when calculating the design thickness of boiler pressure parts. Experience has shown such allowances to be unnecessary under normal service conditions. Ordinarily, the boiler water treatment controls oxygen levels, dissolved solids, and pH so that corrosion on the waterside is insignificant. Similarly, fireside corrosion has not generally been a problem, except under unusual circumstances attributable to certain temperature ranges and some particularly corrosive constituents in the fuel. Even in these cases, it is not usually practical to solve the problem by a corrosion allowance. In the unusual circumstance of severe local corrosion or erosion, any extra wall thickness allowance might at best serve to prolong component life for only a short time. Furthermore, extra thickness added to the wall of a heated tube raises the temperature of the outside surface, tending to increase the rate of corrosion. Some other solution to the problem is usually necessary, such as elimination of the corrosive condition, a change in material, or the provision of shielding against fly-ash erosion.

It is interesting to compare similar provisions for corrosion or erosion found in Pressure Vessels, Section VIII of the ASME Code and in the Power Piping Code, B31.1. In UG-25 and UCS-25, Section VIII provides similar but somewhat more specific guidance on corrosion and erosion than Section I. With some exceptions for very thin vessels, allowances for corrosion or erosion are left to the designer's judgment. Section VIII also states that extra material provided for this purpose need not be of the same thickness for all parts of the vessel if different rates of attack are expected for the various parts. Finally, Section VIII states that no additional thickness need be provided when previous experience in like service has shown that corrosion does not occur or is only of a superficial nature.

The Power Piping Code, B31.1, 102.4, requires a corrosion or erosion allowance only when corrosion or erosion is expected and in accordance with the designer's judgment. Thus, the same general approach to corrosion and erosion is seen in Section I, Section VIII, and B31.1.

OPENINGS AND COMPENSATION

When an opening is made in a pressure vessel for an attachment, such as a nozzle, or for a handhole or manhole, or for any other purpose, the vessel is weakened to some extent. Accordingly, in most cases, the Code requires the designer to provide what is known as **reinforcement**, or **compensation**, for the material removed to make the opening. (The terms ''reinforcement'' and ''compensation'' are used interchangeably.) The Section I rules covering openings and compensation have evolved over the years in an incremental fashion, and they are replete with cross-references and exceptions, making them somewhat difficult to follow. Two different design approaches are used, one for so-called individual openings in shells, headers, and heads, and the other for what are called openings in a definite pattern. The latter method is often used when banks of tubes are connected to shells and headers. The basic rules covering individual openings and compensation for them are found in PG-32 through PG-39. When groups of openings form a definite pattern, the metal remaining between adjacent holes is called a ligament. Rules for ligament design are found in PG-52 and PG-53.

The principal difference between the two design methods is in the placement of the required compensation. Under the rules for individual openings, the compensation for the metal removed must be replaced in the region immediately around the opening, which is defined by PG-36 as being within the *limits of metal available for compensation*, or the *limits of compensation*. The compensation for a single opening may be provided in the wall of the vessel, in the wall of the nozzle or other connection, or as a reinforcing pad on the vessel, or in attachment welds, or in any combination of those locations. However, when the ligament design method is used for openings in a definite pattern, all the compensation is provided in the wall of the shell and the design is based on the average strength of the shell, expressed in terms of its so-called efficiency E. **Efficiency** is a measure of the strength of a shell with a series of holes in it compared to the strength of the same shell with no holes. For example, a shell with a single longitudinal row of 2-inch holes on 4-inch centers is said to have an efficiency of 0.50, since at the row of holes, on average only half the wall is available to carry hoop stress.

Describing the two methods as individual openings versus groups of openings is perhaps a little misleading, because the design rules of PG-32 to PG-39 for individual openings also apply to multiple openings near each other, even if they are so close that their limits of compensation overlap. Also, although the ligament design method is at first defined in PG-52 as applying to "groups of openings which form a definite pattern in pressure parts," that definition is immediately extended by PG-53 to "groups of openings which do *not* form a definite pattern in pressure parts." PG-52 covers groups of tubes with either uniform spacing or a repeated pattern of spacing, such as shown in Figs. PG-52.3 and PG-52.4 (e.g., 5¼, 6¾, 5¼, 6¾, etc. inches). PG-53 is intended to cover the ligament design method for groups of tubes whose spacing for some reason departs from the regular repeating pattern permitted by PG-52. PG-53 calls these tubes *unsymmetrically spaced.*

Limits of Compensation

The term limits of compensation, or limits of metal available for compensation, describes a zone around a vessel opening where extra material may be placed as compensation for the metal removed to make the opening. The size of this zone is arbitrary, but it is loosely based on experimental and theoretical studies of stresses near openings and how they attenuate with distance. Moreover, many decades of experience have shown the rules for openings and compensation to be safe and satisfactory. PG-36 and Figs. PG-36 and PG-36.4 explain and illustrate the concepts pretty clearly. For convenience, Fig. PG-36 is reproduced here.

The limits of compensation are defined by a rectangle encompassing the opening. The width of the rectangle, x, is measured along the vessel, a certain distance on either side of the centerline of the opening. The height of the rectangle is the sum of the vessel thickness and two times y, a distance measured radially from the inner and outer surfaces of the vessel. The dimensions x and y vary somewhat, depending on particular circumstances. The following terminology is used to define x and y:

d = Diameter of the finished opening in the plane being considered
t = Thickness of the vessel wall (actually furnished, as opposed to minimum required)
t_n = Nominal thickness of the nozzle wall (Note: See discussion of this definition under Interpretations on Openings, later in this chapter.)

The dimension x is defined as the *greater* of $2d$ or $d + 2(t + t_n)$, but in no case greater than the pitch between openings. This last proviso is necessary to prevent any compensation from being credited twice. For example, suppose the limit of compensation on either side of an opening turns out to be d. If d happens to be greater than half the pitch between two adjacent openings, the limits of compensation overlap. This situation is treated in PG-38 on compensation for multiple openings, which contains the stipulation that no portion of the cross section may be considered as applying to more than one opening.

The dimension y is defined as the *smaller* of $2\frac{1}{2}t$ or $2\frac{1}{2}t_n$. However, in this direction, another factor is also considered in the determination of available compensation: the dimension t_e. This dimension is defined as the thickness of an attached reinforcing pad or height of the largest 60° right triangle supported by the vessel and nozzle outside diameter projected surfaces and lying completely within the area of integral reinforcement. The best way to grasp this latter definition is to study the sketches in Fig. PG-36.4.

PG-36.4 explains what metal within the limits of reinforcement may be considered as compensation. There are three such categories:

1. Metal in the wall of the vessel over and above the thickness required to resist pressure.
2. Metal over and above the thickness required to resist pressure in that part of a nozzle wall extending outside the vessel wall. All metal extending inside the vessel wall may also be included.
3. Metal added continuously around the nozzle as compensation, when it is welded to both the vessel and the nozzle. This could be in the form of a reinforcing pad or a buildup of deposited weld metal. The metal included in attachment welds may also be included as compensation.

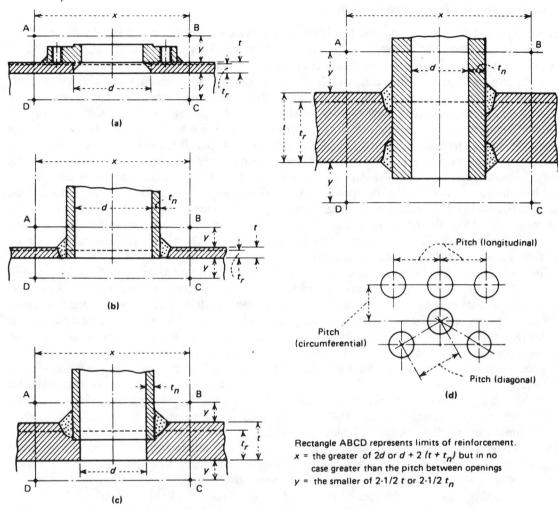

Rectangle ABCD represents limits of reinforcement.
x = the greater of $2d$ or $d + 2$ $(t + t_n)$ but in no
      case greater than the pitch between openings
y = the smaller of 2-1/2 t or 2-1/2 t_n

FIG. PG-36
LIMITS OF REINFORCEMENT FOR TYPICAL OPENINGS
(REPRODUCED FROM ASME SECTION I)

PG-36.4.2 gives formulas for determining the compensation available in the vessel and in the nozzle. Two formulas are given in each case, because the compensation limits may be defined by either of two different formulas, as explained above. The *larger* of the two values is to be used in determining compensation in the vessel wall; the *smaller* of the two values is to be used in determining the compensation available in the nozzle wall. The definition of the diameter of the finished opening used in these formulas is not always the same. Compare, for example, the two inserted nozzles shown in sketches (e-1) and (e-2) of Fig. PW-16.1. When the nozzle is inserted fully into the shell, the diameter, d, is taken as the inside diameter of the nozzle. However, when the nozzle is inserted only partially into the shell, d is taken as the diameter of the unfinished opening in the shell. This change in d affects the width of the limits of compensation and also the required strength of certain load-carrying paths discussed later under the topic Nozzle Attachment Rules. Section I Appendix examples A-65 through A-69 illustrate the application of the rules for calculating compensation.

Compensation for Individual Openings

As just explained, the rules of PG-32 to PG-39 apply to individual openings or several individual openings that are not considered to fall under the rules of PG-52 or PG-53 for the design of ligaments. (Often a group of openings can be designed by either set of rules, as discussed later.)

PG-32 provides a general introduction to the rules for openings and compensation, with advice about the shape of openings and the maximum size permitted. However, it also contains a unique chart, Fig. PG-32, which can be used to determine the maximum size of a single opening for which a shell has sufficient **inherent compensation** so that no further calculations need be made to demonstrate the availability of compensation. The term **single opening**, implicitly defined in PG-32.1.3.1, means one spaced far enough away from any other opening so that the limits of compensation for the two openings do not overlap. Under this definition, a whole series of openings spaced at a little more than twice the diameter of the openings could be treated as a series of single openings. Alternatively, such a series could be designed using the ligament rules of PG-52, provided the diameter of the largest hole in the group doesn't exceed the maximum size permitted by Fig. PG-32. As a general rule, a designer is free to use whichever of the two methods suits the circumstances. (This was affirmed by Interpretation I-92-37 on reinforcement calculations for multiple openings with overlapping limits of compensation.)

The basic concept of inherent compensation or reinforcement is that if a vessel wall is thicker than is required for its design pressure, the extra wall thickness can serve as compensation for a single opening, whose size depends on a number of factors, the most important of which is the amount of extra thickness. The variable along the x-axis of Fig. PG-32 is the product of the vessel diameter and its thickness. Along the y-axis is the maximum size opening for which the vessel is considered to have inherent reinforcement. On the figure are a series of curves called K-lines, plotted for values of K varying from 0.50 to 0.99. The parameter K is defined by the formula $K = PD/1.82St$, in which P is the maximum allowable working pressure, D is the outside diameter of the vessel, t is its actual thickness, and S is the maximum allowable stress. K is thus seen to be a measure of how much extra wall thickness the designer has provided, since $PD/1.82S$ is an approximation of the PG-22.7.2.2 design formula for vessel thickness. K is therefore (approximately) the thickness required divided by the thickness actually provided. A K value of 0.5 indicates that the vessel wall has about twice the thickness needed for pressure alone, and a value of 0.99 would mean that there is very little extra thickness in the wall. Figure PG-32 shows this relationship: as K increases, the size of an opening considered to have inherent compensation diminishes.

Figure PG-32 provides an equation that may be used instead of the figure itself to determine the maximum opening with inherent reinforcement. However, a note on the figure states that the equation may not be used to extend the range shown. The practical effect of this note is to restate the 8-inch diameter maximum opening size already specified on the figure. The K-lines are not plotted for values of Dt less than 10, essentially because they would run together and be difficult to read in that region. However, in Interpretation I-92-65, Subcommittee I advised that it is permissible to use the equation with values of Dt less than 10. (As a matter of fact, an earlier version of the figure, in the mid 1970s, showed the K-lines down to a Dt value of 1.)

Another note on Fig. PG-32 advises that K-lines are shown only for K values of 0.50 and higher, because under the provisions of PG-33, an opening with K less than 0.50 is already fully compensated, as explained above. In other words, if K is 0.50 or below, the wall is twice as thick as it needs to be, and thus the compensation, or area replacement rules, of PG-33 through PG-39 are already satisfied, and no further calculations need be made to demonstrate the availability of compensation. This was reiterated in Interpretation I-92-79, under the heading Openings with Inherent Compensation in Cylindrical Shells.

Figure PG-32 has a long history. Until 1977 the curves on the figure stopped at a Dt value of 200, and the designer was told that for larger values, the opening size given for $Dt = 200$ was to be used, even though it was apparent that the curves were trending upward. In 1973 an inquiry was received by the Boiler & Pressure Vessel Committee noting that current boiler drum designs had Dt values much higher than 200 and asking whether the curves could be extended to take advantage of that upward trend. Since the answer to Code inquiries must always be based on the Code as then written, the answer was no. However, the request seemed quite reasonable, and the task of investigating the feasibility of extending the curves was assigned to the Pressure Vessel Research Council (PVRC) Subcommittee on Reinforced Openings. That committee asked one of its members, Everett C. Rodabaugh, a distinguished researcher who at the time was at the Battelle Laboratories in Columbus, Ohio, to look into the problem. The results of Rodabaugh's study were published in a Batelle report to the PVRC dated July 15, 1975 entitled "Report on Evaluation of Figure PG-32 of ASME Boiler Code, Section I (1974)."

Rodabaugh first asked why Fig. PG-32 was cut off at $Dt = 200$. He tried to find the origin of the figure by reviewing ASME files and questioning senior members of Subcommittee I, but without much success. One member recalled a comparable figure in the 1932 edition of Section VIII and thought it was based on work published in England, probably between 1900 and 1930. With no source document available, Rodabaugh made an independent analysis of the figure, using methods that had recently become available for calculating the limit pressure for a hole in a cylinder and the maximum elastic stress at a hole in a cylindrical shell with internal pressure loading. Using rather conservative assumptions, Rodabaugh concluded that:

> While Figure PG-32 is somewhat inconsistent in the margins of safety, it is on the whole a valid method of limiting the size of openings which have inherent compensation in the cylindrical shell. The evaluation also indicates that extrapolation of Figure PG-32 to Dt greater than 200 is appropriate. Indeed, the extrapolation is such that it forms the most conservative part of Figure PG-32.

Rodabaugh made some interesting comments in his evaluation. One of these was that the family of curves shown in Fig. PG-32 are a representation of the equation $d = 2.75[D(1 - K)]^{1/3}$, an equation shown directly on the curves as published in an old and superseded version of the Ohio Boiler Code. (This equation was therefore included on the figure when it was extended beyond $Dt = 200$ in 1977.) Of the equation he says: "It is immediately apparent that the equation is dimensionally inconsistent; i.e., the dimension of the left-hand side is length, whereas the dimension of the right-hand side is length to the two-thirds power. Accordingly, the equation cannot be justified on any theoretical basis." However, Rodabaugh's evaluations showed the equation to be of adequate accuracy over the range of parameters it covers.

Rodabaugh also points out what he calls another peculiar aspect of Fig. PG-32, which he illustrates by an extreme case, that of a simple hole in a shell, e.g., an inspection opening with some kind of cover to seal the pressure. He assumes a shell with $D = 40$ inches, a required thickness of 2 inches, and an actual thickness of 4 inches. He notes that for this case, PG-32.3.2 would permit an opening up to 20 inches in diameter if the normal compensation rules were used. However, Fig. PG-32 applied to the same shell would indicate that the maximum size opening considered to have inherent reinforcement is only 8 inches. If there is a lesson for the designer here, it is that he or she should be aware of the limitations of the two methods and use the method best suited to any particular situation.

Compensation Required

The basic rules for determining the amount of compensation required for openings in shells and heads are given in PG-33. Although the title of this paragraph doesn't mention headers, this is believed to be an inadvertent omission, since headers are specifically mentioned in the title of PG-32. PG-33 starts with a list of openings it *doesn't* cover, because they are covered elsewhere. The openings not covered are these: flanged-in openings in formed heads, openings in flat heads, openings in a definite pattern covered by the rules for ligaments, openings for connections whose size doesn't exceed NPS 2, and openings that have inherent compensation. That leaves individual connections that fail to meet one of the exemptions or for which other rules are provided.

As explained more fully under the heading Design Philosophy of Nozzle Attachment later in this chapter, a nozzle opening in the wall of a cylinder removes metal that would otherwise be carrying its share of the membrane stress caused by pressure in the cylinder. The highest membrane stress is in the hoop, or circumferential, direction. The area that must be replaced is based on t_r, the required thickness of a seamless shell or head calculated by the design rules of PG-27 or PG-29. Note that the area to be replaced is based only on t_r, the minimum thickness required, even if the actual thickness is greater.

Consider now a plane that passes through the center of an opening and that also contains the longitudinal axis of the vessel. In that plane the area that needs to be replaced is the diameter of the finished opening, d, times the required thickness, t_r, times a correction factor, F. Thus the general equation for the area, A, to be replaced by compensation elsewhere is

$$A = d \times t_r \times F$$

PG-33 introduces the correction factor F to account for the variation in membrane pressure stress that occurs on different planes through the opening, depending on the angle between any plane and the longitudinal axis of the cylinder. The chart in Fig. PG-33 shows this variation. Although no equation is provided for F, the expression $F = \frac{1}{4}\{\text{Cos } 2\theta\} + \frac{3}{4}$ fits the curve shown in PG-33. F is equal to 1.0 when the angle is zero and 0.5 when the angle is 90°. These values reflect the fact that the circumferential stress in a cylinder is approximately twice as much as the axial, or longitudinal, stress. (The expression for F can be derived using Mohr's circle.) However, there is a constraint on the use of the factor F. Its definition stipulates that a value of 1.0 is to be used, except for integrally reinforced openings in cylinders. Although not explicitly defined, ''integrally reinforced'' would appear to mean that the reinforcement—which might be a large solid forging— has to be provided as extra thickness in the wall of the cylinder, by some configuration such as that shown in Fig. PW-16.1(q-1). Only then could reduced values of F be used, which would enable the forging to be elliptical in shape, since a smaller area of compensation is required as the direction of stress changes from hoop to axial. This constraint on the use of F values other than 1.0 does not seem soundly based, since F merely describes the actual variation of stress in the cylinder, which has nothing to do with whether the reinforcing is integral or not. As a matter of fact, example A-69 in the Appendix appears to violate this rule (values of F less than 1.0 are used when the compensation is provided partly by weld metal and partly by extra wall thickness in the nozzles). The Subgroup on Design is presently (1998) considering whether the rules for the use of factor F should be changed. To assure that a certain minimum portion of the reinforcing is provided in the shell, factor F is used also when a drum or shell has a series of holes in a definite pattern. This rule is found in PG-38.4.

Other Openings Exempt from Compensation Calculations

Openings for certain small connections remove so little material that the vessel is presumed to have sufficient inherent compensation and no compensation calculations are required. PG-32.1.3.1 gives the details: First, the diameter of the opening may not exceed ¼ of the inside diameter of the vessel (it usually wouldn't). If that condition is met, welded connections are permitted, up to NPS 2. Threaded, studded, or expanded connections also qualify for the exemption if the diameter of the hole is not greater than NPS 2.

Shape of Openings

PG-32.2 gives advice about the shape of openings and expresses a preference for openings that are round, elliptical, or obround. Other shapes are also permitted, but in such cases PG-32.2.2 advises that all corners should be provided with a ''suitable radius.'' These two paragraphs conclude their advice with the admonition to subject the vessel to a hydrostatic proof test, in accordance with the provisions of PG-100, when doubt exists about the strength or safety of a vessel with odd-shaped openings. Since these paragraphs offer no specific criteria by which to judge suitability of radii or the strength and safety of the vessel, this advice can only serve to warn the designer that he or she must proceed carefully, using past experience, engineering judgment, and if deemed necessary, a hydrostatic proof test.

Size of Openings

In PG-32, Openings and Compensation, is a sometimes overlooked subparagraph on size of openings, PG-32.3. The latter states that properly reinforced openings in cylindrical and spherical shells are not limited as to size, but must comply with the next two subparagraphs. The first of those does impose a size limit of sorts, by serving notice that Section I's area replacement, or compensation, rules are intended to apply to openings not exceeding the following dimensions:

> *For vessels 60 inches in diameter and less:* $\frac{1}{2}$ *the vessel diameter, but not over 20 inches.*
> *For vessels over 60 inches in diameter:* $\frac{1}{3}$ *the vessel diameter, but not over 40 inches.*

Since diameter, *D*, is defined as outside diameter in the pipe formulas of PG-27 and the shell formulas of PG-32, the 60-inch diameter mentioned in these rules is generally construed to mean outside diameter (at least in Section I; Section VIII may have a different tradition). Thus, the largest opening permitted by the above rules may not exceed half the outside diameter for vessels up to 40 inches in outside diameter; then the maximum permitted opening size stays constant at 20 inches until the outside diameter exceeds 60 inches, when the second rule applies, allowing a maximum opening size of only 1/3 the outside diameter. While this sort of design rule does not seem unreasonable, by today's standards it would seem arbitrary and perhaps hard to justify, other than as an example of the ASME Code's conservative, experience-based methods that have long served the industry well.

The next question is what to do if larger openings are needed? Some advice is found in PG-32.3.3, which warns the designer that larger openings should be given special attention and may be reinforced in any suitable manner that complies with the intent of the Code rules. Since that intent is far from clear, it is fortunate that some nonmandatory guidance is then offered:

- Put about two-thirds of the required reinforcing in the inner half of the limits of compensation, adjacent to the opening.
- Give special consideration to the fabrication details used and inspection employed.
- Reduce stress concentration by grinding welds to a concave contour and rounding inside corners of the opening to a generous radius.
- Finally, proof testing may be advisable for large openings approaching the full vessel diameter and openings of unusual shape.

Again, the burden is placed on the designer to base the design on experience, engineering judgment, and if necessary, hydrostatic proof testing.

At first thought, manhole openings in the drum seem to be the only major openings in a boiler, and they usually fall within the half-diameter size limit of PG-32.3.2. However, in the design of large superheater and reheater headers and piping, occasionally a branch connection can be as large, or larger, than the header itself. Designing such a connection becomes a problem in the analysis and design of a tee. Over the years, manufacturers of these large fittings and some boiler manufacturers have developed design methods for them, including experimental testing with strain gages, proof testing, and more recently, three-dimensional finite element stress analysis, confirmed by experiment.

Occasionally, a manufacturer is caught off guard by a demand from a reviewing inspection agency for a proof test to be conducted on construction with an opening that exceeds the half-diameter size limit. Just such a situation occurred involving the connection of a 12-inch saturated steam line from the drum to a 14-inch diameter superheater inlet header. The designers had more than met the recommendations of PG-32.3.3, by using an extra thick header that effectively provided more than two-thirds of the required reinforcing close to the opening. However, the inspector who reviewed the design for the jurisdiction was concerned that this was a large opening, approaching the full-vessel diameter, that might fail in service and insisted that Section I required a full-size proof test of the header with its large branch connection. Since the header had been completed and partially covered with refractory, such a proof test would have entailed either construction of a second prototype for testing or removal of the refractory from the original. Either alternative would have caused a major delay in an already tight delivery schedule and significant additional expense. Under such circumstances, a manufacturer's thorough knowledge of the Code and confidence in a design can help in obtaining prompt approval of the design from the inspection agency.

In the above-noted case, the manufacturer presented a strong case to the inspection agency, making the following points:

- PG-32.3.3 does not mandate proof tests; it merely says such a test may be advisable in extreme cases.
- More reinforcing had been provided near the opening than Section I recommended.
- A three-dimensional finite element analysis of the pipe-to-header connection showed unremarkable stresses.
- The header had passed a hydrostatic test at 1.5 times MAWP.
- The header is intended to operate at a temperature below the creep range, so the hydrostatic test suffices to demonstrate a substantial design margin.
- Several identical headers had been in service in the same jurisdiction for more than 20 years, with no problems.

After some further weeks of delay, during which a higher authority at the authorized inspection agency may have intervened, the design was approved and construction of the boiler proceeded on schedule.

Different Rules for Size of Openings in Section VIII

Section VIII vessels come in a greater variety of sizes and shapes than do Section I vessels, since they serve so many different industries. For some applications, Section VIII vessels do indeed have large openings or openings of unusual shape. Section VIII used to rely on the same type of nonmandatory cautionary language about large openings in UG-36 as Section I uses in PG-32.3.3. However, sometime in the mid-1980s, Section VIII changed its rules so that the recommendation to concentrate the reinforcing adjacent to the opening became mandatory. Later, based on PVRC-sponsored studies of large openings in relatively thin vessels, Section VIII added to Appendix 1–7 in the 1995 Addenda additional rules and restrictions covering large openings. For radial nozzles exceeding the one-half or one-third vessel diameter limit, formulas were provided for calculating membrane and bending stresses near the opening. Membrane stress is limited to its usual design allowable stress S, and the combination of membrane and bending stress is limited to $1.5S$. For openings for radial nozzles whose ratio of nozzle radius to shell radius exceeds 0.7, a design in accordance with U-2(g) is required. That paragraph is the equivalent of the second paragraph of the Section I Preamble, which in effect tells the designer to prove to the Authorized Inspector that a proposed new design is just as safe as otherwise provided by the rules in the Code. (See further discussion of this Preamble paragraph in the Introduction and in Chapter 8.)

It turned out that the new Appendix 1–7 rules were too conservative and were causing a problem for some manufacturers whose vessels had openings that did not meet the new requirements. On the basis of the long satisfactory service experience with those vessels, Code Case 2236 was approved at the September 1996 meeting of the Main Committee to grant relief from the new rules while they were being reconsidered. The case permits large openings to be designed using alternatives to the rules in Appendix 1–7. Acceptable alternatives include successful experience with a specific design or any method that satisfies the requirements of U-2(g).

Flanged-in Openings in Formed Heads

PG-34 is a short paragraph with advice about flanged-in openings in formed heads and could be considerably clearer than it is. First, let it be said that the various head thickness formulas in PG-29 are generally fairly conservative, so that there is likely to be some extra strength in the wall to act as inherent compensation. Moreover, the calculated thickness of heads with flanged-in openings usually must be increased by 15% (see PG-29.3, 29.5, 29.7, and 29.12). Finally, the flange itself, the equivalent of a short nozzle pointing into the head, acts as a reinforcing element. Thus the PG-33 compensation rules do not apply to flanged-in openings in heads that meet all the requirements stipulated, including the requirements of PG-29.13 for knuckle radius and knuckle wall thinning.

There are two other provisions of PG-34 worthy of mention. The first is that compensation for a manhole opening may be provided by a manhole frame or other attachment as an alternative to flanging, in accordance with the compensation requirements of PG-33. More important is that any openings (other than flanged-in openings) in torispherical, ellipsoidal, or hemispherical heads must be provided with reinforcement in accordance with PG-33.

Compensation Required for Openings in Flat Heads

PG-35 provides rules for flat heads with openings whose diameter doesn't exceed half the head diameter or shortest span. If the size of the opening does exceed that limit, the head must be designed in accordance with rules for bolted flange connections, since a flat head with a large opening resembles an inwardly turned flange. Section I doesn't have rules for bolted flange connections, so the designer would have to look elsewhere. Such rules can be found in Appendix 2 of Section VIII, Division 1.

When the size of an opening is less than half of the diameter of a flat head, the compensation required is half of that required for a cylinder. This is due to the fact that a structural member such as a flat head carries load chiefly in bending rather than as a membrane force, although it may carry some membrane force due to the fact that the cylinder to which it is attached expands under pressure and tends to pull the head with it. Also, since the head is in bending, any connection attached to the opening acts as a stiffener. As an alternative to providing calculated compensation, presumably within the normal limits for placing compensation, the designer may modify the design formulas for the thickness of flat heads in PG-31. Different coefficients are used in those formulas, increasing the required head thickness by about 40%, effectively doubling the bending strength of the head.

Strength of Compensation

This subject is discussed under the heading Design Philosophy of Nozzle Attachment later in this chapter.

Compensation for Multiple Openings

As explained above under Limits of Compensation, when adjacent openings are close enough to each other that their limits of compensation overlap, special rules apply to ensure that no portion of the compensation is applied to more than one opening. The condition of overlapping compensation is covered by the rules of PG-38. That paragraph also establishes the minimum spacing permitted. The center-to-center distance between two adjacent openings may not be less than $1\frac{1}{3}$ times their average diameter. The designer must provide a total area of compensation equal to the sum of the areas of compensation required by PG-33 for the separate openings. Since the width of each zone of compensation is restricted by the overlap, the required compensation in the vessel wall is likely to become thicker.

Another rule comes into effect when a vessel has a series of holes whose limits of compensation overlap. In such cases PG-38.4 requires that the net cross-sectional area within the shell wall between any two finished openings must equal at least $0.7F$ times the cross-sectional area obtained by multiplying the center-to-center distance between the openings by the required thickness t_r of a seamless shell. F is the previously mentioned correction factor from Fig. PG-33, which accounts for the variation in stress on different planes through the opening. PG-38.4 states that a cross-sectional area of metal equal to at least 70% of that required for a seamless vessel, over a length equal to the centerline spacing of the openings, must be provided as net cross-sectional area in the shell between the openings. Moreover, any compensation not fused to the wall of the shell may not be credited toward the 70%. This rule is a little hard to grasp without studying the examples shown in Fig. PG-38, which illustrate what is meant by compensation fused to the wall of the shell. Note that the fillet welds shown don't qualify for the 70% rule, because that rule considers only net cross-sectional area within the actual shell wall.

LIGAMENT DESIGN

The metal remaining between adjacent holes in a vessel is called a **ligament**. When groups of openings form a definite pattern, a series of ligaments is created. These ligaments may be designed using rules found in PG-52 and PG-53, so long as the largest hole in the group doesn't exceed the size permitted by the chart in Fig. PG-32. As explained in the introduction to Openings and Compensation, ligaments are designed on the basis of their so-called efficiency. Efficiency is a measure of the strength of a shell with a series of holes in it compared to the strength of the same shell with no holes. If there is a longitudinal row of holes along the length of a vessel, the ligaments between those holes are called **longitudinal ligaments**, and their efficiency is called longitudinal efficiency. Similarly if there is a circumferential row of holes spaced around the shell, the ligaments between those holes are called **circumferential ligaments**, and their efficiency is called circumferential efficiency. Similar terminology applies to a row of holes on a diagonal line, resulting in **diagonal ligaments**. The definition of longitudinal ligament efficiency is given in PG-52.2 for several examples of definite, repeating patterns of openings. In the simplest such example, a longitudinal tube row comprises a series of holes of diameter d at a uniform spacing (longitudinal pitch) equal to p. In that case, the longitudinal efficiency of the ligaments is: $E = (p - d)/p$. Pitch is measured either on the flat plate before rolling, or on the median surface after rolling. One of the difficulties of designing ligaments is that potentially all three types just defined have to be checked to see which is the weakest (has the lowest efficiency) and thus controls the choice of design thickness.

One notable aspect of the efficiency approach is that it establishes an average ligament efficiency for a series of openings that may not be spaced uniformly. Note again that in this design method, all compensation for the openings is placed in the shell wall. As explained at the beginning of this chapter, if a shell has a longitudinal row of 2-inch diameter holes on 4-inch centers, its ligament efficiency, E, is found from the above formula to be 0.50. In other words, at the row of holes, on average only half of the wall is available to carry hoop stress. Although it is nowhere mentioned in PG-52 or PG-53, the required wall thickness of a cylinder with a group of openings forming ligaments is determined from the cylinder design formula of PG-27.2.2, where the ligament efficiency E appears in the denominator of the formula. Because the cylinder formula is not linear with respect to efficiency (due to the extra term in the denominator), an efficiency of 0.5 doesn't quite cause the shell to double in thickness.

The ligament design rules become more complicated when a series of openings in a definite pattern departs from the simple example of a longitudinal row of evenly spaced holes. Figures PG-52.3 and PG-52.4 show examples of nonuniform hole spacing, but each has a repeating pattern defined by p_l, the pitch between corresponding openings in a series of so-called symmetrical groups of openings. Symmetrical groups is a misnomer; they would be better described as repeating patterns of openings. Once the pitch, or length, of the repeating pattern is established, the average ligament efficiency of the group can be determined from the pitch minus all the holes in that pitch, divided by the pitch. PG-52.2.2 gives examples of the calculation of longitudinal efficiency for the patterns shown in Figs. PG-52.3 and PG-52.4. The formula is just $E = (p_l - nd)/p_l$, where d is the diameter of the holes, and n is the number of holes in the repeating pattern length p_l. It may be of interest to know that the examples presently in Section I are unchanged from those in the first edition of 1914, except that the explanation of ligament efficiency was perhaps a little clearer then.

Holes along a diagonal are treated in PG-52.4 and the illustration in Fig. PG-52.5. (Note that this treatment applies only to holes arranged in a triangular or diamond pattern, such as that illustrated, and not to rectangular patterns. For the latter, which are a simpler design problem, the designer needs to worry only about the longitudinal and circumferential ligament efficiencies; the diagonal efficiency does not control.) Diagonal ligaments are designed using Fig. PG-52.1, which provides a means of determining what

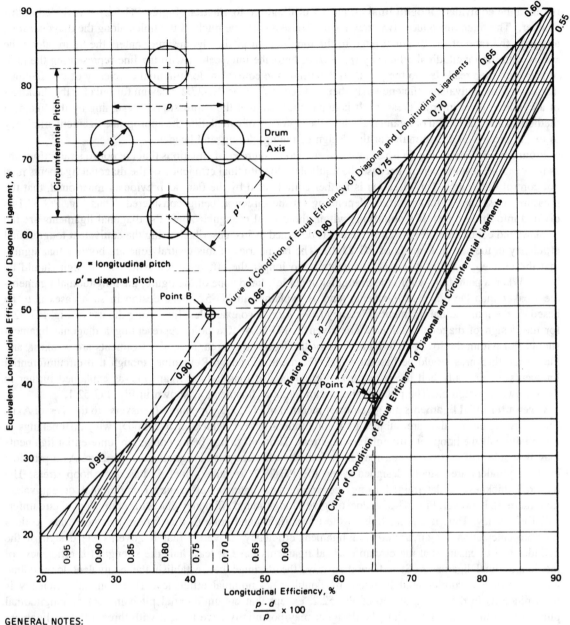

$$\frac{p \cdot d}{p} \times 100$$

GENERAL NOTES:

(a) Equations are provided for the user's option in Notes (b), (c), and (d) below. The use of these equations is permitted for values beyond those provided by Fig. PG-53.1.

(b) Diagonal efficiency, % $= \dfrac{J + 0.25 - (1 - 0.01E_{\text{long.}}) \sqrt{0.75 + J}}{0.00375 + 0.005J}$, where $J = (p'/p_1)^2$.

(c) Curve of condition of equal efficiency of diagonal and circumferential ligaments,

diagonal efficiency, % $= \dfrac{200M + 100 - 2(100 - E_{\text{long.}}) \sqrt{1 + M}}{(1 + M)}$, where $M = [(100 - E_{\text{long.}})/(200 - 0.5E_{\text{long.}})]^2$.

(d) Longitudinal efficiency, % $= E_{\text{long.}} = [(p_1 - d)/p_1]\,100$.

FIG. PG-52.1
DIAGRAM FOR DETERMINING THE EFFICIENCY OF LONGITUDINAL AND DIAGONAL LIGAMENTS BETWEEN OPENINGS IN CYLINDRICAL SHELLS
(REPRODUCED FROM ASME SECTION I)

is called the **equivalent longitudinal efficiency of diagonal ligaments**. (Figure PG-52.1 is included in this chapter.) The figure introduces two new terms. These are p', the pitch of the holes along the diagonal row, and p'/p, the ratio of the diagonal pitch to the longitudinal pitch. The designer enters the figure along the x-axis with the longitudinal efficiency $(p - d)/p$, finds the intersection with the line representing the ratio of p'/p, and then reads left to the y-axis to determine the equivalent longitudinal efficiency of the diagonal ligament. This equivalent efficiency may then be used as E in the PG-27.2.2 design formula for the thickness of a cylinder under internal pressure. It is apparent, then, that the expressions "designing by the ligament method" and "the design of ligaments" describe a process of calculating the appropriate ligament efficiency to use in the PG-27.2.2 formula for the design of piping, drums, and headers.

From a brief study of Fig. PG-52.1, we can envision a number of situations that are not so straightforward as the example just explained, because the equivalent longitudinal efficiency of the diagonal ligaments may not control the design. This situation is further complicated by the fact, as previously mentioned, that the pressure stress in a cylinder varies, depending on the direction being considered. Thus Fig. PG-52.1 is divided into three zones. If a point falls within the central triangular zone, the diagonal ligaments are the weakest and control the design. The cylinder is designed as just described, using the equivalent longitudinal efficiency in the design formula of PG-27.2.2. The boundaries of this central zone are borders that signify that the efficiency of the diagonal ligaments is no longer the efficiency on which the design should be based. When a point falls within the triangular upper left-hand zone of the figure, the longitudinal ligaments are weaker than the diagonal ligaments, and as explained in PG-52.4, the design in such cases is to be based on the longitudinal ligament efficiency $(p - d)/p$. Two examples illustrating the use of Fig. PG-52.1 for the design of diagonal ligaments are given in PG-52.4. If a point representing a diagonal ligament should fall within the lower right-hand zone of the figure, the circumferential ligaments are weakest, and the design thickness would be based on their ligament efficiency. Remember though that circumferential ligaments require only half the strength and efficiency of longitudinal ligaments, as explained further in the following paragraph. The pitch of the circumferential ligaments is shown in Fig. PG-52.1.

Circumferential ligaments are given brief mention in PG-52.3, Openings Transverse to the Vessel Axis. These are ligaments that are subject to a longitudinal stress, which in a cylinder with no openings is essentially half the hoop, or circumferential, stress. Thus, all things being equal, circumferential ligaments may be made half as strong (i.e., may have half the efficiency) as longitudinal ligaments. As a practical matter, cylinders are usually designed using the PG-27.2 formula, which is based on hoop stress. The required thickness is determined using the appropriate ligament efficiency (longitudinal or equivalent longitudinal). However, Fig. PG-52.1 doesn't give a longitudinal efficiency that is the equivalent of circumferential efficiency. Thus if a point falls in the lower right-hand zone of that figure, it is necessary to do a separate calculation for circumferential ligament efficiency, to verify that it governs the design of the cylinder and to ensure that the circumferential ligaments have at least half the strength (or efficiency) of the longitudinal ligaments. Stated another way, the designer can establish the equivalent longitudinal efficiency of a circumferential ligament by doubling its nominal efficiency. That nominal efficiency is determined as in the first example of PG-52.2, except that circumferential pitch instead of longitudinal pitch is used in the formula. Thus the designer may potentially have to deal with three different ligament efficiencies for use in the design formula of PG-27.2: longitudinal efficiency, a longitudinal efficiency that is the equivalent of the diagonal efficiency, and a longitudinal efficiency that is the equivalent of the circumferential efficiency (i.e., is twice the circumferential efficiency). The lowest value controls. However, it would be unusual for circumferential ligament efficiency to govern the design thickness of a cylinder.

PG-52.5 contains some confusing advice about symmetrical groups of tubes arranged in lines parallel to the axis. This advice becomes clearer when turned around and read to say: When tubes or holes are arranged in a drum or shell in symmetrical groups along lines parallel to the axis and the same spacing is used for each group, *the efficiency on which the maximum allowable working pressure is based shall be not less than the efficiency of any of those groups.*

The next complication in the design of ligaments occurs when for some reason the groups of openings do not form a regular repeating pattern. This situation is covered in PG-53 in the description of openings that do not form a definite pattern. These ligaments are designed by using average ligament efficiencies calculated over certain lengths. The vessel design is based primarily on the lowest average longitudinal ligament efficiency over a length equal to the inside diameter of the drum. That lowest efficiency is found by studying the hole pattern and deliberately choosing the position of the diametral length that encompasses the most openings. A secondary design condition is then imposed to prevent a significantly smaller local ligament efficiency within the one diameter length. For a length equal to the inside radius of the vessel and for the position that gives the lowest efficiency, the efficiency may not be less than 80% of that on which the design of the vessel is based.

The last ligament situation (covered by PG-53.3) is that in which rows of holes are placed longitudinally along a drum, but the holes don't line up in the circumferential direction. All of the previously described PG-53 rules would apply, except that diagonal ligaments would be formed between holes in adjacent rows that are offset from each other. The equivalent longitudinal efficiency of these diagonal ligaments is found by using Fig. PG-52.6. (Fig. PG-52.1 and Fig. 52.6 are different representations of the same underlying equations.) From the arrangement of the two diagonal holes shown in the latter figure the designer can easily determine θ, the angle between the diagonal and the longitudinal axis, and the ratio of the diagonal pitch to the diameter of the hole, p'/d. With those values, the equivalent longitudinal efficiency of the diagonal ligament can be determined. An equation in Fig. PG-52.6 is provided as an alternative means of determining that efficiency. The longitudinal efficiency is multiplied by the *longitudinal* pitch of the two holes having the diagonal ligament to get an equivalent (reduced) longitudinal pitch to use in the PG-53 procedures for determining the average ligament efficiency over a length of one inside diameter or one inside radius. Partly because of all these tedious complications, manufacturers generally don't use the irregular patterns that would involve the rules of PG-53. Section IV has evolved a simpler method of ligament design, using the factor F of Fig. PG-33 to account for the variation in pressure stress on diagonal ligaments. Section I may eventually adopt a similar approach.

The rules for ligament design in Section VIII came from Section I originally, and they are almost identical. However, the explanation accompanying the Section VIII rules is a little different and a little more comprehensive. A review of those rules is recommended to gain a better understanding of the rules in Section I.

INTERPRETATIONS ON OPENINGS, COMPENSATION, AND LIGAMENTS

Because this subject can be quite daunting a number of past interpretations may prove helpful. One of these, I-95-06, deals with the question of using individual compensation versus designing by the rules for ligaments and also illustrates what is meant by a single opening:

> Question: *If openings have reinforcement limits which do not overlap, and each opening has sufficient reinforcement within those limits to satisfy the individual requirements, may each opening be considered a single opening for which ligament efficiency calculations are unnecessary?*
> Reply: *Yes.*

Among the PG-36 definitions of metal available for compensation is the nominal thickness of the nozzle wall t_n. Many nozzles are made of pipe, a product form that is usually specified by its nominal wall thickness. However, the governing pipe material specifications permit a manufacturing tolerance on wall thickness of −12.5%, and even though the definition calls for nominal thickness, Subcommittee I has by interpretation insisted that the minimum thickness be used in compensation calculations. That Interpretation is I-86-09, Definition of Nominal Wall Thickness:

> Question: *Does the term* t_n*, as defined in PG-36.4.4, pertain to the nominal wall thickness defined for the schedule rating of the pipe used for the nozzle?*
> Reply: *No. The term* t_n *pertains to the actual thickness of the nozzle wall. See PG-27.4, Note 7.*

Unfortunately, this interpretation puts Section I at odds with Section VIII, which defines t_n as nothing more or less than the nominal wall thickness. Section VIII's reasoning to support this view is that the manufacturing tolerance merely permits an eccentricity between the inner and outer surfaces of the pipe. Consequently, if one side of the pipe is thin, the other side will make up for it; the nominal thickness is actually present and available when the whole circumference of the pipe is considered. The Power Piping Code, B31.1, uses yet another definition for what it calls T_b, the thickness of a branch. This is defined as the actual thickness of the wall (by measurement) or the minimum wall thickness permissible under the purchase specification. While there is probably some justification for the use of each of these definitions, presently an effort is under way to reach agreement on a common definition of nominal thickness for all three codes, in keeping with a general ASME policy that the various Code sections should have consistent rules and definitions, unless there is good reason not to. In 1997 the Subgroup on Design decided to recommend to Subcommittee I that Section I interpret nominal thickness in the same way that Section VIII does, namely, that nominal means nominal. Subcommittee I accepted this recommendation and it is expected that Interpretation I-86-09 will be revised and reissued. Until an agreement on a consistent interpretation of the term "nominal thickness" is reached, it is important that the designer use the definition in the particular code governing the construction involved. Otherwise, difficulties might ensue should an Authorized Inspector discover that the wrong definition had been used.

In 1992 an inquiry was received asking about a nozzle-to-header connection, not specifically addressed in Section I, in which the inside diameter of the nozzle is larger than the opening in the header. The pertinent question from Interpretation I-92-29 is:

Question: *When a nozzle is attached to a header and the inside nozzle diameter is larger than the opening in the header, may the diameter of the opening in the header be taken as* d *in the ligament efficiency calculation of PG-52?*

Reply: *Yes; also please direct your attention to Interpretation I-89-48.*

Interpretation I-89-48, described below, permits the use of as-machined dimensions of the opening in the header to be used for calculating an equivalent opening of uniform diameter.

There have been relatively few inquiries about the rules for ligaments, perhaps because they have been in the Code since there first was a Code, and therefore there is general agreement on how to use them. A few of the inquiries worth mentioning include I-89-47, I-89-48, and I-89-49, and deal with holes that for one reason or other don't have a constant diameter. One of the questions was, essentially, whether the **equivalent average diameter of the hole** could be used in calculating the ligament efficiency. In the case of nozzle-to-header connections, the committee ruled that an average diameter equivalent to the as-machined opening in the header could be used. However, in the case of tube or nozzle connections, such as those shown in Fig. PW-16.1, sketches (y) and (z), the committee ruled that the use of equivalent opening diameter is not permitted, even were credit to be taken only for that portion of the tube or wall actually welded to the shell. Yet, in the determination of whether the limits of compensation overlap, Interpretation I-89-49 does permit the use of an equivalent hole diameter if the variation in diameter is based on the fact that only some portions of the nozzle wall are welded to the vessel wall, as described in PG-38.4 and illustrated in Fig. PG-38. Thus some of these interpretations appear to be inconsistent, and the designer should be forewarned.

In Interpretation I-92-34, an inquirer asked the following question about placing a fully reinforced connection within a ligament.

Question: *For Section I construction, where the shell thickness has been determined by means of a ligament efficiency in accordance with PG-52, is it permissible to locate a fully reinforced connection within the ligament?*

Reply: *Yes; however, t_r used in the reinforcement calculation must be calculated using the ligament efficiency used to determine the shell thickness.*

NOZZLE ATTACHMENT RULES

Design Philosophy of Nozzle Attachment

Rules for the welded attachment of nozzles and other connections to vessels or headers are found in PW-15, PW-16, and PG-37. These rules are intertwined with several related topics: openings, compensation, and load-carrying paths, as further explained below. The rules are directed primarily at the welding of a cylindrical nozzle to a cylindrical vessel (although they also apply to nozzles on vessel heads of various shapes).

Recall from the earlier discussion of membrane stresses that the walls of cylinders such as boiler drums are designed to carry hoop (circumferential) stresses caused by the primary loading of concern, namely, internal design pressure. A nozzle opening in the wall of a cylinder removes metal that would otherwise be carrying its share of this hoop membrane stress. Under Section I design rules (PG-32 through PG-39), the metal removed for the opening usually must be replaced by reinforcement, or compensation, in a region around the opening. That region, described by the term "limits of metal available for compensation" is defined in PW-36. In theory, when compensation is placed around the opening in this manner, the hoop membrane forces are carried around the opening without significantly raising the average local stress. (In the 1930s, this theory was validated by photoelastic stress studies. More recently, three-dimensional finite element analyses have provided further confirmation.)

Several options are available for the placement of the compensation, many of which are illustrated in Fig. PW-16.1. Compensation may be provided by extra wall thickness in the vessel or the nozzle, or by a reinforcing pad around the nozzle, or by deposited weld metal. Any combination of these methods may be used, so long as the compensation is placed within the prescribed limits.

Consider now the case in which the compensation is provided not by extra thickness in the vessel wall, but by extra thickness in the nozzle wall and/or a reinforcing pad on the cylinder. According to the compensation theory, the membrane forces that were formerly carried by the metal removed from the shell must now be carried by compensation that is not integral with the shell. These membrane forces must be transferred from the shell to the compensation, pass around the opening, and then return to the shell. This transfer of load must be accomplished by the weld metal attaching the compensation. Thus, the basis of the nozzle attachment rules, and what is intended by the second sentence of PW-15.1, is the following: *Sufficient weld shall be provided, on either side of the opening, to develop the required strength of the nonintegral compensation through shear or tension in the weld, whichever is applicable.*

Current Section I Rules

In the mid-1970s, Subcommittee I sought some relatively simple equations to help the designer apply the nozzle attachment rules, which until that time had been expressed in words only, as described above. What was needed were equations that would permit the designer to determine the loading to be carried by the nonintegral compensation, and a convenient way to verify that there was sufficient weld to transfer that loading to the compensation. The assistance of the Subcommittee on Design was enlisted, and that committee developed rules for adoption by several book sections. In Section I those rules were placed in PG-37 and PW-15 and illustrated in Fig. PW-16, which is included in this chapter for ready reference. The rules require a two-step approach. The first step is to determine the load, W, to be carried, using the three expressions for W found in PG-37.2. (The lowest value of W determined by those three expressions is the load to be used.) The second step is then to verify that the welds attaching the nozzle are able to carry the load.

To verify the adequacy of these welds the committee introduced the concept of **load-carrying paths**, as shown on the various configurations of nozzles, welds, and reinforcing pads in Fig. PW-16. These paths are intended to represent all possible surfaces where load W could be transferred from a vessel needing reinforcing to its nonintegral reinforcement, such as a welded pad or extra wall thickness in the nozzle. Load W was defined as the *minimum required strength of each load-carrying path*. W may also be thought of as the load required to be carried by any nonintegral compensation. The idea is to check the appropriate

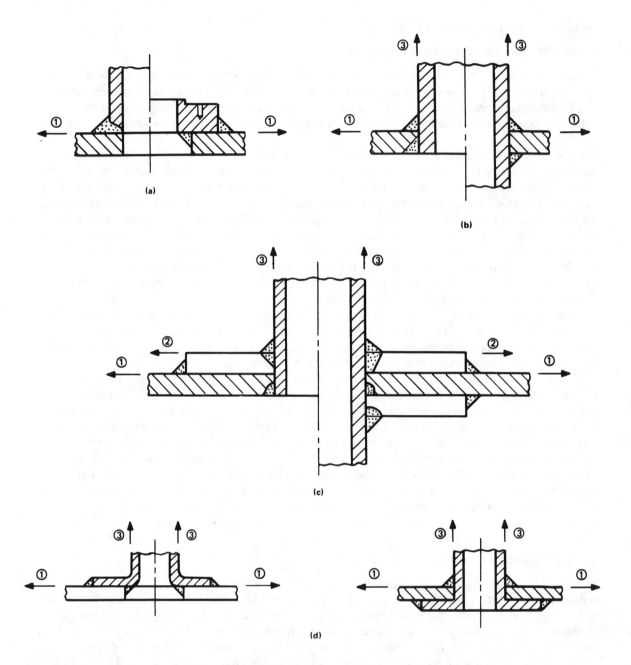

(a)

(b)

(c)

(d)

① Denotes the load-carrying path acting perpendicular to the nozzle centerline about the nozzle at the face of the vessel.
② Denotes the load-carrying path acting perpendicular to the nozzle centerline about the nozzle at the face of the external pad.
③ Denotes the load-carrying path about the nozzle acting parallel to the nozzle centerline.

**FIG. PW-16
LOAD-CARRYING PATHS IN WELDED NOZZLE ATTACHMENTS
(REPRODUCED FROM ASME SECTION I)**

paths to verify that the welds along those paths could carry load W, either in shear or in tension, or a combination of the two, depending on the direction of the load and the type and configuration of the weld. Rules for calculating the strength of individual elements of the path are explained in PW-15.1.1 through PW-15.1.5. Guidance as to which path to check for any particular configuration is provided in PW-15.1.6. Some configurations are exempt because the weld is a full penetration weld in the vessel itself and is thus considered to have ample strength without further verification. Other full penetration welds are not exempt from load-carrying path calculations, except by Code Case 2191, which is discussed below.

To recap, the general approach in designing nozzle attachment welds is first to determine the load to be carried and then to check each of the applicable load-carrying paths to see if the combined strength of the load-carrying elements on that path is adequate to carry that load. This relatively simple design approach seems reasonable enough, but in some cases it does not work so well, as will be explained later.

Discussion of Current Design Rules Determining the Load to Be Carried

The first of the formulas for W in PG-37.2 is seen to be an upper limit on the load to be carried by the compensation; it represents the full area of metal removed for the finished opening multiplied by the allowable stress. That value of W would be appropriate in the case of a vessel whose wall was designed to have the minimum required thickness and thus would have no extra metal available to serve as compensation. In that case, all the compensation would be nonintegral. The second and third formulas for W are supposed to represent the load carried by that portion of the compensation that is not integral with the vessel wall, multiplied by the allowable stress. These latter formulas derive from the two possible limits of reinforcement in the direction of the vessel axis. These two formulas for W can sometimes exaggerate the load to be used, because the full area of metal removed is based on dl, the diameter of the unfinished opening in the vessel prior to the nozzle installation, rather than d, the diameter of the finished opening.

For example, consider sketch (c) of Fig. PW-16.1, which shows a nozzle set into the vessel and attached by a full penetration weld (refer to PG-36, PG-37, and Fig. PG-36 for terminology). In this case, the area of metal removed from the vessel that really ought to be replaced by compensation is dt_r (the product of the inside diameter of the nozzle, d, times the required thickness of the vessel, t_r). However, the second and third formulas for W use the term dlt_r for this product, and dl is seen to be significantly larger than d. The use of the larger diameter of the unfinished opening, dl, was intended for configurations such as the ones shown in sketches (e-2), (k), (s), and (t), where the wall of the inserted nozzle is not attached by full penetration welds and is otherwise not seen to be the full equivalent of the vessel wall that has been removed. However, the use of the larger diameter, dl, overestimates the load to be carried by nonintegral compensation for configurations such as those shown in sketches (c), (g), (h), (m), and (o) of Fig. PW-16.1.

Determining the Strength of Attachment

Having determined load W, the designer's next problem is to determine how and where it is to be carried. Remember that W is supposed to be the load that the nonintegral compensation must carry, and sufficient weld must be provided to transfer that load to whatever compensation will carry it. That compensation may consist of any of the various elements described above.

The designer must now check the applicable load-carrying paths (identified in PW-15.1.6) from among those shown in Fig. PW-16. As explained in PW-15, the strength attributed to the various paths is the sum of three elements: the strength of any fillet welds in shear, the strength of any groove welds in either shear or tension as may be applicable, and the strength of certain pressure parts (such as nozzle necks) in shear. (The concept of including the strength of these pressure parts as part of the total strength of a particular load-carrying path will be discussed further, below.)

The designer is told to base the strength of each of these elements on only half the area in shear or tension. Using half the area is just a convenient device to avoid dividing the areas along the load-carrying path by two to account for the fact that the load must be transferred from the vessel wall to the compensation on both sides of the plane through the center of the opening. An example shown in the appendix makes this clear. Figure A-67 of example A-67 shows a studded flange welded to a vessel opening by means of internal and external fillet welds. Load-carrying path number one of Fig. PW-16.1 is being checked. In this example, one of the elements on this path is the external fillet weld. The strength of this weld is based on half the total area of contact between the bottom leg of the fillet weld and the outer surface of the vessel. That total area is determined by taking the product of the mean circumference of the weld leg times its radial dimension, i.e., the size of the leg.

The allowable stresses for these various elements are given in PW-15.2 as a percentage of the allowable stress of the vessel material. The percentages vary from 49% for fillet weld legs (not throats) in shear, to 74% for groove welds in tension. A note explains that these rather conservative values have been reduced to account for joint efficiency and the effect of combined end and side loading. Unlike Section VIII, Section I does not provide explicit rules to account for the possibility that some portion of the compensation may have a lower allowable stress than the vessel material it is supposed to be replacing. However, the designer is told in PG-37.1 that material of lower strength may be used for compensation only if its area is increased in proportion to the relative strength of the two materials. Note though that no extra credit may be taken for compensation that happens to be stronger than the vessel wall.

Acceptable Nozzle Configurations

It is difficult to generalize about nozzle attachment welds, since there are so many configurations possible. Many are shown in Fig. PW-16.1 (included in this chapter), which is deliberately entitled Some Acceptable Types of Welded Nozzles and Other Connections to Shells, Drums, and Headers. This title clearly implies that other connection details could be used. The use of the phrase ''some acceptable types'' is a Code tradition, and it applies to several illustrations in Sections I and VIII, such as those depicting flat heads and covers (Fig. PG-31), reinforcement dimensions (Fig. PG-36.4), lug weld details (Fig. PW-16.2), and tube attachment details (Fig. PFT-12.1). This tradition is based on the statement in the Foreword to all sections of the Code that the rules are not to be interpreted as limiting in any way the manufacturer's freedom to choose any method of design or any form of construction that conforms to the Code rules. While this statement does imply some flexibility in the rules, as a practical matter, a manufacturer might find it difficult to devise details or configurations different from those already illustrated. Even if the manufacturer were able to develop some new form of construction, its acceptance would have to be obtained from the Authorized Inspector, as noted in the second paragraph of the Preamble.

The extensive variety of nozzle attachment configurations shown in Fig. PW-16.1 includes quite a few where the nozzles are attached by full penetration welds. Examples include sketches (a), (b), (c), (g), (h), (o), (q-1) through (q-4), (u-1), (v-1), and (w-1). If we accept the principle that a weld is as strong as the base metal, it is clear that some of these welds do not represent any kind of potential weak point in the vessel. As explained above, the configurations shown in sketches (q-1) through (q-4) are obviously so strong that they don't need any consideration of load-carrying paths, and PW-15.1.6 exempts them from such consideration. For the rest of these full penetration welds, the designer is told by PW-15.1.6 to check various paths deemed applicable. This requirement has caused some problems, which led the committee to reconsider the rules governing nozzle attachment welds, as explained below.

Inquiries Regarding Nozzle Attachment Strength

Minimum Attachment Welds.   Several unusual aspects of the application of these nozzle attachment rules have been a continuing source of confusion and inquiries to Subcommittee I. The first of these is the fact, duly noted in PG-37.2, that the required weld strength, W, sometimes turns out to be negative, in which case it is to be taken as zero. As explained above, this can happen when there is enough compensation within the vessel wall that none is needed external to it and thus no weld is needed to transfer load from the vessel to the compensation. Therefore, no check of load-carrying paths should be required (see next

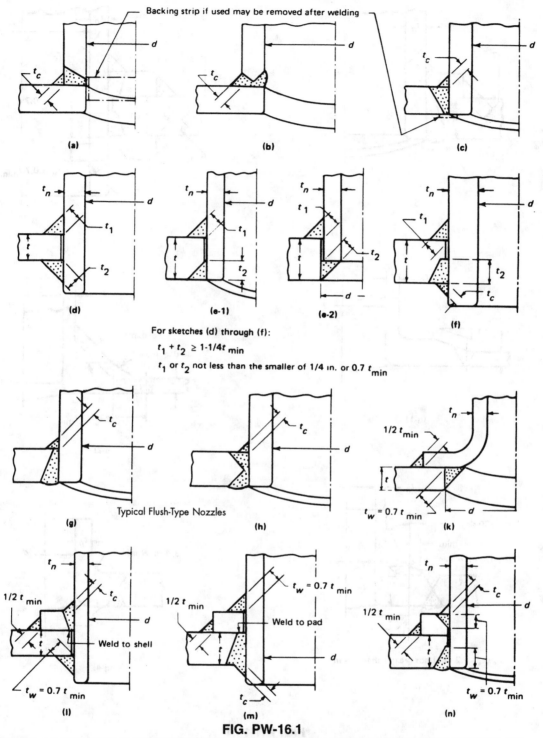

FIG. PW-16.1
SOME ACCEPTABLE TYPES OF WELDED NOZZLES AND OTHER CONNECTIONS TO
SHELLS, DRUMS, AND HEADERS
(REPRODUCED FROM ASME SECTION I)

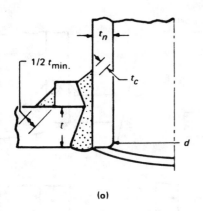

(o)

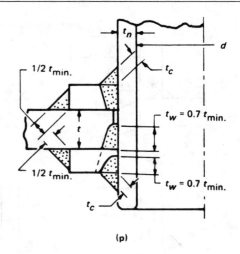

(p)

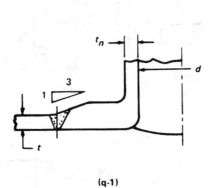

(q-1)

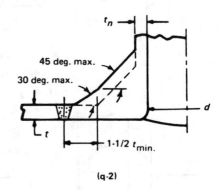

(q-2)

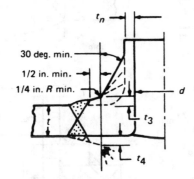

$t_1 + t_4 \leqslant 0.2\, t$ but not greater than 1/4 in.

(q-3)

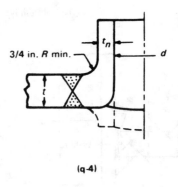

(q-4)

FIG. PW-16.1
SOME ACCEPTABLE TYPES OF WELDED NOZZLES AND OTHER CONNECTIONS TO SHELLS, DRUMS, AND HEADERS (CONT'D)

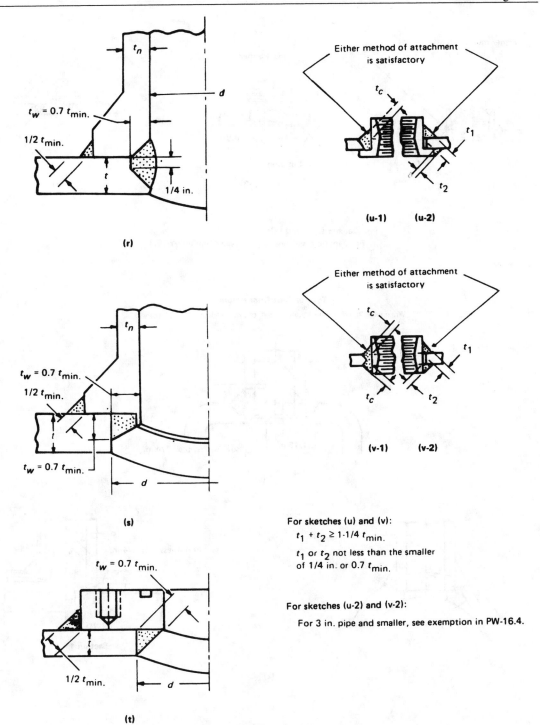

For sketches (u) and (v):

$t_1 + t_2 \geq 1\text{-}1/4\ t_{min.}$

t_1 or t_2 not less than the smaller of 1/4 in. or $0.7\ t_{min.}$

For sketches (u-2) and (v-2):

For 3 in. pipe and smaller, see exemption in PW-16.4.

FIG. PW-16.1
SOME ACCEPTABLE TYPES OF WELDED NOZZLES AND OTHER CONNECTIONS TO SHELLS, DRUMS, AND HEADERS (CONT'D)

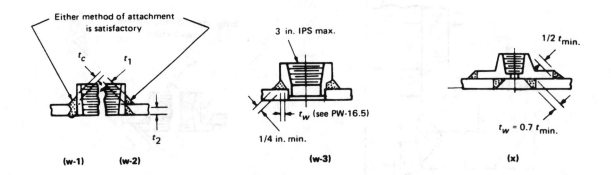

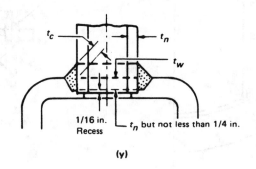

Typical Tube Connections

When used for other than square, round, or oval headers, round off corners.

(y)

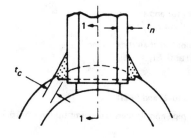

(z)

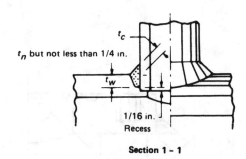

Section 1 – 1

FIG. PW-16.1

SOME ACCEPTABLE TYPES OF WELDED NOZZLES AND OTHER CONNECTIONS TO SHELLS, DRUMS, AND HEADERS (CONT'D)

paragraph). In that case, how much weld is really needed to attach the nozzle? The answer is found in PW-16, which gives minimum requirements for attachment welds and illustrates them in Fig. PW-16.1. Thus if the PG-37.2 rules show that W is small or zero, the minimum requirements of PW-16 apply by default. In the opinion of the authors, these minimum attachment welds are adequate, as has been demonstrated by many decades of safe service using both Section I and Section VIII construction.

Unfortunately, two interpretations by Subcommittee I have confused the issue of whether load-carrying path calculations are required when PG-32 says that no compensation calculations need be made. These interpretations, I-86-78 and I-89-71, stated that load-carrying path calculations were necessary even when no compensation is needed, although the latter interpretation advised that the committee was reconsidering this position. Subsequently, the committee did reexamine the subject and effectively reversed the two interpretations just mentioned. This is explained below, under the headings Problems with Small Nozzles and Consideration of New Rules.

Designers sometimes confuse load-carrying paths with failure paths, thinking of them as potential failure surfaces if very large forces were to be applied to the nozzle. They look at load paths such as path 3 on the right side of sketch (b) in Fig. PW-16 and investigate whether the pressure loading on the nozzle could cause a shear failure through the throats of the fillet welds and pull the nozzle out. This is *not* how these load-carrying path rules are intended to be used; they are only intended to verify whether enough weld has been provided to transfer load from the vessel to any nonintegral compensation. Checking fillet welds to see if they can sustain the pull of a nozzle due to internal pressure is an entirely different problem. In that situation the so-called *nozzle end load* due to internal pressure is carried in shear by the combined throat thickness of the fillet welds, on conical surfaces all around the perimeter of the nozzle. (There is no need to take half the area in this case, because the pressure load is resisted by the full area of the weld.)

A little study of the definitions in PW-16 and the sketches of the minimum attachment welds in Fig. PW-16.1 shows that even in the case of nozzles attached by partial penetration welds, the minimum combined throat thickness of those welds is in most cases equal to 1.25 times the wall thickness of the thinner of the two parts being joined. The allowable axial load on those welds along path 3, based on the combined shear strength of their throat areas, can be shown to be much greater than the maximum axial membrane force developed in the nozzle by internal pressure. Accordingly, minimum attachment welds are more than adequate to resist the tendency of nozzles to blow out of the vessel because of internal pressure.

A word of caution is in order. The design rules for nozzle attachment welds are intended to develop the strength of any nonintegral compensation and to account for pressure loads. They do not necessarily assure adequate strength to resist large external loads such as might occur due to thermal expansion or the weight of attached piping. Footnote 13 to PG-32 cautions that such loads should be considered in unusual designs or under conditions of cyclic loading. This is one of many instances where Section I expects the designer to exercise good engineering judgment. Note also that the nozzle attachment welds in many cases are subject to combined loading: They may have to carry load from the vessel wall to nonintegral compensation and at the same time have to resist the end load from the nozzle due to internal pressure. In all likelihood, they may also be carrying at least some mechanical loads (forces and moments) from the attached piping. The effect of this combined loading may be provided for to some extent by the reduction factors, noted in PW-15.2, which are applied to the allowable stresses. However, if there are substantial external loads on the nozzle, the designer should evaluate the effect of those loads.

Problems with Small Nozzles. In the early 1990s Subcommittee I received a number of inquiries from designers who complained that when they applied the rules of PG-37.2 and PW-15 to small-diameter nozzles, they sometimes found that excessively large attachment welds were required. The problem derives from the fundamental assumption that the load formerly carried by metal removed from a vessel wall must be transfered to external reinforcing if no extra material is available in that wall. An unseen variable implicit in these calculations has a very strong influence on the results and can sometimes make a design look unreasonable. That variable is the ratio of the nozzle diameter to the vessel diameter. If that ratio is fairly large, say about 0.5, the compensation that might be required external to the vessel wall, and the weld necessary to develop its strength, are likely to be quite reasonable for the size of the nozzle. However, consider what happens when the nozzle-to-vessel diameter ratio is small, as in the case of an NPS 2.5 pipe welded to a high-pressure boiler drum whose outside diameter is 66 inches. The minimum thickness of such a drum might be on the order of 5 inches or more. If the vessel wall at the location in question is designed for minimum thickness, the area of nonintegral reinforcing that must be provided is the product of that minimum vessel thickness times the diameter of the hole (~2+ inches) through the vessel wall. In this case, the product is perhaps 10 or 12 square inches. Trying to put that much reinforcing around an NPS 2.5 pipe outside the vessel wall with sufficient weld to develop its strength was resulting in what were perceived to be grossly excessive welds.

This perception was based on the not unreasonable first impression of a designer that a small nozzle has no need of a gigantic weld to attach it. Although that impression is correct so far as it goes, it fails to recognize that the gigantic weld is there not just to attach the small nozzle; its supposed primary function is to transfer load W to external compensation, while at the same time it (the weld itself) may comprise much of that compensation. Another more practical problem in applying these compensation and load-carrying path rules to small nozzles is that it may prove difficult to provide so much compensation outside the vessel within the limits normally permitted for that compensation. Thus some of it would have to be placed within the vessel wall, as integral compensation.

Consideration of New Rules

Subcommittee I decided that the best way to solve the problem of small nozzle connections that appeared to require excessively large welds was to write a Code Case that would provide interim relief while the committee reconsidered the nozzle attachment rules. In writing the Code Case, the committee took advantage of some exemptions implicit in Section I rules and also accepted a precedent, established in Section VIII, of exempting all nozzles attached by full penetration welds from load-carrying path calculations for pressure loading. (The authors think that exemption is fully justified, because in their opinion a full penetration weld is strong enough to transfer whatever load is needed to develop the strength of compensation placed in the nozzle.)

The Section I exemptions, found in PG-32, are based on a concept called **inherent compensation**. Under this concept, a vessel is assumed to have sufficient inherent compensation so that for certain relatively small openings, no additional reinforcing, or compensation, is needed to replace the metal removed to make the opening. The rules of PG-32 exempt certain nozzles from the calculations normally required to show compliance with the rules for openings and compensation. It follows that if no nonintegral compensation is needed in such cases, there is no need to consider the strength of load-carrying paths that are intended to develop the strength of that compensation.

Accordingly, Subcommittee I developed Code Case 2191, which exempts a broad category of welded connections from load-carrying path calculations. The exempt configurations are those attached by full penetration welds, nozzles smaller than NPS 2 (although it is generally understood that these were already exempt under the provisions of PG-32), and openings that meet the ligament rules of PG-52. A number of committee members, including the authors, wanted to extend the exemption to all openings that met the rules of PG-32 for inherent compensation, but others expressed some concern about nozzles not attached by full penetration welds. In the interest of granting prompt relief for many commonly used nozzle configurations, the committee accepted and approved the more restrictive version of the case. For the longer term, Subcommittee I decided to reexamine the rules for designing nozzle attachment welds to see how they might be improved. At this writing (1998), the Subgroup on Design of Subcommittee I is reviewing those rules with the goal of revising any inappropriate provisions, and aligning the rules more closely with those of Section VIII. In the normal course of events, this action would include the incorporation of the provisions of Code Case 2191 into Section I.

Further Discussion of Current Nozzle Attachment Rules

In one form or another the nozzle attachment rules have been in use for many years, and experience has shown them to be safe. The equations in PG-37.2 for W, the minimum required strength of the load-carrying paths, and the instructions in PW-15.1 and Fig. PW-16 for determining the strength of those paths, first appeared in Section I in 1978. They were originally intended to provide a simple design approach that would cover a large variety of nozzle configurations. It is therefore hardly surprising that the rules may have shortcomings in certain circumstances. For example, sometimes the welds used to attach the nozzle and reinforcing pads comprise multiple parallel load-transfer paths, so that it is difficult to determine exactly how the load is shared among the paths. Sketches (n), (o), and (p) of Fig. PW-16.1 are examples of such a situation. Consider sketch (n), which shows a set-through nozzle with a reinforcing pad. Both the nozzle and pad are attached to the vessel by partial penetration welds. Assume for purposes of discussion that there is no extra material in the vessel wall and that all compensation is nonintegral. According to the advice in PW-15.1.6.3 consideration of load paths 1, 2, and 3 of Fig. PW-16(c) is required. Let's examine some of the assumptions implicit in this process, and consider their validity.

Load path 1 is along the surface of the vessel. It passes under the reinforcing pad and through the nozzle. Since there is no integral reinforcing in the vessel, the minimum required strength of the load-carrying path, W, is determined by the first equation in PG-37.2. As previously explained, that equation gives the largest possible value of W, the full strength of the material removed for the opening. A little study of sketch (n) shows that load path 1 does not have to carry all of load W out of the vessel and into the nonintegral compensation; some of that load can obviously be transferred by the partial penetration groove weld on the inside surface of the vessel, connecting the nozzle to the vessel. Unfortunately, there is no simple way to determine just how these load paths share the transfer of load from the vessel to its nonintegral compensation. However, it is clear that in this instance designing load path 1 for the full value of W would appear to be a very conservative approach. The use of the verb "would appear to be" is deliberate, because the rules for determining the strength of the load-carrying path 1 count the strength of the nozzle itself (see PW-15.1.4 and Appendix example A-69). As explained above under Design Philosophy of Nozzle Attachment the designer is supposed to provide "sufficient weld to develop the required strength of the nonintegral compensation through shear or tension in the weld, whichever is applicable." It seems to the authors that the weld alone should provide this required strength, and no credit should be allowed for the strength of the nozzle wall itself. Otherwise, this would be an example of "lifting oneself up by the bootstraps." How can the nozzle help the weld transfer load from the vessel to the nozzle itself?

Fortunately, there are several other factors and rules at work that ultimately result in a safe nozzle attachment, despite the design inconsistency just explained. Once again, the default rules for the minimum size of attachment welds are helpful. These sizes, shown on sketch (n) of PW-16.1, are seen to provide a fairly sturdy attachment. Further assurance of adequacy would result from the mandated consideration of load path 3, the path along the outer surface of the nozzle. Again load W is a conservative upper limit for the load to be transferred to the nozzle, since the nozzle doesn't provide all of the nonintegral compensation (some of it is in the reinforcing pad). Thus the three welds on load path 3 are likely to be overdesigned, lessening any concern that calculations for load paths 1 and 2 may have been an inappropriate way to verify the adequacy of the fillet welds attaching the outer edge of the reinforcing pad to the vessel and its inner edge to the nozzle.

The principle of crediting some part of the nozzle itself as contributing to the strength of a load-carrying path seems somewhat more plausible for path 3 of the flanged connection shown on the right side of sketch (d) of Fig. PW-16. In that situation, the nozzle presumably could not be pulled out of the vessel without shearing off the internal flange. However, the designer is supposed to be checking the ability of the weld to transfer load to nonintegral compensation; he or she is not supposed to be looking for potential failure paths should the nozzle be subjected to extremely large external loads. As explained earlier, that is an entirely different design problem. Accordingly, in the opinion of the authors, verifying the adequacy of load paths 1 and 3 under the present rules is not a justifiable design approach. A better approach would be to count on the strength of the welds only, including both the inside and outside welds, because together they must transfer any load from the vessel to nonintegral compensation provided by extra thickness in the nozzle wall.

Some general conclusions can be drawn from this rather long dissertation on the Section I rules for welding nozzles to vessels. When no question of reinforcing the vessel by nonintegral compensation is involved, Section I (and Section VIII as well) call for certain minimum size attachment welds, which are stronger than the nozzle itself, unless the wall of the nozzle is thicker than 0.75 inches. This is because PW-16.2 doesn't require the minimum throat thickness of nozzle attachment welds to be larger than 0.75 inches. When nonintegral compensation for openings in the vessel is required, the strength of load-carrying path rules of PW-15 and PG-37.2 provide a relatively simple method of assuring that enough weld is provided to develop the strength of that compensation. Because there are so many possible nozzle, reinforcing pad, and weld configurations, the rules sometimes don't work well, but they generally err on the conservative side. So far as the authors are aware, the rules have provided for the safe attachment of nozzles. Subcommittee I is at this time reexamining the rules for designing nozzle attachment welds to see how they might be improved. The goal is to revise any inappropriate provisions, and align the rules more closely with those of Section VIII.

DISHED HEADS

Heads are used as end-closure members for drums and headers. They may vary in shape from flat to hemispherical. Section I uses the term **dished heads** to describe several types of rounded heads for which design formulas are provided in PG-29. These include:

- Heads whose shape is described as **torispherical**, consisting of a segment of a sphere surrounded by a portion of a torus called the knuckle. These heads are designed using the formula of PG-29.1 for a head described as a **blank unstayed dished head**, which is a segment of a sphere, with pressure on the concave side. The corner radius of the dish, also known as the knuckle radius, and the amount of wall thinning permitted during forming of the knuckle are governed by the rules of PG-29.13.
- Heads whose shape is described as **semiellipsoidal**, typically the shape of a two-to-one ellipse, i.e., a head in which half the minor axis (the depth of the head) is one-quarter the diameter of the shell. Design rules for such heads are given in PG-29.7. They must be fabricated to the shape tolerances of PG-29.8 so that their shape closely approximates that of a true ellipse. The thickness of a blank ellipsoidal head (one with no openings) is required to be at least equal to that of a seamless shell of the same diameter, as provided in PG-27.2.2.
- Heads that are full **hemispheres**. These are designed in accordance with PG-29.11 and may be made thinner than any other type of head.
- Heads may be made thinner than called for in the design formulas of PG-29 only if they are stayed in accordance with the rules of PG-29.
- Section I also permits the use of flat heads and covers, in accordance with the rules of PG-31.

It happens that the most efficient shape for containing pressure is the sphere. When a sphere is subject to internal pressure, it develops only membrane stresses, which are essentially uniform across the thickness, so that all of the material in the wall is able to carry the full allowable stress. As the shape of a head departs from a spherical shape, the stresses are no longer just membrane stresses; bending stresses develop also. The greater the departure from a spherical shape, the more bending stresses develop. In a flat head, the stresses are almost all bending stresses. Carrying pressure loads by developing bending stresses is less efficient than carrying them by membrane stresses, because bending stresses vary through the head thickness, changing from tensile stress on one surface to compressive stress on the other. The stress can reach the allowable stress only at the surfaces of the material; the rest of the material is not used at its full potential. This behavior is reflected in the head design formulas. The formula for a full hemispherical head requires the thinnest wall; the other formulas call for much heavier walls. Unfortunately, PG-29 is a poorly organized hodgepodge of rules that are far from clear, particularly in distinguishing one type of head from another. Accordingly, the following commentary is offered on several different aspects of these rules.

One of the first constraints on the shape permitted for a dished head is found in PG-29.2, which describes a head formed by two radii, one for the crown and one for the knuckle. The crown radius, L, is limited to the outside diameter of the flanged portion of the head. (PW-13 requires a short flange on all but hemispherical heads, to facilitate the welding of the head to the cylinder. A full hemisphere doesn't need such an extra flange, because where it meets the cylinder, it is already essentially cylindrical in shape.) This limit on L ensures that the head will not be so shallow that excessively high bending stresses would develop, since with a large radius a head begins to resemble a flat head.

Various types of heads with flanged-in manholes are described in PG-29.3, PG-29.7, and PG-29.12. In each case, the head is required to have extra thickness to compensate for the flanged-in opening, as specified in PG-29.3 and PG-34. Because they are thus provided with what experience has shown to be sufficient compensation for the opening, PG-29.4 exempts these flanged-in openings from having to comply with the PG-33 compensation rules. All other openings in any type of head must be compensated in the usual manner, following the rules of PG-33, unless they are exempt from compensation calculations by the rules of PG-32 for heads having inherent compensation. PG-29.3 also specifies the minimum permitted distance between two manholes placed in the same head (one-quarter of the outside diameter of the head).

The design rules of PG-29 are intended primarily for internal pressure on heads that are concave to the pressure, so that the stresses are tensile and there is no need to be concerned about buckling or instability. Usually, heads are concave to the pressure, but once in a while this may not be the case. PG-29.9 gives some limited advice about such situations: it arbitrarily limits the maximum allowable working pressure to 60% of that for heads of the same dimensions with pressure on the concave side. However, this rule may not be applied to either the semiellipsoidal heads of PG-29.7 or the full hemispherical heads of PG-29.11. Presumably, this is because those heads are relatively thin and would be more likely to buckle under compressive stress. Thus the 60% rule can be applied only to torispherical heads. If a designer wanted to use a hemispherical or ellipsoidal head with pressure on the convex side, he or she would have to do a special buckling analysis to meet the intent of the second paragraph of the Preamble or perhaps do a proof test, as described in PG-18.

Full hemispherical heads are designed by the provisions of PG-29.11. This paragraph has two formulas. The second of the two requires thinner walls for any realistic design conditions, and as a practical matter under the provisions of PG-29.11, it may be used for all heads thicker than ½ inch. This paragraph also has formulas for use at high pressures when the usual formulas (based on so-called thin shell theory) are no longer accurate. This threshold occurs when the required thickness of the head exceeds 35.6% of the inside radius, an unlikely occurrence in the boiler industry.

Also included in PG-29.11 is some last advice about joints in full hemispherical heads. They must meet all requirements for longitudinal joints in cylindrical shells, except that when a head is butt welded to a shell, the middle lines of the plate thicknesses need not be in alignment.

FLAT HEADS

Dished heads are usually more economical than flat heads as end closures on drums, but flat heads and covers are often used on smaller vessels and headers, particularly at lower pressures, when a slightly thicker flat plate costs less than a forged head. Rules for the design of unstayed flat heads, cover plates, and blind flanges are given in PG-31. The rules cover both circular and noncircular heads and covers. Some of the many types of flat heads available are illustrated in Fig. PG-31. (For convenience, Fig. PG-31 is included in this chapter.) As usual when Section I provides illustrations, it notes that other forms of construction that meet the rules are also acceptable. The heads shown in Fig. PG-31 are welded, threaded, bolted, and crimped. The head thickness design formulas vary somewhat, depending on the method of attachment to the shell and the shape of the head. Design advice and limitations are given with the thickness formulas for the various types of heads shown.

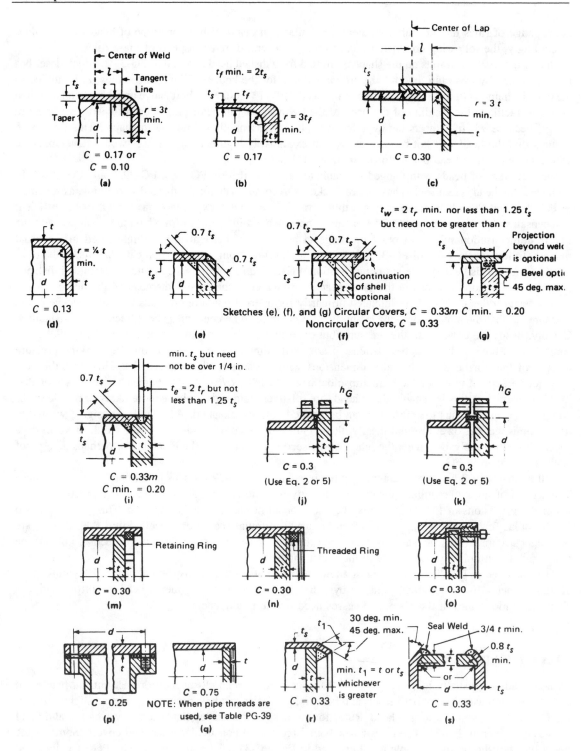

FIG. PG-31
SOME ACCEPTABLE TYPES OF UNSTAYED FLAT HEADS AND COVERS
(REPRODUCED FROM ASME SECTION I)

The design formulas of PG-31 are all based on the bending stress caused by pressure acting on a flat plate. This is an exception to the general rule that Section I design is based on hoop membrane stress in a cylinder. Since flat head design is based on maximum bending stress, the allowable stress for welded flat heads has effectively been increased to 1.5S, as noted in the definition of the factor C that appears in the design formulas (S is the normal allowable membrane stress). The bolts of some bolted covers, such as those shown in sketches (j) and (k) of Fig. PG-31, introduce a bending moment in addition to that caused by pressure. This extra bending moment is accounted for by the second term under the square-root sign in formulas (2) and (5).

Radiography of most of the types of welds used to attach the various heads shown does not produce meaningful results, because the configurations don't present a uniform thickness to the passage of X rays. Consequently, it is difficult to detect and interpret the kind of discontinuity that might indicate a defective weld. Accordingly, footnote 12 exempts welds shown in sketches (e), (f), (g), (i), (r), and (s) from radiography.

INTERPRETATIONS ON HEADS

There are a few Interpretations that clarify the application of PG-29 rules on the design of heads.

Interpretation I-86-19 addressed the thickness of the flue or flange in flanged-in openings:

> Question: *When using ellipsoidal dished heads with flanged-in (flued) manways meeting the requirements of PG-29.3 and PG-29.7, does the minimum thickness apply to both the gage (main portion) and the flanged-in (flued) section?*
> Reply: *Yes.*
> Question: *May the rules of Section VIII Division 1, UG-32, for flued-in openings in ellipsoidal heads be used in lieu of the rules in Section I?*
> Reply: *No.*

This is another case in which Section I and Section VIII rules have drifted apart, and in such cases, the Section I rules apply.

Although there seems to be no good reason why a flued-in opening could not be designed under the rules of PG-33 like any other opening if the designer so chose, Interpretation I-81-26 stated otherwise:

> Question: *May a flued-in (flanged-in) opening be designed per the rules for welded nozzle attachment in PG-33?*
> Reply: *No. The rules of PG-33 apply to welded construction as illustrated in Fig. PW-16.1 and are not intended for use with flued-in (flanged-in) constructions.*

A related Interpretation, I-80-31, issued not long before I-81-26, discussed the question of flanged-in versus compensated openings:

> Question: *Are there valid reasons for penalizing flanged-in openings in hemispherical heads as compared to built-up and welded internal projecting necks, or compared to cut openings through weld overlaid reinforcing metal?*
> Reply: *The added thickness required by the Code for flued-in (flanged-in) openings helps compensate for the material removed in the forming operation. Other designs are permitted but require full replacement of the metal removed.*

This last statement provides a reasonable basis to question the validity of the statement in I-81-26, that the rules of PG-33 may not be applied to flanged-in openings. In other words, if a flued-in opening happened to provide full replacement of the metal removed, there is no theoretical need to require an extra 15% wall thickness. However, the extra 15% rule was imposed as a good rule of thumb on the reasonable assumption that a flued-in opening would not provide full replacement of the metal removed for the opening.

Interpretation I-86-41 dealt with the design of a hemispherical head that wasn't quite a full 180° hemisphere, for reasons of economy and ease of fabrication:

Question: *May the head design formula of PG-29.1 be used for a head which is less than a full hemisphere provided the finished configuration of the vessel encompasses a full hemispherical head and all other rules of Section I are met?*
Reply: *Yes.*

Interpretation I-86-64 explained that PG-29.7 should be used for the design of semiellipsoidal heads, not the formula of PG-29.1, which applies to heads whose central portion forms a segment of a sphere.

Interpretation I-92-88 asked whether the PG-29.7 rules for a semiellipsoidal head with a flanged-in manhole apply to a welded manhole or access opening in a semiellipsoidal head:

Question: *May compensation for welded manholes or access openings in an ellipsoidal head be provided only by increasing the head thickness by 15% in accordance with PG-29.3, and with no consideration of the rules of PG-32 through PG-39?*
Reply: *No.*

There were two reasons for this reply. The flanged-in portion of the manway would not be available as compensation, and it was not clear that the extra head thickness imposed by using the radius specified by PG-29.7 in the formula of PG-29.1 would have been provided.

There have been a few Interpretations on flat heads. One of these, I-86-27, dealt with the proof testing of bolted head flanges:

Question: *For Section I construction, is proof testing permitted for the design of flanges other than those covered by ANSI standards listed in PG-42?*
Reply: *Yes, or as an alternative, the rules of Section VIII, Division 1, Appendix 2, as permitted by the Preamble, may be used.*

Ordinarily, if Section I has rules for the design of some component, proof testing is not permitted, in keeping with a policy described in the Appendix paragraph on proof testing, A-22. However, in this case, the rules are elsewhere, so proof testing is permitted.

The fact that other types of heads than those shown in Fig. PG-31 may be used was affirmed in Interpretation I-86-86:

Question: *Is it permissible to use an alternative to the types of flat heads shown in Fig. PG-31?*
Reply: *Yes, provided the type of flat head is not expressly prohibited by Section I and the concurrence of the Authorized Inspector is obtained.*

Interpretation I-89-46 asked about the omission of one of the welds attaching the head to the vessel:

Question: *May an unstayed flat head similar to the type shown in Fig. PG-31, sketch (i), be used without the fillet weld if equivalent strength is provided?*
Reply: *Yes. Fig. PG-31 is not all-inclusive and only some of the acceptable types that may be used are shown.*

Subcommittee I gave its approval to this particular arrangement, because the remaining weld appeared adequate by itself. The answer would probably have been different for the configurations shown in sketches (e) and (f), because omission of the inside fillet weld would have left only a single fillet weld holding the head.

STAYED SURFACES

The cylindrical, dished, or flat pressure-retaining elements comprising a boiler are usually designed to be self-supporting. However, certain flat and curved surfaces may sometimes be designed more economically if they are designed as stayed surfaces, which receive some or all of their support from an array of tension members called stays, staybolts, or stay tubes.

Stays may be **through-stays**, which carry loads from one stayed surface to another, parallel surface. **Diagonal stays** are sometimes used when through-stays would be much longer or would interfere with access into the boiler through a manway. Diagonal stays between a tubesheet of a firetube boiler and the boiler shell transfer some of the pressure load on the tubesheet to the shell, where it is balanced by axial membrane loads in the shell coming from the other end of the shell. In a firetube boiler, the tubes also act as stays between the tubesheets at either end of the cylindrical shell. These tubesheets can therefore be much thinner than if they were designed by the formulas for flat heads in PG-31. Stay tubes, stays, staybolts, and diagonal stays are illustrated in Figs. A-8, PW-19.4(a), PFT-25, PFT-27, and PFT-32.

Stayed Flat Plates

Section I rules for stayed construction are found in PG-46 through PG-49, PW-19, PWT-12, and PWT-13, and many paragraphs of Part PFT, which covers construction of firetube boilers. The design of flat stayed surfaces rests on a few simple principles. The stays are usually laid out with uniform spacing called the pitch, p, in rows that may be described as horizontal and vertical or radial and circumferential. If the pitch between vertical and horizontal rows is not equal, the larger value is used in determining the required thickness of the stayed surface. The design may be approached in either of two ways. The designer can choose the spacing of the staying members and then determine the plate thickness required to span the distance between stays and the size of stays required to carry the load brought to them by the plate they support. Alternatively, the thickness of the plate may be chosen first and the stay pattern then laid out so that the maximum allowable pitch for that thickness of plate is not exceeded.

Consider first the problem of determining the required thickness of a flat stayed surface, using the formula of PG-46 that establishes thickness as a function of pressure and pitch of the stays. This formula is seen to be nothing more than a bending stress formula of the same type as that used to design flat heads in PG-31, except that the coefficient C reflecting the particular type of construction happens to be in the denominator instead of the numerator and its values have been adjusted accordingly. A bending stress formula in its essence is simply a statement that the bending stress is the bending moment divided by the section modulus of the bending element. The bending moment is usually expressed as the product of some appropriate coefficient k, the unit load (the pressure, P), and the square of the span of the element (the pitch, p, squared), or kPp^2. The section modulus assumed for the plate is that traditionally assumed for a 1-inch wide section of thickness t, namely $t^2/6$. Using this formulation, we can show that the bending stress is $S = 6kPp^2/t^2$. Rearranging to solve for t, we find that $t = p\sqrt{6kP/S}$. This result is seen to be the same as formula (1) in PG-46, if we let $6k = 1/C$. Thus $k = 1/6C$.

The values of C given for different types of stays vary from 2.1 for threaded or welded stays in plates whose thickness is not over 7/16 inch to 3.2 for a threaded stay that effectively strengthens the plate it supports with large, thick washers on the outside and nuts both inside and out. We can find the value of the coefficient k for the first type of stay just described to be $k = 1/6C = 1/6(2.1) = 0.079$. This value approximates the coefficient appropriate for the bending moment in a uniformly loaded square plate of dimension p and thickness t, supported at its corners by stays, and by symmetry, having no rotation at the edges. Thus the stayed surface is in effect being designed as a series of small squares with edges fixed against rotation (but not deflection), supported by stays at each corner. (Timoshenko discusses this problem in his book *Theory of Plates and Shells*, under the heading Bending of Plates Supported by Rows of Equidistant Columns.) The coefficient k for the stronger construction (with large, thick washers) is found to be $1/6(3.2) = 0.052$ and would result in a slightly thinner plate if the pitch were kept constant or a slightly greater permissible pitch if the plate thickness were held constant. However, it happens that threaded stays and large, thick washers are not at all common these days, when it is so much easier (and thus cheaper) to use welded stays.

Stayed Curved Surfaces

Most stayed construction today involves the staying of flat (as opposed to curved) surfaces. Rules for the design of stayed curved surfaces are provided in PFT-23. They involve calculating the maximum allowable working pressure for such surfaces as the sum of the pressure that could be carried by the curved surface alone and the pressure that can be carried by the stays. For most of today's designs, the use of stayed curved surfaces is no longer economical. Many details of stayed construction, such as the locomotive boiler shown in Fig. PFT-23.1, hark back to the early 1900s and are now little used.

Direct Stays

The second problem of stayed construction is the proper design of the stays themselves. Although Section I is somewhat imprecise in its use of the terms ''stays'' and ''staybolts,'' the design method is the same for either. The load carried by a **direct stay** (one perpendicular to the surface it supports) is the product of the pressure times the area supported by the stay. Here again, the basic approach is very simple. By PFT-26, the area supported by a stay is a rectangle whose dimensions are the pitch of the stays in each direction. In other words, each stay supports its tributary area, half way to the next stay in each direction. PFT-26 stipulates that the area occupied by the stay is to be deducted from the area it is assumed to carry, since by considerations of force equilibrium the area occupied by the stay cannot contribute to the net tensile load it carries. The actual cross section of a direct stay or a staybolt is determined by following the rules of PFT-28 and PG-49. The load on the stay is divided by the allowable stress for the material of the stay, found in Table 1A of Section II, Part D, and the result is multiplied by 1.10. The extra 10% provides a degree of conservatism and may compensate for any inaccuracy in the assumption that the load on a given stay is a function only of its so-called tributary area.

Diagonal Stays

PFT-32 advises that for **diagonal stays** the required area of a so-called direct stay is to be scaled up by a length ratio, the ratio of the diagonal length, L, shown in Fig. PFT-32 to its straight projected length, l. This increase reflects the higher load in a stay not perpendicular to the plate it supports. Applying the length ratio is seen to be the equivalent of dividing the area of a direct stay by the cosine of the angle that the stay makes with the shell. As an alternative design method for diagonal stays, PFT-32.2 allows the use of 90% of the regular allowable stress. This alternative is permitted only when the length ratio mentioned above does not exceed 1.15, i.e., when the angle between the diagonal stay and the shell doesn't exceed about 30°. For shallower angles, the reduced allowable stress design method may be used. For steeper angles, the stay area required is that of a direct stay scaled up by the length ratio of PFT-32.1.

Stay Tubes

The design of stay tubes supporting tubesheets in firetube boilers is quite different from the design of watertubes used in other types of boilers. The stay tubes are under external pressure and must be designed for stability under compressive hoop stresses, using the special rules of PFT-50 and PFT-51, which have been borrowed from Section VIII. In addition, of course, these stay tubes are axially loaded, in tension. The designer must verify that the tube is adequate to withstand both of these loadings.

Code Changes

Design of Stay Tubes.   In 1995 Subcommittee I approved changes to the PFT-31 rules for stay tubes used to support the tubesheets in a firetube boiler. The revised rules were issued with the A96 Addenda.

The old PFT-31 formula had treated the tubes in the aggregate, requiring them to have a total metal cross-sectional area equal to the following: the pressure load on the net area of the tubesheet containing the tubes (after deducting the aggregate area of the holes) divided by S, the maximum allowable stress value for the tubes. S was not allowed to exceed 7000 psi. The pitch of the stay tubes was based on the formula in PG-46 using values of C specified in PFT-31.2, which depended on tube arrangement and whether or not the tube ends were shielded from the action of flame or radiant heat from the firebox. The old method had a number of shortcomings: It did not consider the loading on individual stay tubes; only the average tube load was considered. The allowable stress for the tubes was a constant value of 7000 psi regardless of the tube material or temperature. Also, the practice of measuring the pitch of stay tubes between centers was resulting in considerable overdesign of the tubesheet because it ignored the fact that large-diameter tubes occupy a considerable portion of the area being stayed, effectively reducing the actual span of the tubesheet between tubes. Another problem was that there was no clear definition by which the designer could distinguish between a so-called stay tube and a regular tube. (There actually is none; any tube rolled into the two tube sheets of a firetube boiler cannot help but act as a stay tube.)

In the 1996 Addenda, PFT-31 was revised to define when a tube needs to be checked as a stay for its ability to support an axial load as well as external pressure. When tubes are used as stays supporting tubesheets, their required cross section is to be determined by the PG-49 method of taking 1.1 times the load carried, divided by the allowable stress of the material. However, the new rules state that if tubes fall within or on the perimeter of a group of tubes that are spaced at less than twice their average diameter, they are not considered to be stay tubes, and no calculations need be made to check their cross-sectional area. That is because experience shows that tubes spaced so closely can easily carry the pressure load on the area tributary to them, and the span between the outside diameter of the tubes is small enough that no check of bending stress in the tubesheet is needed. Whatever thickness is chosen for the stayed portion of the tubesheet in a location where the tubes or stays are not closely spaced, will suffice where the tubes are closely spaced. The last change was to include in PFT-31.2 a modified version of the tubesheet design formulas of PG-46 that establish tubesheet thickness as a function of pressure and tube or stay pitch. A new term under the square-root sign in these formulas ($\pi d^2/4$) recognizes some beneficial effect of the reduced span between the outside diameter of the tubes compared to the center-to-center pitch. Finally, only two values of the coefficient C are permitted, 2.1 and 2.2, depending on whether the tubesheet is over 7/16 inches thick. The old penalty of 20% on the C value when the ends of the tubes are not shielded from the action of flame or radiant heat was eliminated; it was not considered to be justified.

Design Temperature.   In the firetube boiler the hottest components are the firetubes, the furnace, and possibly the tubesheets, since they have hot gases on one side, and water on the other. Until changed by the 1997 Addenda the PFT-50 rules for the choice of design temperature to be used for the design of firetube boiler components subject to external pressure were not entirely clear. For the furnace and firetubes, both of which are under external pressure, PFT-50 used to call for a design temperature which was ''the greater of that specified in PG-27.4, Note 2, or PFT-17.7 (when applicable).'' The former rule is for tubes under internal pressure, and the latter rule (now deleted) applied to ring-reinforced furnaces under external pressure. Thus it was not clear which rule was applicable to tubes under external pressure. The PG-27.4 rule is that the design temperature shall be not less than the maximum expected mean wall temperature, but in no case less than 700°F for tubes absorbing heat. PFT-17.7 had called for a furnace design temperature 100°F higher than the water temperature. For tubes absorbing heat it is probably conservative to use a minimum design temperature of 700°F. However, such a choice would probably have little economic impact since these tubes are usually selected to be thicker than the design formula requires, just to provide adequate strength for handling and expanding. For furnaces, a design temperature of 100°F above the water temperature is adequate unless the furnace is over 1 in. thick or has a high heat input, in which case the designer should calculate the maximum expected mean wall temperature. The choice of design temperature for other parts of a firetube boiler, those subject to internal pressure, such as the shell and tubesheets, is left to the designer.

Taking all of these factors into consideration, in 1997 the Subgroup on Design revised and clarified the rules for choosing the design temperature for components under external pressure and consolidated them in a single paragraph, PFT-50. That paragraph now permits the designer to calculate or measure the maximum expected mean wall temperature or, as an alternative, to use 700°F. A footnote warns that a higher temperature may be appropriate for furnaces thicker than 1 in. or subject to high heat input.

Areas of Heads to Be Stayed.   Rules for calculating the areas of segments of heads or tubesheets to be stayed have been included in one form or another in every edition of Section I since the first. Strangely enough, however, no instructions are provided as to how this area is to be used, other than to record it on the Manufacturer's Data Report Form. The area might conceivably be used to determine the total number of stays required by dividing the area supported by a single stay into the total area required to be stayed. However, such advice is not provided by Section I and, if it were, would usually result in an excess number of stays because of special rules that increase the allowable pitch of stays adjacent to the shell. In 1993 the Subgroup on Design of Subcommittee I decided to review the rules for calculating the area of segments of heads to be stayed, for both firetube and watertube boilers, with the object of deleting them. Letters were sent out by the secretary of Subcommittee I to about a dozen manufacturers of firetube boilers asking them if they saw any purpose or need in continuing to require the calculation of the areas to be stayed. The responses received confirmed that no one was using the calculated area in designing the stays or the tubesheets. The only apparent use of the area to be stayed was to list it on the P-2 Manufacturer's Data Report Form. Accordingly, in late 1996, the Subgroup on Design voted to delete the rules requiring the calculation of areas to be stayed. These deletions included PWT-13, the first part of PFT-24, Figs. PFT-24.1 and PFT-24.2, notes on Appendix Fig. A-8, sketches (q) and (r), and Appendix Table A-4, Net Areas of Flanged Heads. Revisions were also made to PFT-27.9, the P-2 Manufacturer's Data Report, its instruction guide (A-351), and the Section I index, to remove references to the area to be stayed. At the same time, the Subgroup on Design recommended clarifications of the PFT-50 rules for choosing the design temperature of furnaces and tubes under external pressure. These changes were approved by Subcommittee I and the Main Committee in 1997 and appeared in the 1997 Addenda to Section I.

SUPPORTS AND ATTACHMENT LUGS

This subject is dealt with rather briefly in PG-55. That paragraph gives advice on several aspects of lug design and fabrication. The first requirement is that the lugs or hangers must be properly fitted to the surfaces to which they will be attached. The lugs may be attached by welding that meets the requirements of Part PW, including postweld heat treatment. No radiography is required because the welds are of the type for which radiography doesn't yield meaningful results. Three types of welds are permitted, full and partial penetration welds, as well as fillet welds along the entire periphery of the lug. Acceptable forms are shown in Fig. PW-16.2. Loads on lugs and hangers are to be determined by the designer, taking into account the weight of the component and its contents. Advice is provided on how to calculate the allowable load on fillet welds, but not on so-called full penetration welds. Guidance on the strength of the latter may be found in PW-15.2. Finally, materials for lugs are not restricted to the Section I pressure part materials listed in Section II, Part D; any structural material of weldable quality will do. A material of weldable quality is one whose carbon content is less than 0.35% and which has been proven weldable by the weld procedure qualification requirements of Part PW.

LOADING ON STRUCTURAL ATTACHMENTS

In many boilers the tubes are supported in some fashion or other by lugs welded to them. The loading from the lugs causes bending stresses in the tube walls that add to the stresses already present due to internal pressure. Some means is needed to limit the combined stress due to pressure and mechanical loading to a safe value. To help designers arrive at a safe limit, for decades Section I provided Fig. PW-43.1, Chart for Determination of Allowable Loading on Structural Attachments to Tubes.

The chart had a series of curves for a limited number of tube sizes for use when the tube lug was pulling on the tube and a single curve for use when the lug was compressing the tube. Along the x-axis of the chart was a variable, the outside diameter of the tube divided by the square of the wall thickness. The maximum allowable loading in pounds per inch of lug length was given along the y-axis. The designer entered the chart with the appropriate value of OD/t^2, moved vertically to the tube size being checked, and then horizontally to the y-axis to determine the allowable load per inch of lug. (The lug is welded along the tube in the axial direction, as shown in Fig. PW-43.2.) Several examples were provided in Appendix A-71 through A-74, illustrating how to determine the allowable lug load for several loading conditions: tension, bending, combined compression plus bending, and loading on a tube bend.

One drawback of the chart in Fig. PW-43.1 was the limited number of tube sizes that it covered; another was the fact that no allowance was made for tube materials of different allowable stress. A manufacturer submitted an inquiry asking whether the chart could be expanded to include more tube sizes, and perhaps include the effect of different allowable tube stresses. Since the origin of the existing chart could not be ascertained, the task of developing an expanded version of it was turned over to the Subcommittee on Design. A major function of that subcommittee is to assist other subcommittees in the development of special analysis or design methods.

The development of the new chart and new design method took a long time, with the results finally appearing in the 1992 Addenda. The new method handles any size tube, any allowable tube stress, and accounts for a new factor, the so-called lug attachment angle (the angle subtended by the tube lug, as shown in Table PW-43.1). Also, any extra tube wall thickness above that needed for pressure alone is now accounted for; it increases the allowable loading on the tube lug. There are only two curves on the new chart, one for compression loading on the tube, the other for tension loading. The designer can read load factor values from the plotted curves or, alternatively, can calculate load factors from equations provided. The equations also extend the range of the plotted curves. The load factor equations are merely curve fits of the underlying ring equations. The x-axis of the new chart in Fig. PW-43.1 is essentially unchanged from the pre-1992 chart, but the variable measured on the y-axis of the chart is now called a load factor, L_f. The maximum allowable unit load in pounds/linear inch of attachment, L_a, is derived from L_f, as explained further below.

Very briefly, the new design approach is based on combining the tube stress due to internal pressure with the local circumferential bending stress caused by the lug load. The tube is modeled as a ring with a concentrated zone of loading acting radially on an arc whose length is equal to the thickness of the lug. The allowable combined stress was based on the design-by-analysis methods of Section VIII, Division 2, considering the combination to be in the category of primary plus secondary stress, but with a lower stress limit than that permitted by Section VIII. The reason there are different curves for tension and compression is that the ring is treated as a curved beam. It happens that when a curved beam is subjected to a bending moment, the bending stresses on the opposite surfaces are not equal and opposite, as they are in a straight beam. They are higher on the inside radius.

A compression load from the lug causes a tensile bending stress on the inside surface of the tube, which adds to the pressure stress, which is also a tensile stress. A tensile load from the lug again causes a higher stress on the inside than the outside, but it is a compressive stress and is not additive to the pressure stress. The governing combination in this case occurs on the outside surface, where the slightly lower bending stress adds to the pressure stress. The fact that the bending stress is slightly lower on the outside explains why the chart shows slightly higher allowable load factors (and thus higher allowable lug loads) for tension loading. Notice that the two curves converge as the D/t^2 ratio approaches 500, because as that ratio increases, the curvature of the tube grows smaller and smaller (i.e., locally it becomes straighter and straighter) so that the curved beam effect that results in higher stresses on the inside surface disappears.

When PW-43 was revised in 1992, the sample problems in Appendix A-71 to A-74 illustrating the determination of allowable loading on tube lugs were revised to show the new method. Somehow, the explanation of the examples as printed was not as clear as the subcommittee had intended, and the examples are a little hard to follow. Accordingly, to serve as a guide to the method, the first of the sample problems, shown in A-71, is reworked and explained as Design Exercise No. 4 in Appendix IV.

DESIGN OF SAFETY VALVE NOZZLES

Section I (and Sections IV and VIII also) are silent on the subject of designing safety valve nozzles to resist the forces that develop when a safety valve discharges. Historically, this is probably due to the fact that the manufacturers of safety valves and the designers of the relatively short nozzles on which they are mounted are well aware of the potentially high reaction forces and moments the safety valves can generate and have designed the valves and nozzles to accommodate those loads.

As previously explained, under Section I's design philosophy, most pressure-containing components are designed on the basis of a design stress that is a membrane stress. Bending stresses are not explicitly covered (even though they are implicit in some of the design formulas for flat heads and stayed surfaces, such as those found in PG-31 and PG-46). Safety valves are usually equipped with a discharge elbow, permitting the discharge to go vertically into an exhaust stack. The reaction force from the discharge causes a large bending moment on the nozzle supporting the safety valve. The designer of the safety valve nozzle thus must deal with a combination of normal pressure loading plus the bending moment and possibly an axial thrust downward on the nozzle. Since Section I provides no design rules for this situation, the designer must rely on engineering judgment in determining how best to proceed.

There are several aspects to this design problem. One is the determination of the reaction force on the valve while it is discharging. The safety valve manufacturers know what these forces are, from theory and experiment, and often provide this information in their literature. Alternatively, B31.1, Power Piping Code, has an Appendix II, Nonmandatory Rules for Design of Safety Valve Installations, which is a comprehensive guide containing advice on determination of discharge loads for both open and closed systems. Open systems discharge to atmosphere, often through exhaust stacks (called vent pipes in this appendix). An elbow on the discharge of the safety valve directs the flow into the stack, which is independently mounted on cold steel. The safety valve itself moves with the piping on which it is mounted as that pipe undergoes thermal expansion. In closed systems, not often used for steam, the discharge is piped to some kind of receiving vessel.

Appendix II also cautions that all parts of the piping system (including the exhaust stacks) must be designed not only for the effects of the safety valve discharge forces, but also for thermal expansion, dead weight, earthquake, and other mechanical loads (such as vibration or the effects of more than one valve opening at the same time). Other topics covered include dynamic amplification of reaction forces due to the sudden nature of their application, the various load combinations to be considered, and the allowable stress criteria for each load combination.

Note that the design of safety valve installations on a boiler differs depending on their location. If the safety valves are mounted on the drum or directly on a superheater or reheater header, their nozzles are considered part of the boiler proper, subject only to Section I design rules. On the other hand, if they are mounted on main steam or reheat piping, the nozzles are considered to be boiler external piping, subject to the design rules of B31.1. This turns out to make a significant difference, as will now be explained.

Around 1980 designers of safety valves began to complain that designing safety valve nozzles using the design stresses of Section I (which are membrane allowable stresses) was resulting in very massive nozzles, seemingly all out of proportion to the valves mounted on top of them. Code Case 1876 was developed to overcome the two major causes of the problem: The first was that at elevated temperature, the governing criterion that set the allowable stress at a relatively low value was "two thirds of the average stress to cause creep rupture in 100,000 hours." Since the safety valves can never blow for that length of time, this was considered an inappropriate criterion. The Code Case therefore was based on the design philosophy pioneered in Section III's elevated temperature design rules, which permit some allowable stresses to be based on the length of time they are expected to be applied. Using this philosophy the 1000-hour time-dependent rupture stress was chosen for the case, thereby giving the designer an allowable stress higher than ordinarily permitted. A thousand hours was considered to be 10 times the cumulative time that a safety valve was likely to be blowing during the life of a boiler.

The second cause of the problem was the allowable primary membrane stress criterion of two-thirds of the yield stress. This is a reasonable criterion for primary membrane stress, but it appears inappropriate for a primary bending stress, which is what the safety valve nozzle is subjected to when the valve is discharging. In the NB-3200 design-by-analysis rules for Class I Nuclear Power Plant Components, the allowable primary bending stress limit is 1.5 times the basic allowable membrane stress, i.e., it is $1.5S_m$, in recognition of the margin remaining before full yielding would take place in a rectangular cross section subject to bending. In a cylinder subject to internal pressure, a 50% increase in a membrane stress of two-thirds yield would bring the cylinder to incipient yielding, with all fibers at yield stress. However, the onset of yielding at the extreme fiber of the same cylinder subjected to bending stress is not so serious, because most of the cross section is still well below yield stress. Depending on the shape of the cross section, the design margin between the moment at initial yielding and the fully plastic moment can range from 20% to more than 50%. Moreover, the axial stress in a heavy safety valve nozzle is so low that it was found to have very little effect in reducing the limit moment on the cross section.

Based on these principles Code Case 1876 incorporated allowable stresses for the design of safety valve nozzles significantly higher than those normally used for design of other pressure parts. These stresses are provided for a relatively small number of materials commonly used for safety valve nozzles and may only be used for the combined short-term loading condition of design pressure plus any forces and bending moments due to the discharge of the safety valve. The nozzle and the vessel or header on which it is mounted must also be designed for design pressure alone, using the normal allowable stress given in Section II, Part D. As soon as Case 1876 was adopted, Subcommittee I asked the B31.1 Section Committee to adopt a similar Code Case so that the same relief could be granted for the design of safety valve nozzles mounted on boiler external piping. However, members of that committee expressed a number of concerns, which could never be resolved to their satisfaction, despite an ongoing dialog over more than a dozen years. Consequently, the design of safety valve nozzles on main steam or reheat piping may utilize only the basic allowable membrane stress, increased by 20% for what B31.1 calls occasional loading. Such low design stresses sometimes result in very massive, thick-walled nozzles that present a number of difficulties when it comes to welding the much thinner safety valve to the nozzle.

The members of Subcommittee I's Subgroup on Design who have been involved in the development of Code Case 1876 believe that B31.1's rules are overly conservative. They cite the long record of satisfactory service of safety valves with relatively thin walls and note that those walls are subject to the same bending and axial loads as the nozzles supporting the valves. The lesson here for the designer is to be very careful to use the nozzle design rules applicable to the location—boiler proper or boiler external piping—of the safety valves. For the future, efforts to convince the B31.1 Section Committee to relax their rules for the design of safety valve nozzles continue. The incorporation of Code Case 1876 into the text of Section I would require a special category of short-term allowable stresses in Section II, Part D, so Subcommittee I may simply keep the case as a permanent one, renewing it every three years.

THERMOCOUPLE INSTALLATIONS

In large boilers, thermocouples are often used to measure temperatures of tubes and other pressure parts, and of water, steam, and flue gas. At some locations, such as the fireside of furnace tubes, it is so hot that a thermocouple installation on the surface would not long survive. To measure the metal temperature at or near the outside surface of such tubes, manufacturers resort to the use of a so-called chordal thermocouple, which is installed within the tube wall in a small hole drilled along a chord of the circle forming the outside surface of the tube. By using two such drilled holes, it is possible to install a thermocouple just at or below the hottest surface of a furnace wall tube and have the wires come out on the back face of the tube outside of the furnace. There, the temperature is low enough to permit installation of wiring to take the thermocouple signals to appropriate instrumentation in the control room or elsewhere.

The ends of drilled thermocouple holes on the furnace side of the tubes can be sealed by means of a silver solder designed for high-temperature service. Thermocouple assemblies consisting of a short length of tube with one or more chordal thermocouples installed some four to six inches apart are usually prefabricated for welding into furnace wall panels. A slightly different type of thermocouple installation is used to measure metal temperature near the inside surface of tubes or pipe. In such cases the thermocouples can be placed in blind holes (drilled from the outside, but not all the way through the wall). For measuring fluid temperatures in headers or pipe, thermocouples can be placed in what are called thermowells—short, small-diameter closed-end tubes or pipes projecting radially into the pipe or header. The thermowells can be attached to the pipe by welding or by bolted flanges. In all of the thermocouple installations just described, holes have been drilled in the pressure parts. Since Section I is silent on the subject, designers may wonder whether any reinforcing or compensation is required for the affected components. In 1983 in response to just such a question, Subcommittee I issued Interpretation I-83-60, Blind Drilled Holes for Thermocouple Wires:

> Question: *For Section I construction, is it permissible to drill isolated external holes not exceeding 0.13 in. (3.3 mm) in diameter to a depth of not more than half the wall thickness in fired superheater coils for the attachment of thermocouple lead wires?*
> Reply: *Section I does not provide rules for blind drilled holes; see the second paragraph of the Section I Preamble.*

As will be recalled from the discussion in the Introduction, the second paragraph of the Preamble places on the designer the burden of proof that any proposed new details of design and construction are as safe as those otherwise provided by the rules in the Code.

Let's consider how the designer might justify the various forms of thermocouple installation without providing any compensation. For chordal thermocouple installation, the diameter of the hole is very small. Although in theory this small hole has a slight weakening effect locally, analysis has shown that the hoop stresses due to internal pressure normally carried by the material removed for the thermocouple can easily be carried by adjacent material. If necessary, an Authorized Inspector might be persuaded that the installation was safe by the solution to a closely related but axisymmetric problem, which can be easily solved using finite element analysis. The analytical model used for the axisymmetric problem is a so-called long cylinder under internal pressure, with an annular hole whose diameter matches that of the hole drilled for the chordal thermocouple installation. The hole could be at midthickness of the cylinder wall or close to the outer surface without much change in the results. (Note: A long cylinder is one whose length is such that axisymmetric rotations or deflections at one end have little or no effect on the other end. An approximation for such a length is $l \geq 2.33\sqrt{rt}$, where r is the radius and t is the thickness of the cylinder. In the analytical model just described, the annular hole should be placed at midlength of a cylinder whose length is twice a long cylinder, so that end effects will have died out at midlength.)

If the average membrane stresses resulting from this model exceeded the allowable Section I design stress, they could be justified by showing them to be lower than stresses similarly determined for other configurations permitted by Section I, such as the 2-inch diameter unreinforced opening in a shell permitted by PG-32.1.3.1.2. That configuration can be solved without much difficulty by using a 3D finite element program. Note also the concept of inherent reinforcement, i.e., rules for establishing the maximum size of an opening in a cylindrical shell for which no reinforcement is needed, given in Fig. PG-32. That figure shows, for example, that a fully stressed 12-inch OD by 2-inch wall header is permitted to have an unreinforced opening greater than 1 inch. Thus a blind drilled thermocouple hole of perhaps an eighth of an inch in diameter is surely acceptable. (For further discussion of inherent reinforcement and Fig. PG-32 see Openings and Compensation earlier in this chapter.)

A similar approach could be used to justify a so-called blind drilled hole for a thermocouple. Again, note that the hole diameter is very small. For this situation, we can again invoke the Section I rule in PG-32.1.3.1.2, which permits unreinforced openings for threaded, studded, or expanded connections for which the diameter of the hole in the vessel wall does not exceed NPS 2. By comparison, the thermocouple hole is much smaller. Moreover, it doesn't fully penetrate the wall. Another justification for a blind drilled hole can be found in Code Case 1998, Drilled Holes Not Penetrating through Vessel Wall, which permits blind holes up to 2 inches in diameter without reinforcing if the remaining thickness at the bottom of the hole satisfies the PG-31 design rules for a flat head. As discussed in Chapter 15: Determination of Allowable Stresses, the design philosophy of Section I is based on the use of conservative values of average hoop membrane stress. Bending stresses and the effects of stress concentrations are not explicitly recognized, but the allowable stresses, design formulas, and details of construction permitted limit bending and peak stresses to a safe level. This has been demonstrated by a long, satisfactory record of experience. Thus the effect of small holes drilled for thermocouples is seen to be little different from the effect of other permitted details that increase stress above the average membrane stress. If any further justification is needed, it can be found in the long and successful record of typical thermocouple installation by boiler manufacturers.

CHAPTER

5

PIPING DESIGN

INTRODUCTION

As explained in Chapter 4 under the heading Distinction Between Boiler Proper Piping and Boiler External Piping, Section I's rules for the design of boiler proper piping are somewhat limited. They actually cover the design of pipe for internal or external pressure only. More extensive rules are provided in B31.1, Power Piping Code, applicable to the design of boiler external piping. Those rules cover, in addition to pressure, many other loads that piping might be subject to, such as mechanical loads that may be developed due to thermal expansion and contraction of the piping, impact loading, gravity loads, and seismic loads. Although such comprehensive rules are not provided by Section I for the design of the boiler proper piping, the boiler manufacturers over the years have managed to design this piping so that it serves its purpose in a satisfactory manner, often using design methods of B31.1. Designers of boiler piping, whether boiler proper piping or boiler external piping, are thus faced with a mixed bag of interrelated rules and should be aware of a number of special provisions and potential pitfalls that need to be considered. These are now reviewed.

SHOCK SERVICE FOR FEEDWATER AND BLOWOFF PIPING

There are two types of boiler external piping that are in a type of service called ''shock service''—**feedwater piping** and **blowoff piping**. Both types must be designed for a pressure higher than the MAWP of the boiler itself. The feedwater piping is close to the pump and is potentially subject to pressure surges from it, especially if flow can be restricted by a valve somewhere in the flow path between the feed stop valve and the boiler. The blowoff piping can be subject to shock and vibration when the blowoff valves are opened, causing a flow-induced pressure drop in the blowoff line. Since the boiler water being discharged is at saturation temperature or close to it, the pressure drop can induce boiling within the blowoff line, with considerable vibration and shaking. The discharge of flashing liquid can also generate significant reaction forces on the blowoff pipe. To cover this shock service, the committee thought it prudent to add some safety margin to the design pressure used for feedwater and blowoff piping, thereby forcing the designer to make the walls somewhat heavier than would otherwise be required. The margin chosen was 25% of the MAWP, but not more than 225 psi (this design rule is now found in B31.1, 122.1.3, and 122.1.4). Thus for boilers whose MAWP exceeds 900 psi, the extra margin is a constant 225 psi. When the MAWP is 2900 psi, approximately the highest MAWP of a natural circulation boiler, this represents a margin of only about 8%. However, this high-pressure piping is so strong (to resist the design pressure) that it doesn't need much additional margin for shock service.

Note that the zone of higher design pressure extends only for the length of the feedwater piping or the blowoff piping. For feedwater piping, this is typically from the upstream boundary of the boiler external piping at the feed stop or check valve to the end of the feedwater pipe where it is welded to an inlet nozzle on the economizer or, in the absence of an economizer, to an inlet nozzle on the drum. For blowoff lines, the zone of higher design pressure would extend from the end of the blowoff nozzle (which might be on a header at a low point on the boiler) to and including both valves normally required on a blowoff line. The result is somewhat unusual in that design pressure undergoes a step change at the junction of boiler external piping and the boiler proper, in the absence of any valve. A similar series of step changes in design pressure occurs in the case of forced-flow steam generators with no fixed steam and waterline that are designed for different pressure levels along the path of water-steam flow (see PG-21.2, PG-16.2, and Fig. PG-67.4).

PUMP DISCHARGE HEAD

The pressure developed by a boiler feed pump follows a typical pump curve, a head versus flow relationship in which for any given speed the pump discharge head varies inversely with feedwater flow. The pump designers choose a point along the curve as a design point, which for design pump speed establishes the design flow at the desired design pressure. At greater flow, the pump discharge pressure decreases. At reduced flow, the pressure increases, usually reaching a maximum—called the **pump shutoff head**—at or near zero flow, as would occur with a closed valve in the feedwater line. Pump pressure can be controlled by various means, such as varying pump speed or providing bypass/recirculation lines. Depending on the type of pump and its controls, the pump shutoff head can be considerably higher, perhaps 30% higher, than the pressure developed at full design flow. Such a pressure excursion could represent a significant overstress condition for the piping, valves, or other components between the pump and a closed block valve downstream.

The Power Piping Code, B31.1, deals with this potentially hazardous situation in 122.13, Pump Discharge Piping. That paragraph requires that the piping from the pump up to and including the valve normally used for isolation or flow control be designed for the maximum sustained pressure exerted by the pump and for the highest coincident fluid temperature, as a minimum. This is a logical and reasonable design approach. However, B31.1 relaxes this approach to some extent, by invoking the short-term overstress rules in 102.2.4, Ratings: Allowance for Variation from Normal Operation, which permit certain pressure and temperature excursions above their design values. These pressure or temperature excursions as applied to pump discharge piping are described in 122.13 as occasional and inadvertent, but they are permitted under three conditions, which cover almost anything that could happen:

A. During operation of overpressure relief devices designed to protect the piping system and the attached equipment;

B. During a short period of abnormal operation, such as pump overspeed; or

C. During uncontrolled transients of pressure or temperature.

Note that it may not always be possible to take advantage of the so-called allowance for variation from normal operation; see discussion under the heading Short-Term Overstress in Boiler External Piping later in this chapter.

UNUSUALLY HIGH PIPING DESIGN PRESSURE

There are certain situations in which piping design pressure can significantly exceed the MAWP of the boiler. One which Section I fails to mention explicitly (although B31.1 does address the subject in 122.13, Pump Discharge Piping) is the presence of a stop valve somewhere in the flow path between the feed stop and check valves (which establish the upstream boundary of the boiler) and the boiler drum. If there are no valves between the feed stop valve and the drum, the safety valves on the drum provide overpressure protection for all upstream piping and components between the drum and that stop valve, such as feedwater heaters, or the economizer itself. However, sometimes a stop valve is installed downstream of an economizer, between the economizer and the drum, or an economizer is equipped with shutoff valves and bypass piping. (Such economizers are called **isolable economizers**, since they can be shut off from the boiler.)

With the valves on either side closed, an isolable economizer can become a fired pressure vessel, and Section I, PG-67.2.6 requires installation of a certain minimum safety relief valve capacity to relieve the pressure caused by heating any water trapped within it. That relieving capacity, in pounds/hour, is calculated from the maximum expected heat absorption, in Btu/hour, as determined by the manufacturer, divided by 1000. That capacity is relatively modest compared to the full capacity of the boiler feed pump. Consider now what would happen if the pump were started when the feed stop valve was open but the stop valve downstream of the economizer was inadvertently left closed. The pressure in the feedwater line would quickly approach the pump shutoff head, because the relieving capacity of the economizer safety relief valve is so small compared to full flow. Pump shutoff head can exceed the MAWP of the boiler by 30% or more, depending on how much margin the pump designers have provided. This pressure would prevail from the pump to the stop valve downstream of the economizer. Components affected would include the feed stop and check valves, the feedwater piping, the economizer, and piping downstream of it up to and including the stop valve. Various means of dealing with this overpressure condition are discussed below.

A somewhat similar situation that could lead to an overpressure condition in the feedwater piping is described in PG-58.3.3: if feedwater heaters are installed between the feedwater stop valve and the boiler and those heaters are equipped with isolation and bypass valves, "provisions must be made to prevent the feedwater pressure from exceeding the maximum allowable working pressure of the piping or feedwater heater, whichever is less." In this situation, if the wrong combination of valves were closed, all flow to the boiler could be blocked. The pump could then develop its full shutoff head, which would quite probably exceed the design pressure of both the feedwater piping and the feedwater heater. PG-58.3.3 continues with advice on how to solve this problem: "Control and interlock systems are permitted in order to prevent overpressure." Such a method of providing overpressure protection is unusual (the Code in most cases requires pressure relief valves), but it is simply not feasible to provide enough pressure relief valves on the head of a feedwater heater to relieve the full flow of the boiler feed pump, which is somewhat greater than the full steam-generating capacity of the boiler. In fact, typical industry practice is to provide the tube side of feedwater heaters with only a small pressure relief valve on the head, intended to relieve the thermal expansion of trapped water. (Of course, in the absence of any isolation and bypass valves between the feedwater heaters and the drum, the safety valves on the drum protect the heaters.) Thus other means of protecting against overpressure are needed and permitted. An interlock system to prevent overpressure could involve a bypass with valves so arranged that the bypass valve must be open before the valve on the normal path can be closed. A control system might include one or more pressure-sensing controls that can reduce pump output as required, by reducing speed or by opening bypass or recirculation lines. Multiple controls could provide extra reliability. This design approach would also seem appropriate when feedwater heaters equipped with isolation and bypass valves are installed upstream of the feed stop and check valves, and thus outside of Section I's jurisdiction. In such a case, the above-mentioned pump discharge piping rules of 122.13 of B31.1, which permit short-term overstress conditions, could reasonably be applied. The basis for these rules is explained in the following section.

SHORT-TERM OVERSTRESS IN BOILER EXTERNAL PIPING

An unusual and now somewhat confusing aspect of piping design is a design approach now covered only in B31.1, 102.2.4, Ratings: Allowance for Variation From Normal Operation. This approach was formerly found in Section I also, but was deleted in 1972, when the B31.1 committee was given responsibility for the construction rules governing materials, design, fabrication, installation, and testing of boiler external piping. The B31.1 committee subsequently modified that design approach, but it can sometimes still be used to overcome the pump overpressure problem just described. The essence of this approach is to permit short-term overstress in the piping system caused by limited excursions in pressure and/or temperature above their nominal design values, i.e., above the maximum design capability of the pipe. The Section I version of this approach last appeared in PG-58.1.1.1 of the 1971 edition, as follows:

> *58.1.1.1 Variation in pressure and temperature of piping systems shall be considered safe for operation if the maximum sustained pressure and temperature on any part does not exceed the design pressure and temperature under this Section of the Code for all component parts of the system. It is recognized that variations in pressure and temperature inevitably occur and therefore the piping system shall be considered safe for occasional operation for short periods at higher than the design pressure and temperature.*
>
> *Either pressure or temperature or both may exceed the design values if the stress in the pipe wall calculated by the formulas using the maximum expected pressure during the variation does not exceed the allowable S value for the maximum expected temperature during the variation by more than the following allowances for the periods of duration indicated:*
> *58.1.1.1.1 Up to 15 percent increase above the S value during 10 percent of the operating period.*
> *58.1.1.1.2 Up to 20 percent of the S value during 1 percent of operating period.*

Until 1973 these Section I piping rules allowing short-term overstress applied explicitly to steam, feedwater, and blowoff piping, but it's not clear that they couldn't have been applied to any boiler piping. When the rules for design of boiler external piping were transferred to B31.1, the short-term overstress rules (which had long been in B31.1 also) became applicable to boiler external piping only, and not to the boiler proper piping. As explained earlier, boiler external piping is a category that includes feedwater, main steam, vent, drain, blowoff, and chemical feed piping.

Some time in the 1970s, the B31.1 committee became aware that designers were taking advantage of the permitted short-term overstress for long periods of time. In one case, a designer argued that because a piping system had been designed for a 40-year life, a 15% overstress condition was acceptable for 10% of 40 years, or 4 years, and no immediate correction in operating conditions was required. This was never the intent of the committee; the intent was to permit relatively short-term excursions for unusual overload or upset conditions, perhaps for a few hours or even days when power demand was very high, but not for so long that significant creep damage would accumulate if the piping (such as main steam piping) were operating in the creep range.

Accordingly, to close this loophole, or rather to reduce its size, the B31.1 committee added a time limit to these permitted excursions in pressure and temperature. Excursion time limits described vaguely as 10% of the operating period and 1% of the operating period were changed to 10% or 1% of *any 24-hour operating period*. (It happens that the committee at this time, 1998, has given preliminary approval of a proposal to increase this arbitrary 24-hour-based rule to something more reasonable. The proposal would permit a *15% overstress for no more than 8 hours at any one time and no more than 800 hours per year or a 20% overstress for no more than 1 hour at a time and no more than 80 hours per year*. Those times represent a maximum of about 9% and 0.9% of the year, not very different from the 10% and 1% of the previous rule.)

Still another modification of the rule permitting short-term temperature and pressure excursions occurred after some manufacturers of standard pressure parts (such as valves and fittings) realized that the Code was, in effect, saying that it was permissible to exceed the manufacturer's design stresses (inherent in the published pressure and temperature ratings of these standard components) by as much as 20%. The manufacturers objected, arguing with some justification that the short-term overstress rule reduced the manufacturer's design margins and imposed on the manufacturers implied warranties and liabilities they were unwilling to accept. The B31.1 committee was apparently persuaded by these arguments and modified the rules again. Thus the 1997 B31.1 version of the old Section I paragraph reads as follows:

> *102.2.4 Ratings: Allowance for Variation From Normal Operation. The maximum internal pressure and temperature allowed shall include considerations for occasional loads and transients of pressure and temperature.*
>
> *It is recognized that variations in pressure and temperature inevitably occur, and therefore the piping system, except as limited by component standards referred to in Paragraph 102.2.1*

or by manufacturers of components referred to in Paragraph 102.2.2, shall be considered safe for occasional short operating periods at higher than design pressure or temperature. The calculated stress resulting from such a variation in pressure and/or temperature may exceed the maximum allowable stress from Appendix A by:

(A) 15% if the event duration occurs less than 10% of any 24-hr operating period; or

(B) 20% if the event duration occurs less than 1% of any 24-hr operating period.

Let's consider the two categories of component standards covered by the added clause. The first covers those piping component standards listed in Table 126.1 that do provide pressure and temperature ratings, and the second covers those listed in Table 126.1 that don't provide pressure and temperature ratings. (Ratings of the latter, e.g., buttwelding fittings such as tees and elbows covered by ASME/ANSI B16.9, are based on the rating of matching seamless pipe of the same material.) In other words, the inserted clause is intended to cover all standard piping components of Table 126.1. It is a warning that either the standards themselves or the manufacturers of those components whose standards don't have specific pressure and temperature ratings may limit or prohibit the use of the short-term overstress provisions of 102.2.4. In effect B31.1 is saying don't exceed the ratings of the piping component standards.

Two very important and frequently used component standards among those found in Table 126.1 are ASME/ANSI B16.5, Pipe Flanges and Flanged Fittings, and ASME/ANSI B16.34, Valves—Flanged, Threaded, and Welding End. They are among several standards that provide pressure and temperature ratings. However, of all the component standards in Table 126.1 that provide pressure/temperature ratings, only one, the B16.34 valve standard in 2.5 Variances, addresses and limits any short-term overstress conditions caused by exceeding the published ratings. According to 2.5.1, the operating pressure may exceed the rated pressure of a valve by no more than 10% when pressure-relieving devices are in operation. It then notes that "such conditions are necessarily of limited duration" and warns that "Pressure excursions in excess of the aforementioned are solely the responsibility of the user." The next paragraph, 2.5.2 Other Variances, makes it clear that the valve manufacturers don't approve of any other pressure or temperature excursions above the established ratings, by making this statement: "Subjecting a valve to operating variances (transients) in excess of its rating is solely the responsibility of the user."

The component standards that don't provide pressure and temperature ratings (but rather are designed to be as strong as seamless pipe of the same size and material) do not address the question of variances in the pressure–temperature ratings. This leaves unresolved the question of what limits, if any, the manufacturers of such components might impose on short-term excursions. However, it would appear from the specific language of B16.34 that the user who subjects valves or fittings to pressure or temperature excursions above those (if any) permitted by the governing standards assumes the burden and responsibility of ensuring that the magnitude and duration of any excursions will not cause a significant reduction in safety or an excessive accumulation of creep damage (for components operating at elevated temperature).

Since virtually all boiler external piping includes standard valves and fittings, it would at first glance seem that the changes to B31.1, 102.2.4 since 1973 have effectively blocked its use to permit any short-term excursions causing 15% or 20% overstress. However, this is not necessarily the case; the standard valves and fittings are chosen by pressure class, as explained in Chapter 11. These pressure classes (e.g., classes 600, 900, 1500, and 2500) are typically available in very large design pressure increments. For example (see Table 2-1.1 of B16.34, the ASME/ANSI standard covering flanged, threaded, and welding end valves, included in Appendix IV, Design Exercises), at 800°F, each successive pressure class has a pressure rating from 30% to 60% higher than the next lower rating. It is likely, therefore, that valves with the lowest pressure class capable of sustaining any given design conditions may have a considerable design margin (either pressure, temperature, or both) above those conditions. Thus a short-term excursion of pressure or temperature resulting in a permitted 15% or 20% overstress of the piping itself might not exceed the pressure/temperature rating of the particular pressure class of the valve(s) chosen for the boiler external piping.

As explained in the Introduction, Section I rules apply to the construction of new boilers. The same is true for the piping design rules of B31.1, that is, they too are new-construction rules. Thus it seems odd that either code should contain any provisions that deal with the actual operation of the boiler, even to the point of saying how much overstress is permitted during short-term transients. However, the committees who develop the new-construction rules over the years have occasionally crossed the line that separates new construction from ongoing operation. We can speculate that these exceptions were justified as good advice based on experience or on a desire to warn about potentially unsafe conditions. Moreover, the rule writers of yesteryear worked in a less formal time, when apparently fewer objections were raised about including an occasional rule dealing with boiler operation. A Section I or Section VIII rule permitting short-term excursions above design conditions would probably not be approved today, on the grounds that such a rule has nothing to do with new construction, and also that it would be virtually unenforceable, since it is often impractical to obtain and monitor a continuous record of the operating temperature and pressure for a large number of pressure parts. Moreover, Section I and Section VIII presently provide no role for the Authorized Inspector once construction of the boiler or pressure vessel is completed, so there would be no third-party assurance that any pressure/temperature excursions stayed within the limits prescribed. Yet there are some B31.1 committee members who justifiably believe that the designer of a new piping system should give some recognition to the likelihood that it will inevitably be subject to temperature excursions and that the limits prescribed are useful in preventing premature creep damage.

HYDROSTATIC TEST OF BOILER EXTERNAL PIPING

The boiler external piping is given a hydrostatic test with the boiler, in accordance with PG-99 of Section I and 137.3 of B31.1, Power Piping, which simply invokes the rules of PG-99 and repeats the Section I requirement that an Authorized Inspector be present during the test. Figure PG-58.3.1 of Section I illustrates the Code jurisdictional limits for piping. That figure shows that the ordinary arrangement of the boiler feed shutoff valve (also called the feed stop valve) and the check valve is to have the latter upstream of the shutoff valve. However in the case of a single boiler-turbine unit, PG-58.3.3 permits the order of these valves to be reversed, with the shutoff valve located upstream. During hydrostatic testing, that arrangement poses some minor difficulties that can be avoided by appropriate planning. When the feed stop valve is upstream of the check valve, it establishes the upstream boundary of the complete boiler unit. During the hydrostatic test, water is pumped into the system using some convenient connection, usually at a low point on the boiler, such as a drain line from a waterwall header. Since the check valve in the feedwater line will prevent any flow from the boiler toward the upstream feed stop valve, some provision must be made to introduce the water upstream of the check valve, to ensure that all the feedwater piping is tested. This might be accomplished by removing the disk from the check valve during the test or by use of a temporary connection to bypass the check valve.

During the hydrostatic test, the closed feed stop valve defining the upstream boundary of the boiler and all other valves that establish the pressure boundary of the boiler have to be capable of withstanding a hydrostatic pressure of 1½ times the maximum allowable working pressure of the boiler. Occasionally, this presents a problem. These stop valves are designed in accordance with the ASME/ANSI B16.34 standard for valves. That standard's hydrostatic test requirements for valves differ from Section I's requirements. B16.34 provides for two test conditions: one with the valve open, and one with the valve closed. In the open position, B16.34 requires the valve body (or **shell**, as the standard calls it) to be tested at 1½ times its 100°F pressure rating with no visible leakage through the pressure boundary walls. A valve body that can sustain a hydrostatic test at 1½ times its 100°F pressure rating is more than adequate to pass a hydrostatic test of the boiler, since valves on the boiler are designed for pressures at least as high or higher than the MAWP of the boiler, but at temperatures much higher than 100°F. Since the pressure rating of a valve declines with temperature (because the allowable stress declines with temperature), the 100°F pressure ratings of all these valves are higher than their pressure ratings when the boiler is in operation at design temperature. Accordingly, the body of an open valve has no problem sustaining a pressure of 1½ times the MAWP of the boiler.

When a valve such as a stop valve or check valve is designed for shutoff or isolation service, B16.34 requires a closure test. In the closed position the valve must be able to sustain a pressure equal to only 110% of its 100°F pressure rating without significant leakage, as a test of the closure member. (Actual leak tightness requirements vary with the intended service and are not covered by B16.34, which advises that guidance on the subject may be found in MSS SP-61 or API-598.) Sometimes the hydrostatic test pressure of 1½ times the MAWP is too much for a valve in its closed position; that is, 1½ times the MAWP is greater than 110% of the 100°F pressure rating of the valve. For example, consider a feed stop valve for a boiler whose MAWP is 2900 psi. As described earlier under Shock Service, the design pressure for such a valve as part of the feedwater piping is 225 psi greater than the MAWP of the boiler, and the design temperature is the saturation temperature corresponding to the MAWP of the boiler. Accordingly, design conditions would be 3125 psi at 691°F. For convenience in this discussion, the design temperature can be rounded up to 700°F.

Feed stop valves are normally made of the plain carbon steels that B16.34, the governing valve standard, designates as group 1.1 materials. Welding end valves are readily available in two pressure classes—standard class and special class. Table 2-1.1 of B16.34 (see Appendix IV) lists the pressure ratings (the maximum design pressure) of these two classes as a function of temperature. The table shows that either a 2500 standard class valve or a 1500 special class valve would meet the required design conditions (3125 psi at 700°F). The 1500 special class, with its thinner walls, could be the design choice because it might cost less than the standard class valve. It happens though that the hydrostatic test pressure of the boiler in this example (1.5 × MAWP or 4350 psi) exceeds the nominal capability of a 1500 special class valve in its closed position, which according to B16.34 need be only 110% of its 100°F pressure rating, or 4125 psi. This is a rather small difference, only about 5%, and in all probability the valve closure member and operating mechanism could sustain such a pressure without significant leakage. However, unless the valve manufacturer's agreement were obtained ahead of time, by the purchase specification, the manufacturer would be likely to disclaim any responsibility for the consequences of exceeding the normal limit of 110% of the 100°F rating for a closed valve. Some further discussion on hydrostatic testing of boiler external piping is found in Chapter 7.

ALLOWABLE STRESS WHEN SAFETY VALVES ARE DISCHARGING

A basic principle in the Code design of pressure parts is that during operation of a boiler or pressure vessel, stresses due to internal design pressure are not allowed to exceed the maximum allowable stress values (design stresses) listed in Section II, Part D. There are exceptions to this principle, such as when the safety valves are discharging. Since the pressure in a natural circulation boiler is permitted to rise 6% above the maximum allowable working pressure when the safety valves are discharging (see PG-21.1 and PG-67.2), it follows that actual stress in components designed to the minimum required thickness could exceed the design stress by 6% under those circumstances. For forced-flow steam generators with no fixed steam and waterline that are equipped with automatic controls and protective interlocks responsive to steam pressure, the special overpressure protection rules of PG-67.4 apply. Those rules limit the pressure rise to 20% above the maximum allowable working pressure of any part of the boiler (any part other than the main steam piping, whose design pressure is sometimes lower than the rest of the boiler, as explained in the next section). In this type of boiler, groups of pressure parts are designed for different pressure levels along the path of water-steam flow, with the highest pressure at the economizer inlet and the lowest at the superheater outlet. Often, stop valves are installed in the water-steam flow path (see PG-67.4.4). These valves and any others used on the boiler external piping are usually selected to meet the ASME/ANSI B16.34 valve standard, which provides tables of temperature and pressure ratings for various pressure classes and material groups, as explained in Chapter 11.

That valve standard contains provisions which permit an overpressure of only 10% when safety valves are discharging, a value that is consistent with the overpressure permitted by Section VIII of the Code for vessels with a single pressure relief valve when that valve is discharging. The designer must therefore be careful that any valves chosen for use in a forced-flow steam generator with no fixed steam and waterline have sufficient margin in pressure rating to take care of the difference in overpressure capability provided by the valve standard (10% of its rating) and that potentially required by Section I (20% of the MAWP at the superheater outlet). However, as explained above under Short-Term Overstress in Boiler External Piping, the valve pressure classes usually available have large differences in their pressure ratings, so that the necessary margin to take care of the overpressure mismatch may already have been provided. That is, the lowest pressure class valve that meets the design pressure and temperature requirements may be able to sustain not only the design pressure required, but 1.2 times the so-called master stamping pressure, the design pressure at the superheater outlet (see PG-106.3).

FEEDWATER PIPING

Rules covering this piping are found both in Section I and B31.1, Power Piping. It's best to read both sets of rules carefully, since neither code covers all the requirements. Section I requirements for feedwater piping are stated in PG-58 and PG-59 and are illustrated in Figs. PG-58.3.1 and PG-58.3.2. Figures 100.1.2(B) and (A) of B31.1 are equivalent. Figure PG-58.3.1 is included in this chapter for convenience. It shows the different valve arrangements required for a single boiler fed from a single source and for two or more boilers fed from a common source, with and without bypass valves around the required regulating valve. The required boiler feedwater check valve is typically placed upstream of the feed stop valve, except that on a single boiler-turbine unit installation, the stop valve may be located upstream of the check valve.

Note that the upstream boundary of the scope of Section I varies slightly depending on feedwater valve arrangements. This is shown by solid and dotted lines in Fig. PG-58.3.1 of Section I or its equivalent, Fig. 100.1.2(B) in B31.1. Changes in the Section I boundary can have some significance in the choice of design pressure for the feedwater piping and valves, which is governed by the rules found in 122.1.3 of B31.1. The rules call for a design pressure higher than the MAWP of the boiler, as described under the heading Shock Service for Feedwater and Blowoff Piping at the beginning of this chapter. They also stipulate, indirectly, a design temperature no less than saturation temperature corresponding to the MAWP of the boiler. The application of these rules is explained in Appendix IV, Design Exercise No. 3, Choice of Feedwater Stop Valve. Note also that a closed stop valve in the feedwater piping can cause the pressure upstream to rise to pump shutoff head. Design for this situation is addressed in 122.13 of B31.1 and is also discussed earlier in this chapter under Pump Discharge Head and Unusually High Piping Design Pressure.

One last requirement that is sometimes overlooked is found in B31.1, 122.1.3(C), which stipulates that the size of the feedwater piping between the boiler and the first valve upstream may be no smaller than the connection to the boiler.

MAIN STEAM PIPING

Main steam piping from the steam drum or from the superheater falls under Section I's jurisdiction, but as explained in Chapter 4, this piping is *boiler external piping* (BEP), and the design rules for it are found in ASME B31.1, 122.1.2, Steam Piping. There, five different situations are covered in A.1 through A.5, with these subparagraphs providing slightly different rules for each situation. It turns out that main steam piping can sometimes be designed for a pressure somewhat less than the MAWP of the boiler.

Subparagraph A.1 describes steam piping connected to the steam drum or to the superheater inlet header up to the first stop valve in each connection. Such piping is boiler external piping. An auxiliary steam line from the drum is also in that category. All of these lines contain saturated steam. The design pressure specified for such lines is not less than the lowest pressure at which any drum safety valve is set to blow.

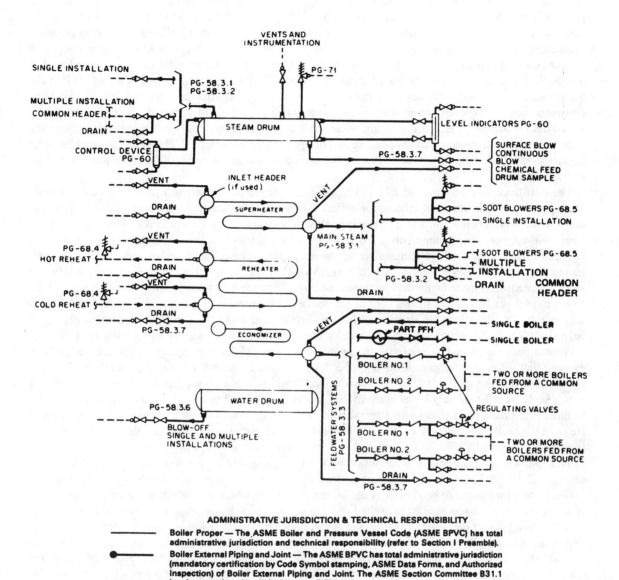

ADMINISTRATIVE JURISDICTION & TECHNICAL RESPONSIBILITY

—————— Boiler Proper — The ASME Boiler and Pressure Vessel Code (ASME BPVC) has total administrative jurisdiction and technical responsibility (refer to Section I Preamble).

●————— Boiler External Piping and Joint — The ASME BPVC has total administrative jurisdiction (mandatory certification by Code Symbol stamping, ASME Data Forms, and Authorized Inspection) of Boiler External Piping and Joint. The ASME Section Committee B31.1 has been assigned technical responsibility.

○— — — Non-Boiler External Piping and Joint — Not Section I jurisdiction (see applicable ASME B31 Code).

FIG. PG-58.3.1
CODE JURISDICTIONAL LIMITS FOR PIPING—DRUM-TYPE BOILERS
(REPRODUCED FROM ASME SECTION I)

The design temperature for this piping is taken as the saturated steam temperature corresponding to the lowest pressure at which any drum safety valve is set to blow. Although the low set drum safety valve is ordinarily set at the MAWP of the boiler, this is not always the case; it may be set lower, for example, to provide steam for a system not designed for the full MAWP of the boiler. Note that PG-67.3 of Section I requires only that one or more safety valves be set at or below the maximum allowable working pressure.

Recall also that a steam line leading from a drum to an integral superheater (one connected directly to the boiler without intervening valves) is *boiler proper piping*. Piping design rules in that case are found in Section I. The design formulas of PG-27 would be used, with design pressure taken as the MAWP of the boiler. Section I actually provides no guidance as to choice of design temperature for such piping. Most designers would choose saturation temperature corresponding to the MAWP of the boiler, if the piping would not be absorbing heat. If the piping would be in a hot gas stream, the designer would calculate the maximum anticipated mean wall temperature and design for that temperature. The same design philosophy is seen in PG-27.4, Note 2, which advises on the choice of design temperature for tubes that do, and do not, absorb heat.

The next three subparagraphs of B31.1, 122.1.2 deal with the design of main steam lines coming from the superheater. This boiler external piping differs somewhat from steam piping connected directly to the drum, especially in being much hotter. For typical high-pressure utility boilers in the United States, main steam outlet temperature is nominally 1010°F, compared to saturated steam temperature of about 690°F in the drum. Because of the high temperature of the main steam piping, plain carbon steel cannot be used; more expensive low alloy materials such as SA-335 Grade P-22 (2¼Cr-1Mo), or even austenitic stainless steel, are needed. At these higher temperatures, the pipe is operating in what is known as the creep regime, where time-dependent failure modes control the choice of allowable design stress. This allowable stress is less than half of the allowable stress used for the carbon steel drum plate, whose design temperature is only about 700°F. As a consequence, main steam piping is relatively expensive and thick-walled. In what may have been an attempt to minimize the required thickness of this hot piping, design rules have been evolved that in some cases permit the use of a design pressure somewhat lower than the MAWP of the boiler, in recognition of the fact that the steady, long-term operating pressure in the main steam line of a power boiler at full load may be as much as 10% below the MAWP of the boiler. Two factors contribute to this lower pressure. First, operating pressure in the drum is usually about 5% below MAWP, to avoid frequent lifting of the safety valves. Second, the pressure drop between the drum and the superheater outlet is often 5% of the MAWP.

Thus a lower design pressure is permitted by the next paragraph of the B31.1 design rules for steam piping, 122.1.2 (A.2), which covers piping connected to the superheater outlet header up to the first stop valve. Design pressure is to be chosen as the lowest pressure at which any safety valve on the superheater is set to blow, but not less than 85% of the lowest pressure at which any drum safety valve is set to blow. It is customary to set the safety valve on the superheater at a pressure low enough so that it will be the first to lift, thus assuring continued cooling flow of at least some steam through the superheater in the event of a turbine trip. The superheater safety valve is usually set lower than the MAWP by slightly more than the pressure drop through the superheater, which is often about 5% of the MAWP for large utility boilers. Consequently, the design pressure of the main steam piping could be reduced by that amount, to the safety valve set pressure. Design temperature is the expected steam temperature. This somewhat imprecise definition of design temperature was inherited when these paragraphs were moved from Section I in 1972. (By comparison, the definition of design temperature found in B31.1, 101.3.2 is a better one; it describes design temperature as the metal temperature representing the maximum sustained condition expected. The difference between these two definitions is probably not significant, but note that 101.3.2(c) requires design temperature of piping leading from fired equipment to be based on the expected continuous operating condition plus the equipment manufacturer's guaranteed temperature tolerance.

The next steam piping arrangement, described by B31.1, 122.1.2 (A.3), is a main steam line with two stop valves, used when two or more boilers having manhole openings are connected to a common header (see Section I, PG-58.3.2). A slightly lower design pressure is permitted for this most downstream portion of the main steam line, the piping between the two stop valves: the design pressure may be taken as the maximum sustained operating pressure, but not less than 85% of the lowest pressure at which any drum safety valve is set to blow. Again, the design temperature is described as the expected steam temperature.

The last of the three subparagraphs dealing with main steam lines, 122.1.2 (A.4), is probably the most recent of the three, since it describes large boilers installed on the unit system (i.e., one boiler and one turbine or other prime mover) and provided with automatic combustion control equipment responsive to steam header pressure. In this case, the design pressure is chosen as the highest of the following:

- the design pressure at the turbine throttle inlet plus 5%
- 85% of the lowest pressure at which any drum safety valve is set to blow
- the expected maximum sustained operating pressure at any point in the piping system

Design temperature is the expected steam temperature at the superheater outlet. Tacked on at the end of this paragraph is advice that the design pressure for steam piping for *forced-flow steam generators with no fixed steam and waterline* is required to be no less than the expected maximum sustained operating pressure.

The last of the B31.1, 122.1.2 subparagraphs dealing with steam piping sets a minimum design pressure of 100 psi for any steam piping that is boiler external piping. The purpose of this provision is to ensure some minimum mechanical strength in piping that might otherwise have paper-thin walls, were it to be designed for the MAWP of a very low pressure boiler.

Note that by using stipulations such as expected steam temperatures and expected maximum sustained operating pressure, B31.1 places the responsibility for establishing and justifying design temperature and pressure for steam piping on the designer. It's up to the designer to meet this responsibility by using recognized engineering principles and methods, and prudent judgment.

BLOWOFF AND BLOWDOWN PIPING

Some of the requirements for blowoff and blowdown piping are found in Section I, PG-58 and PG-59, dealing with piping and piping connections to the boiler. Other requirements are found in B31.1, 122, Design Requirements Pertaining to Specific Piping Systems, and specifically in subparagraphs 122.1.4, Blowoff and Blowdown Piping; 122.1.7(C), Blowoff Valves; and 122.2, Blowoff and Blowdown Piping in Nonboiler External Piping. This rather odd state of affairs, having requirements divided between the two codes, is the result of the transfer of the boiler external piping rules from Section I to B31.1 in 1972. Although there is some overlap among these requirements, certain of them are found only in Section I, and others are found only in B31.1. Thus it is necessary to study both codes carefully to ensure full compliance with the rather special rules that apply to blowoff and blowdown systems.

There is some confusion between the terms blowoff and blowdown, since they both refer to a process of discharging water from the boiler. PG-59.3.1 defines a **blowoff** as a pipe connection provided with valves located in the external piping through which the water in the boiler may be blown out under pressure. However, certain pipe connections that would apparently meet this definition are excluded from it, namely, "drains such as are used on water columns, gage glasses, or piping to feedwater regulators, etc., used for the purpose of determining the operating condition of such equipment." B31.1 in its definition adds drain piping to this list of pipe connections not considered blowoffs, but imposes additional rules to help ensure that a drain line won't be used as a blowoff unless special valving is provided (see B31.1, 122.1.5, Boiler Drains, and the next section of this chapter).

The subtle distinction between a blowoff system and a blowdown system is explained in B31.1, 122.1.4: "**Blowoff systems** are operated intermittently to remove accumulated sediment from equipment and/or piping, or to lower boiler water level in a rapid manner. **Blowdown systems** are primarily operated continuously to control the concentrations of dissolved solids in the boiler water." Note that the definition of a blowdown system has some flexibility; such a system doesn't have to operate continuously, so long as that is its primary, or usual, mode of operation.

The importance of distinguishing between a blowoff and a blowdown system is that their design pressures and valving differ. There is also a maximum size of NPS 2½ imposed on blowoff lines. Blowdown systems require only a single shutoff valve and are designed for the MAWP of the boiler. Blowoff systems have unique double valving requirements and a design pressure higher than the MAWP, because they are designed for so-called shock service, described at the beginning of this chapter. As explained there, boiler water being discharged through a blowoff is at saturation temperature or very close to it, and the flow-induced pressure drop in the blowoff line can induce boiling within that line, resulting in considerable vibration and shaking. According to Subcommittee I folklore, in the early days an ordinary manually operated valve was used as a block valve on the blowoff line. Workers attempting to blow sediment out of the boiler might open the block valve a small amount, in an attempt to keep the process under control. However, there were instances when the vibration caused by the flashing discharge would suddenly shake the valve open, increasing the flow to such an extent that there was a danger of scalding any nearby workers.

To overcome this problem, rules were instituted in Section I long ago, rules now found in B31.1, 122.1.7(C.4) that require two valves on blowoff lines for all but a few small electric boilers and other boilers whose MAWP doesn't exceed 100 psi. One of these two valves must be a so-called slow-opening valve. A **slow-opening valve** is defined as one "which requires at least five 360° turns of the operating mechanism to change from fully closed to fully open." Such a valve is less susceptible to being shaken open. The designer may furnish either of the following two valve arrangements: two slow-opening valves anywhere in the blowoff line or one ordinary stop valve (described as a quick-opening valve) at the nozzle connection on the boiler, followed by a slow-opening valve. The latter could be downstream at some location more convenient for the operator. Paragraph 122.1.7(C.10) also permits another valve arrangement, with the two required independent blowoff valves combined into a single body, but says nothing about which of the two (slow- or quick-opening) should be upstream. (It's hard to see what difference this choice would make.) Note that the boiler external piping, and thus Section I's jurisdiction, extends out to and includes the second valve.

Further information on what Section I requires in the way of blowoff and blowdown systems is found in PG-58 and PG-59, which deal with boiler piping requirements. With the exception of forced-flow steam generators with no fixed steam and waterline and high-temperature water boilers, all boilers must have a bottom blowoff outlet in direct connection with the lowest water space practicable. There are maximum and minimum size limits imposed on the size of blowoffs: The maximum size is NPS 2½, and the minimum size is NPS 1, except for boilers with 100 square feet of heating surface or less, in which case the minimum is NPS 3/4. Miniature boilers may have blowoff connections as small as NPS 1/2. The reason for these size limits, as explained by old hands on Subcommittee I, is that any size larger than NPS 2½ would discharge so much water so quickly that the natural circulation of the boiler could be upset. The minimum size was set to ensure that accumulated sediment could indeed be blown out. A smaller size was deemed adequate for miniature boilers. There is no maximum size specified for a blowdown connection, piping, or valves.

A perforated surface blowoff pipe is usually installed in the drum below the normal water level, for a number of purposes. This blowoff can be used to control the level of dissolved and/or entrained solids in the boiler water. (The dirty water discharged through the blowoff is replaced with clean feedwater.) The surface blowoff can also be used as an auxiliary drum water level control if needed. PG-59.3.2 limits the size of this blowoff pipe to NPS 2½.

At most large boilers today, the water discharged from both the blowoff and blowdown systems is sent to a flash tank, where some of its energy is often recovered in the form of low-pressure steam for use in the de-aerators. The dirty water that collects at the bottom of the flash tank is then sent to an atmospheric tank, where more steam flashes off, and the remaining water can then be discharged to a sewer or to a facility for further treatment.

Further detailed requirements for blowoff and blowdown piping are given in the B31.1 paragraphs mentioned above. One of the pitfalls sometimes overlooked is a requirement that the size of the blowoff piping cannot be less than the size of the connection on the boiler. This is also the case for blowdown piping. Moreover, 122.1.7(C.3) requires that the blowoff valves themselves, the pipe between them, and the boiler connection must be the same size in almost all cases. There are also some special provisions in 122.1.7(C) on blowoff valves, prohibiting the use of valves with dams or pockets where sediment could collect.

DESIGN CONDITIONS FOR BLOWOFF AND BLOWDOWN PIPING

B31.1 also provides advice on what design temperature and pressure to use for blowoff and blowdown piping. The design temperature is the same for both. That temperature is the temperature of saturated steam at the maximum allowable working pressure of the boiler. This is a logical choice of design temperature, since the piping in question is attached to water-filled components of the boiler, and it's hard to imagine how the temperature of the water could ever exceed saturation temperature in the drum.

Design pressure for blowdown piping is required by 122.1.4(B.1) to be no less than the lowest set pressure of any safety valve on the drum, which is usually the MAWP of the boiler, although it could be lower. As previously explained, blowoff piping is in shock service, and its design pressure is required by 122.4(A.1) to exceed the maximum allowable working pressure of the boiler by either 25% or 225 psi, whichever is less, but may not be less than 100 psi.

DRAINS

PG-58 and PG-59 provide a great deal of information about piping arrangements and piping connections to the boiler and repeat many of the requirements that were transferred to the B31.1 Power Piping Code in 1972.

Among the various topics covered in these two paragraphs are the Requirements for Drains, in PG-59.4. This paragraph has three subparagraphs; the first of them, PG-59.4.1, lays down a general rule that all components of the boiler shall be provided with drains. That rule is as follows:

> *Ample drain connections shall be provided where required to permit complete drainage of all piping, superheaters, waterwalls, water screens, integral economizers, high-temperature water boilers, and all other boiler components in which water may collect.*

The next two subparagraphs add further particulars about superheaters and high-temperature water boilers. The advice in PG-59.4.1.1 is that

> *Each superheater shall be equipped with at least one drain connection so located as to most effectively provide for the proper operation of the apparatus.*

This is an instance of a Section I rule that either doesn't mean exactly what it says or by interpretation has come to mean something slightly different. For example, many large boilers are equipped with nondrainable superheaters. No one would ever think to raise a question about this arrangement, perhaps because a drain at the bottom of a pendant superheater would be completely impractical. (There is no way to keep such a drain cool enough to survive in the middle of the furnace with no flow through the drain connection.) Thus it could be said that the drains normally provided on superheater headers *are* located so as to provide most effectively for the proper operation of the superheater, even though they can't drain condensate that may collect in pendant loops.

With the above background in mind, consider the following situations which occurred in the 1970s and 1980s and which established a policy that there are exceptions to the rule that all boiler components must be drainable. In the mid 1970s the states began to take a stronger stand on air pollution from industrial plants that had previously managed to fend off such regulation on the grounds that the industry and its jobs might leave if costly pollution control equipment were mandated. Finally, however, a certain state set a deadline for a large copper refinery to clean up the exhaust gases emanating from its smelter.

The most efficient way to effect the cleanup was to install a waste heat boiler to cool down the dirty gases sufficiently to send them to a scrubber. This solution also had the advantage of capturing some of the waste heat in those gases. The refinery engineers chose a type of boiler that had proven itself in similar service in Japan, Canada, and several states in the United States. The boiler was equipped with pendant water cooled surface within the furnace, comprised of vertical U-bend elements. The dirty gases passed over this surface, depositing most of their burden of particulate matter on it. Once during every shift, a rapping device was used to knock the slag deposits off the pendant elements for disposal.

Just as the addition of waste heat boilers and scrubbers was nearing completion, the Authorized Inspector for the organization assembling the boiler, the A stamp holder, challenged the lack of a drain on the pendant water cooled surface in the boiler. The Inspector invoked both the general rule of PG-59.4.1 requiring all components to be drainable and the particular rule of PG-59.3.4, which requires all water screens that do not drain back into the boiler to be equipped with outlet connections for a blowoff or a drain line.

The manufacturer of the boiler was unable to persuade the Inspector that it was both impractical and unnecessary to drain the pendant surface. The manufacturer cited the precedent of the pendant superheater with no drain, the fact that the boiler was to be installed in a heated building where there was no danger of freezing the water in the pendant surface, the successful application and acceptance of this design in other jurisdictions, and the feasibility of draining the boiler by using compressed air to blow the water out if necessary. The Inspector insisted that he would accept the boiler and sign the Data Report Form only if Subcommittee I issued a ruling that a drain wasn't needed for the pendant surface. The manufacturer had no choice but to bring his case to Subcommittee I. Moreover, the inquiry had to be carefully phrased so that the committee would perceive it as a general question and not as a request for a ruling on a particular design. As is often the case, the committee modified the inquiry somewhat, in the interest of making the published reply as broadly useful and informative as possible. The following Interpretation was issued in 1977 as Interpretation I-77-16, under the heading Section I, PG-59, Drain and Blowoff Requirements:

> Question: *The design of a forced circulation waste heat boiler for a copper refining process incorporates vertical U-bend elements in the waterwalls and in pendant heating surface above or within the furnace. The pendant design is specifically desired for this application to facilitate mechanical rapping for slag removal. The individual elements are necessarily undrainable.*
>
> *A continuous blowdown connection from the steam drum is used for control of boiler water concentration. Further, the design provides a bottom blowoff connection from the lowest water space available in the main downcomer (circulating pump suction) piping. In addition, drain valves are provided for the circulating pump discharge header and for the inlet and outlet upper headers from which the nondrainable boiler elements are suspended.*
>
> *Are these provisions for blowoff and drain connections adequate to meet the requirements of Section I?*
>
> Reply: *It is the opinion of the Committee that the blowoff connection as described satisfies the requirements of PG-59.3 provided the size and type of blowoff valves meet the special Section I requirements for such service. The various drain connections as described in the inquiry will meet the intent of PG-59.4 if so arranged as to permit draining all the boiler piping and headers except for the vertical U-bend elements.*

Both authors were participants in the deliberations which lead to the above interpretation and believe it made sense. It was practical. Moreover, it was philosophically consistent with the words in the Preamble referring to forced-circulation boilers, boilers with no fixed steam and waterline, and high-temperature water boilers. The Preamble exempts such boilers from the requirements for ''accessories that are manifestly not needed or used in connection with such boilers.'' Further support for this interpretation also can be found in B31.1, 122.1.5(A), Boiler Drains, which begins: ''Complete drainage of the boiler and attached piping shall be provided *to the extent necessary* to ensure proper operation of the steam supply system.'' (Italics were added by the authors for emphasis.) The drainage arrangements described in the inquiry did meet this criterion, as evidenced by the successful operation of that boiler and many others like it.

The committee again ruled on the issue of drainability on January 3, 1980 in Interpretation I-80-01, under the heading: Section I, Blowoff Requirements, PG-59.3 and PG-59.4. Although ostensibly about blowoffs, this interpretation provides further evidence that the committee does not believe it is necessary to be able to drain the water by gravity from every component. The boiler in question was an unusual one; it generates steam in so-called bayonet tubes, a tube-inside-a-tube arrangement in which water flows down the inner tube and up through the annular space between the inner and outer tube, where the steam is generated. It is clearly not feasible to provide either a drain or a blowoff at the bottom of each individual tube. The inquirer sought guidance on how he might deal with this situation with the following explanation and questions:

> *These inquiries apply to a drum boiler equipped with vertical bayonet tubes suspended from the drum. Water is force-circulated down the inner tube and up through the annular space between the inner and outer tube.*
>
> Question: *May a threaded cap at the bottom of the each boiler tube serve as the drain required in PG-59.4?*
>
> Reply: *No. A drain is defined as a pipe connection with a valve or valves as stipulated in PG-58.3.7. The individual bayonet tubes need not necessarily be drainable, and alternate means of removing the water in these tubes could be employed if the designer so chooses.*
>
> Question: *Must an external blowoff pipe be provided for each boiler tube?*
>
> Reply: *No. PG-59.3.3 requires only a single "bottom blowoff outlet in direct connection with the lowest water space practicable." The blowoff connection and its valve or valves must comply with the provisions of PG-58.3.6.*

In the above Interpretation, the committee extended the policy first enunciated in Interpretation I-77-16 regarding drains and blowoffs, namely, that it is not absolutely necessary to drain by gravity every individual element in the boiler. Moreover, the single blowoff required has always been described as an "outlet in direct connection with the lowest water space practicable." Since it is not practicable to provide a blowoff on individual tubes, a regular blowoff from a water space somewhere low on the boiler, such as a waterwall header, would suffice.

There is another Interpretation dealing with the issue of drainage. That Interpretation is I-83-98, Section I, Drains for Horizontal Axis Helical Coil Forced Flow Steam Generators. It reads as follows:

> Question: *Certain forced flow, no fixed waterline steam generators used for oil field steam flooding service are constructed with helical coils on a horizontal axis. It is impractical to provide complete drainage by gravity of each of these coils. Is such drainage required by PG-59.4.1?*
>
> Reply: *No.*
>
> Question: *What drain provisions are considered necessary to comply with PG-59.4.1?*
>
> Reply: *Ample drains must be provided to permit drainage of all boiler piping and headers. Alternative means of removing the water in the helical tubes may be used.*

Still another inquiry on the subject of drains was answered by Subcommittee I on March 9, 1992 as Interptretation I-92-27, Requirements for Drains, which reads as follows:

> Question: *Do the drain requirements of PG-59.4 require an economizer or superheater to be drainable by gravity?*
>
> Reply: *No. Alternative means of removing the water may be used.*

By these rulings the committee again showed that it is willing to take a practical approach in accepting new boiler designs as they are developed for new applications, so long as the fundamental safety features of the boiler are unimpaired. In the opinion of the authors, drainage by gravity is not a fundamental safety feature. Other means of coping with undrainable water are available if necessary. A start-up procedure known as a boil-out is typically employed to eliminate trapped or condensed water in pendant- or platen-type superheaters and reheaters.

The above discussion dealt with the fundamental issue of drainability. There are other requirements pertaining to drains and drain piping, which are found in Section I, PG-59.3.4 and PG-59.4, and in B31.1, 122.1.5. The latter paragraph stipulates that a drain line intended for use both as a drain and a blowoff must be designed as a blowoff, with the requisite double valving. If the drain is intended for use only when the boiler is not under pressure, a single shutoff valve is acceptable, provided it can be locked closed or can be blanked off on the downstream side. Finally, 122.1.5(A) stipulates that the pipe, fittings, and valves of any drain line may not be smaller than the drain connection on the boiler. This is probably to prevent constrictions downstream of the drain connection which might prevent prompt drainage of the boiler.

DESIGN CONDITIONS FOR DRAIN PIPING

Advice on the design conditions to be used for drain piping is given in 122.1.5(D). This advice is somewhat confusing, in that it introduces a discussion of ''remaining piping'' and how to design it. [That remaining piping is apparently nonboiler external piping (NBEP) beyond (downstream of) the valve on the drain line that defines the boundary of Section I's jurisdiction. Minimum design conditions of pressure and temperature are stipulated for this nonboiler external piping as 100 psi and 220°F.] Paragraph 122.1.5(D) goes on to say that the drain piping and valves that are part of the boiler external piping are to be designed for the same temperature and pressure as the drain connection on the boiler proper. The latter is usually a welded nozzle. Boiler manufacturers design these nozzles for the MAWP of the boiler at the design temperature of the components to which they are attached. If the nozzle were a threaded nozzle and thus part of the boiler external piping, the design pressure and temperature would logically be the same as for the components to which the nozzle was threaded.

PIPE CONNECTIONS TO THE BOILER

Both Section I and B31.1 address the question of the type of connections that may be used to attach piping to the boiler proper, in their respective paragraphs PG-59.1 and 122.1.1(D). Section I's rules are brief: Boiler external piping must be attached by a tapped opening, a bolted flange, or by a butt or socket weld. (The weld or other means, e.g., bolts, by which piping is attached to the boiler nozzle must be designed in accordance with B31.1, since the figures showing the Code jurisdictional limits for piping, e.g., Fig. PG-58.3.1, indicate that the joint between the boiler and the boiler external piping is part of the latter.) PG-59.1.1.4 goes on to say that boiler proper piping may be expanded into grooved holes and seal welded if desired. It also stipulates some special rules for the attachment of blowoff connections for firetube boilers. These old rules were intended to facilitate the replacement of such connections without welding.

B31.1, 122.1.1(D) addresses the connections of boiler external piping to the boiler and echoes the Section I rules with some amplification. Additional B31.1 rules are as follows:

- Nonferrous pipe or tubes are limited in size to NPS 3.
- Hub-type flanges may not be cut from plate material, because of the possibility that any laminations in the plate uncovered by machining might open up when bending stresses are applied to the flange. (Section I actually has an identical provision in PG-42.3.)

Finally, B31.1 paragraph 122.1.1(H) allows the use of American National Standard socket welded flanges in piping or boiler nozzles, provided the dimensions do not exceed NPS 3 for Class 600 and under and NPS 2.5 in Class 1500. This advice is slightly inconsistent with similar provisions for socket welded connections in Section I, PW-41.5, which has no pressure limitations, but provides more detailed design rules and prohibits the use of socket welded joints where they would be in contact with furnace gases. B31.1 is silent on what size Class 900 socket welded flanges may be used; an NPS 2.5 limit is implied.

ADDITION OF SMALL CONNECTIONS TO BOILER PIPING

There is an unusual provision in 100.1.2(A), the Scope paragraph of B31.1, Power Piping, covering boiler external piping. That provision reads as follows:

Pipe connections meeting all other requirements of this Code but not exceeding NPS ½ may be welded to pipe or boiler headers without inspection and stamping required by Section I of the ASME Boiler and Pressure Vessel Code.

First note that any boiler external pipe connections to the boiler do, as a matter of course, meet all other requirements of B31.1, Power Piping (all other than inspection and stamping). Thus the piping should be perfectly satisfactory to serve as power piping, since design, material, and construction are adequately covered by B31.1. The B31.1 waiver of inspection and stamping applies only to small pipe—NPS ½ and smaller, such as might be used for instrumentation connections—that appears to be boiler external piping. The waiver also applies only to small connections welded to pipe or boiler headers (but not explicitly to the drum). Moreover, Section I already waives stamping, but not inspection, for pipe NPS 2 and smaller (see PG-109.2). It thus seems that this special provision might prove useful only for the addition of small instrumentation connections as an afterthought, perhaps after the Authorized Inspector has completed his or her inspection. What seems unusual is that by 100.1.2(A), B31.1 is apparently waiving Section I rules; however there is a comparable provision in Section I, PW-41.3, which states that pipe connections not exceeding NPS ½ may be welded to pipe or headers without the inspection normally required by Section I. Presumably all other Section I rules would still apply. It thus appears that both Section I and B31.1 at one time decided that the rules normally requiring inspection of welded pipe connections could be relaxed for these small lines, and coordinated their respective rules.

FABRICATION AND WELDING

FABRICATION

Fabrication is not specifically defined in Section I, but it is generally understood to mean all those activities by which the manufacturer converts material (plate, tube, pipe, etc.) into completed boiler components. Such activities include welding, bending, forming, rolling, cutting, machining, punching, drilling, broaching, reaming, and others. Section I permits the manufacturer very broad latitude in fabrication because of the wide variations in manufacturing practice, machinery, and trade methods.

Section I provides only a limited number of general rules for fabrication, in PG-75 through PG-82. (Welding is covered extensively in Part PW, and a discussion of those rules is given in the next section of this chapter.) The general rules for fabrication in Part PG specify tolerances for drums and ellipsoidal heads and limit certain cutting, drilling, and punching processes such that subsequent machining will remove all metal whose mechanical and metallurgical properties have been affected by the earlier process.

The manufacturer is permitted to repair defects in material, provided acceptance of the Authorized Inspector is first obtained, both as to method and extent of repairs.

WELDING—SOME HISTORY

An important factor in the evolution of today's efficient high-pressure boilers was the development of welding as a replacement for the riveted construction used in the 19th century and the first half of the 20th century. Although welding was already in use in 1914 when the the first edition of the ASME Code was in preparation, no mention of welding appeared in that edition because the method was not considered sufficiently reliable. The use of gas and electric welding for steel structures and ships was brought before the Boiler Code Committee at its first meeting after the adoption of the Code, in 1915, but the committee, though interested, took no action. In 1916 the Committee responded to a request for consideration of welding from the Union Iron Works of California, stating that it had under observation all sorts of welding of pressure vessels and had asked welding experts to write papers for the ASME to bring out further information on these various processes. Further requests were received for permission to use the rapidly evolving art of welding, and when the first edition of the Code was revised in 1918, paragraph P-186 permitted fusion welding, but only when stresses were carried by other construction and safety did not depend upon the strength of the weld.

An ASME Subcommittee on Welding was appointed in 1918, and in 1920 several societies involved in welding formed the American Welding Society (AWS). Utilizing government-sponsored research on welding that had been motivated by the recent war effort, these groups formulated the first rules for fusion-welded boilers. The first rules for joining pressure boundary materials by welding appeared in 1923, in the Low Pressure Heating Boiler Code, the predecessor of today's Section IV. The first edition of Section VIII, then called Unfired Pressure Vessels, was published two years later, in 1925. It also contained limited rules for welding, but it was not until the late 1920s and early 1930s that fairly comprehensive rules for welding appeared in both Section VIII and Section I. A separate book section on the *qualification* of welding was first published in 1941, entitled Section IX, Welding and Brazing Qualifications.

An important aspect of welding, compared to other means of construction, is the absolute need for careful control of the welding process in order to achieve sound welds. The ASME Code attempts to achieve this control by allowing welds to be made only by qualified welders using qualified procedures. Section IX provides the rules by which the welders and the weld procedures may be qualified. However, there are many other aspects of welded construction besides qualification, and rules covering these other aspects are found in the other ASME book sections covering welded construction (Sections I, III, IV, and VIII).

The ultimate goal of welding procedure and performance qualification is to achieve a weldment whose properties are at least the equivalent of the base material being joined, as demonstrated by certain tests. Qualification of the procedure establishes that the weldment will have the necessary strength and ductility when it is welded by an experienced welder, one who has demonstrated his or her ability to deposit sound metal when following that procedure. Qualification of the welder establishes that he or she can deposit sound weld metal in the positions and joints to be used in production welding.

An assumption implicit in the welding rules of Section I is that a weld is the equivalent of the base material it joins; accordingly, a properly made weld does not weaken the vessel. Consequently, there is no bar to superimposing attachments on welds or having nozzles or other openings placed where they intersect welds (this is stated explicitly in PW-14). There is also no rule that would prevent a manufacturer from making an opening for a nozzle in a vessel, then deciding the nozzle isn't needed, and replacing the material removed for the opening with a properly designed and welded patch.

Note: A number of different terms are used in this chapter to describe various types of welders and welding. The following generally accepted definitions of those terms are provided for general information:

- A **welder** performs a manual or semiautomatic welding operation.
- A **welding operator** operates machine or automatic welding equipment.
- In **manual welding**, the welder controls and performs the entire welding process by hand.
- In **machine welding**, the equipment performs the welding under the constant observation and control of the welding operator.
- In **semiautomatic arc welding**, the equipment controls only the filler metal feed and the advance of the welding is manually controlled.
- In **automatic welding**, the equipment performs the welding without adjustment of the controls by the welding operator.

WELDING RULES

Most of the Section I welding rules are found in Part PW, Requirements for Boilers Fabricated by Welding. Part PW provides rules specifically applicable to boilers and their components and simply refers to Section IX for the qualification of the weld procedures and the welders. The rules for the welding of boiler external piping (as opposed to boiler proper piping) are not found in Section I; they are found in B31.1, Power Piping. Power Piping also invokes Section IX for the qualification of weld procedures and welder performance, but is somewhat more liberal with respect to the transfer of procedures and welders from one organization to another.

Among the many aspects of welding covered by Part PW are these: responsibilities of the manufacturer or other organization doing the welding; the materials that may be welded; the design of welded joints; radiographic and ultrasonic examination of welds and when such examination is required; the welding of nozzles; attachment welds; welding processes permitted; qualification of welding; preparation, alignment, and assembly of parts to be joined; the use of backing strips; advice on preheating; requirements for postweld heat treatment; repair of weld defects; exemptions from radiography; the design of circumferential joints; the design of lug attachments to tubes; duties of the Authorized Inspector related to welding; acceptance standards for the radiography and ultrasonic examination required by Section I; preparation of test plates for tension and bend tests; and welding of attachments after the hydrostatic test. The remainder of this chapter is devoted to the discussion of these topics.

QUALIFICATION OF WELDING PROCEDURES

It is a general rule under any section of the ASME Boiler and Pressure Vessel Code and the ASME Code for Pressure Piping, B31, that all welding must be done by qualified welders, using qualified welding procedures. With some exceptions, every manufacturer (or other organization doing the welding, such as an assembler or parts manufacturer) is responsible for the welding it does, and is also responsible for conducting the tests required by Section IX to qualify the welding procedures used and the performance of the welders who apply those procedures. Section I stipulates these responsibilities in PW-1.2.

The qualification of a welding procedure is a process in which the manufacturer prepares certain documentation, known as a **Welding Procedure Specification (WPS)** and a **Procedure Qualification Record (PQR)**, which documents and supports the WPS. The WPS comprises a set of instructions, primarily for the welder (but also for the Authorized Inspector), regarding how to make a production weld to Code requirements. These instructions are written in terms of certain parameters applicable to the welding process that are called essential, nonessential, and when required, supplemental essential variables (see below). A WPS lists the essential and nonessential variables, and the acceptable ranges of some of these variables when using the WPS. The WPS must also reference the Procedure Qualification Record (PQR) that supports it. Often, more than one PQR is used to support a WPS.

The PQR is a formal record of the welding data (the actual value of the variables recorded during the welding of the test coupons) for each welding process used and the results of mechanical test specimens cut from those coupons. Nonessential variables used during the welding of the test coupon may be recorded at the manufacturer's option. The tests (usually tensile and bend tests) are used to demonstrate that a weldment made using the WPS has the required strength and ductility for its intended application. Various other tests are sometimes used instead of the bend and tensile tests. All of these tests are more fully explained in QW-140 and QW-202 of Section IX. The variables used in production welding may vary somewhat from those used during the welding of the test coupons. Note that the purpose of the Procedure Qualification test is just to establish the properties of the weldment, and not the skill of the welder, although it is presupposed that the welder performing the procedure qualification test is a skilled worker.

QUALIFICATION OF WELDER PERFORMANCE

To complete the qualification process for welded construction, performance tests must be conducted for each welder or welding operator, to establish the welder's ability to deposit sound weld metal. Each manufacturer is responsible for conducting the tests to qualify the performance of its welders in accordance with its qualified welding procedure specifications. (QW-300.2 of Section IX states that this responsibility cannot be delegated to another organization, but that prohibition has been weakened by the series of Section I interpretations described below, under the heading Transfer of Procedures and Welders to Another Organization.) The purpose of making each manufacturer responsible for qualifying its own procedures and welders is to assure that they can produce an acceptable weldment, using those procedures.

In general, the **Welder Performance Qualification (WPQ)** is accomplished by having the welder weld test coupons. The welder may then be qualified by mechanical bend tests of specimens cut from performance test coupons or by radiography of a test coupon or the welder's initial production weld. For certain types of welds, such as fillet welds, stud welds, resistance welds, and corrosion resistant weld metal overlay, radiography is not an appropriate test, and other tests are substituted. These include fracture tests, macroexamination, torsion tests, metallographic examination, and liquid penetrant examination. Particulars of these tests may be found in QW-100 and QW-200 of Section IX.

The WPQ must list the essential variables, the type of test and test results, and the thickness and diameter range qualified for each welder and welding operator. The **range qualified** is a convenient term applicable to some essential variables used in both procedure and performance qualification. For example, a WPS is qualified for welding base materials up to a certain thickness only, depending on thickness, T, of the test coupons welded. Similarly, a welder is qualified to deposit weld metal up to a certain thickness only, depending on the thickness, t, of weld metal he or she deposited in making the test coupons. Long experience has shown that a weld procedure that can be qualified using a test coupon of a given thickness will produce satisfactory weldments for a base metal thickness somewhat greater, or lesser, than that used in the test coupon. Similarly, a welder who has shown by a performance test that he or she is capable of depositing sound metal of a given thickness is also capable of depositing sound weld metal that is either thicker or thinner than the welder deposited in making the test coupon. Based on this concept, Section IX has developed tables QW-451 and QW-452, which establish thickness ranges for production welding based on procedure and performance qualifications.

For procedure qualification, Table QW-451 defines the range of thickness of base metal qualified as a function of the thickness of the test coupon welded and the maximum thickness of deposited weld metal qualified as a function of the thickness of weld deposited in making the test coupon. These tables contain so many different test coupon thicknesses and weld types that any generalization about thickness ranges qualified could be misleading. However, in many cases the range of thickness of base metal qualified by a particular procedure is twice the thickness of the test coupon welded, and the maximum thickness of deposited weld metal qualified when using a particular weld process is twice the thickness of the weld metal deposited in using that process to make the test coupon.

For performance qualification, Table QW-452 defines only the thickness of deposited weld metal qualified as a function of the thickness of the weld metal deposited in the test coupon. Base metal thickness qualified is not a consideration in performance qualification. For groove welds, welding a test coupon up to 3/4 of an inch thick qualifies the welder to weld twice the thickness he or she deposited in making the coupon. If the coupon thickness is 3/4 of an inch or greater, the welder is qualified to deposit weld metal of any thickness. This seems quite reasonable since if a welder can make a satisfactory weld 3/4 of an inch thick, he or she could presumably make a thicker one with no difficulty.

When welders are qualified, identification marks are assigned to them, so that all welded joints can be identified as to who made them. PW-28.4.1 describes ways of identifying welds, subject to the acceptance of the AI. The welder can stamp his or her mark on or adjacent to the welds, or the manufacturer may keep a record of welded joints and the welders who made them. This is discussed further under Welding Documentation, below.

EXPIRATION AND RENEWAL OF WELDING QUALIFICATIONS

As a general rule, once a welding procedure specification is qualified, it remains qualified. In fact, QW-100.2 of Section IX explains that weld procedure specifications, procedure qualification records, and welder performance records made in accordance with the requirements of the 1962 (or later) edition may be used in any construction built to the ASME Boiler and Pressure Vessel Code and the ASME Code for Pressure Piping. These WPS's, PQR's, and WPQ's do not have to be amended to include any variables required by later editions and addenda. However, the qualification of any new weld procedures or the requalification of any existing procedures or welders must be in accordance with the requirements of the edition of Section IX in effect at the time of the qualification.

In contrast to the permanence of the WPS, the skills a welder demonstrates during performance qualification can be diminished or lost if they are not used. Accordingly, QW-322 of Section IX, Expiration and Renewal of Qualification, stipulates that if a welder or welding operator has not welded with a process for six months, his or her qualifications for that process expire. Thus it is important for a manufacturer to maintain careful records for each welder, so that if possible, sufficient welding to maintain the welder's qualifications can be scheduled, thereby avoiding the not inconsiderable cost of requalification. The general rule is this (as usual, there are some exceptions): If during the six-month period prior to his or her expiration of qualification a welder uses a manual or semiautomatic welding process, the welder maintains his or her qualification for that process and extends it for another six months. That same paragraph also provides for the revocation of a welder's qualifications for a particular welding process for cause, that is, when there is a specific reason to question the welder's ability to make welds using that process. (All other qualifications not questioned remain in effect.) Such a revocation might occur if a welding supervisor or the AI noticed obvious evidence of unsatisfactory welding.

In the case of a welder whose qualifications to use a particular process have expired because of a six-month lapse in use, the qualifications can be restored if the welder welds a single test coupon of plate or pipe, of any material, thickness, or diameter, in any position and that coupon passes the qualification tests originally required (by QW-301 and QW-302). Successful testing renews the welder's previous qualifications for that particular welding process for those materials, thicknesses, diameters, positions, and other variables for which the welder was previously qualified. A welder who loses his or her qualifications for cause must requalify, using a test coupon "appropriate to the planned production work." Successful testing of the coupon in accordance with QW-301 and QW-302 restores the revoked qualifications.

TRANSFER OF PROCEDURES AND WELDERS TO ANOTHER ORGANIZATION

For various reasons, procedure and performance qualification tests are ordinarily not transferable from one manufacturer to another. That is, a welder who has been qualified by one manufacturer to use a procedure qualified by that manufacturer cannot switch employers and continue to weld using that same procedure unless the new employer has also qualified the procedure and the performance of the welder. However, in 1948 a modification of this policy was introduced (apparently at the request of the pipefitters union, according to Main Committee minutes of that time). After due consideration, the Main Committee included in the Summer 1947 Addenda to the 1946 Edition of Section I a new paragraph, P-116, and a new reference to P-116 at the end of existing paragraph P-112. The latter stated that the requirements for "welded external piping (not a part of the boiler circulation and beyond the outlets of the boiler)" were now to be found in P-116. When the text of Section I was completely reorganized and its paragraphs renumbered in the 1965 edition (with no change in technical content from the previous version), P-116 became present-day PW-28.5, with only slight editorial change. That paragraph gives the following advice:

> To avoid duplication of qualification tests, it is recommended that procedures, welders, and welding operators qualified as required above be acceptable for any similar welding work on piping using the same procedure (see PW-1.2).

This recommendation has been a source of contention ever since it appeared. First of all, it flies in the face of the PW-1.2 rule that each manufacturer is responsible for qualifying its own weld procedures and the performance of its welders. Also, it raises an interesting question: does a recommendation waive a requirement? (Based on interpretations of this paragraph, the answer is yes.) In February 1978, Subcommittee I issued Interpretation I-78-06, which explained what it had in mind when it wrote PW-28.5. Here is the essence of that interpretation:

> The intent of this paragraph is to provide for the avoidance of duplication of qualification tests of welding procedure specifications, welders, and welding operators as they apply to piping within the scope of Section I.
>
> Manufacturers or contractors are permitted to join together with other manufacturers or contractors into an organization for the establishment of welding procedure specifications and their qualification so that the individual manufacturer or contractor does not have to duplicate this effort. This means that a welding procedure specification adopted by the organization must be qualified in accordance with Section IX of the Code by a member of the organization. Other

members of the organization may subsequently use the welding procedure specification without requalification.

Welding procedure qualification test records shall be available with each welding procedure specification sent to the organization's membership.

No member of the organization may use the organization's qualified welding procedure specifications on ASME Code work unless he holds the applicable ASME Code Stamp and Certificate of Authorization.

Welders and welding operators of a member manufacturer or contractor must pass their performance test on each of the organization's welding procedure specifications they are to weld under, except as otherwise permitted in Section IX. The performance qualification test records are placed on file with the organization. When such welders or welding operators are employed by another member manufacturer or contractor, their performance qualification records are made available to their new employer by the organization and performance requalification is not required for those welding procedure specifications under which they qualified previously.

It is incumbent upon the member manufacturer or contractor using the organization's welding procedure specifications and performance qualification records to assume responsibility for the qualifications by maintaining records, certified by him, and available to the Authorized Inspector.

Note that PW-28.5 recommends that previously qualified procedures and welders be acceptable for "any similar welding work on piping." Yet subsequent Interpretations I-81-17 and I-92-09 extended that recommendation from welding on piping to welding on any product forms that may be used for construction under PW-41, i.e., any kind of circumferential welding on cylindrical product forms.

A further relaxation of the PW-1.2 rule that each manufacturer (or other Certificate holder) must establish and qualify its own weld procedures and welders occurred with the publication of Interpretation I-89-50, entitled Mass Testing of Welders:

Question: *In accordance with PW-28 and PW-1.2, can several contractors or assemblers simultaneously conduct performance qualification test(s) of a welder?*
Reply: *Yes.*
Question: *In accordance with PW-28 and PW-1.2, can a welder simultaneously weld performance qualification test coupons in accordance with Welding Procedure Specifications of several contractors or assemblers when the essential variables are documented as identical?*
Reply: *Yes.*

The use of procedures and welders qualified by other organizations (such as the National Pipe Welding Bureau) under the circumstances covered by Interpretation I-78-06 and the other interpretations quoted above has become fairly common, especially for boiler external piping. However, the practice has also been extended by the later interpretations to boiler proper piping, even though the history of this paragraph shows the committee's original intent was to limit the practice to what is now known as boiler external piping. There are still many committee members who object to this practice (which is related to and sometimes also referred to as welding with standard, or prequalified, procedures using prequalified welders, see below) and insist that each organization should continue to take full responsibility for qualifying its own weld procedures and the performance of its welders. At the same time, they do concede that experience with prequalified procedures has generally been satisfactory when used on the most common materials. However, they argue that an organization that obtains prequalified weld procedures from others may have insufficient knowledge of welding metallurgy to avoid the many potential problems of welded fabrication. Among such problems and concerns are the effects a variation in the composition of the materials to be welded might have on their hardenability, and thus their weldability, and what remedial measures, such as preheat or postweld heat treatment, might then be appropriate. These committee members also believe that requiring welder performance qualification by each new employer would achieve better results than permitting one manufacturer to rely on performance testing supposedly carried out by another organization. Some manufacturers share a related view; if they are to be held responsible for their welding, they want to conduct their weld procedure qualifications and their own welder performance qualifications, rather than relying on others.

In 1996 Subcommittee I tried to delete PW-28.5, on several grounds. The first was that this paragraph was originally aimed at welding of boiler external piping, and rules for such welding since 1973 have been found in B31.1, Power Piping. Accordingly, any recommendation in Section I about qualification of procedures and welders for such piping is inappropriate and redundant. Moreover, Subcommittee I was largely opposed to the idea of weakening the requirements for welding components of the boiler proper, i.e., letting someone other than the manufacturer qualify the procedures and welders to be used. The proposal to delete PW-28.5 was approved by Subcommittee I and the Main Committee. However, it was defeated at the next approval level, the Board of Pressure Technology Codes and Standards. At that level, board members who believed that procedures and welders proven satisfactory for boiler external piping are also appropriate for the boiler proper were able to muster the minimum number of votes to block the action, and PW-28.5 remains in Section I.

On a related issue, the Main Committee has at this time (1998) a task group studying the advisability and feasibility of permitting the use of ANSI/AWS standard (prequalified) welding procedures in the Code. Among the concerns expressed about the use of standard weld procedures is that a Code symbol stamp holder who doesn't qualify his or her own procedures loses the experience and knowledge normally gained thereby. Having at least some minimal welding knowledge and experience is considered necessary to ensure the quality of the weldments an organization produces. Accordingly, one recommendation of the task group is that any organization that wants to use standard weld procedures be required to qualify at least one of its welders on each standard weld procedure specification it intends to use. Should the recommendations of the task group be approved by the Main Committee, Section IX will prepare and publish a list of approved standard welding procedures. It would then be up to the individual book sections to permit their use and to impose any further requirements or restrictions deemed necessary. This is similar to the way base materials are treated within the Code, i.e., Section II adopts material specifications, but they have to be approved by the individual book sections before they can be used. In the meantime, the transfer of weld procedure specifications, their supporting qualification records, and welder performance qualifications, all as described above, will continue to be acceptable practices. Also note that the National Board has at this time already appproved the use of 24 ANSI/AWS standard welding procedures for repairs and alterations.

WELDING DOCUMENTATION

Section IX provides suggested formats for Welding Procedure Specifications and Procedure Qualification Records (Forms QW-482 and QW-483 in Appendix B). These forms are not mandatory; any format that fits the needs of the manufacturer will do, so long as it furnishes the necessary information. In the case of the WPS, paragraph QW-200.1(d) calls for the inclusion of, or reference to, all essential and nonessential variables. For a PQR, paragraph QW-200.2(d) requires a listing of all essential variables, the type and number of tests, and the results of the tests. The PQR includes a signed statement by the responsible manufacturer, certifying that the test welds were prepared, welded, and tested in accordance with the requirements of Section IX.

QW-301.4 requires the record of welder performance qualification tests, the WPQ, to list the essential variables, the type of test and test results, and the range of thicknesses and diameters for which the welder is qualified. Appendix B also provides a suggested format (Form QW-484) for a manufacturer to record the results of a welder's performance qualification test, or WPQ. This form also includes a signed certification by the manufacturer that the preparation, performance, and testing of the welds was done in accordance with the requirements of Section IX.

Section I has its own rules for welding qualification and weld record keeping, in PW-28. These rules essentially invoke all of the Section IX rules just described, but superimpose other rules having to do with maintenance of records of welding procedures and welders employed, showing the date, results of tests, and welding mark assigned to each welder. These records must be certified by the manufacturer and made available to the Authorized Inspector.

PW-28.4.1 requires that the manufacturer have a system that allows welded joints to be identified as to the welder or welding operator who made them. One way this can be accomplished is by having the welder stamp his or her mark on or adjacent to the welds. As an alternative, the manufacturer is permitted to maintain a record of welds and the welders used in making them. Section I has no requirement for maintaining those records after completion of the boiler, and the manufacturer is free to dispose of them at that time. Similarly, the welders marks may soon be lost once the boiler is placed in service. This arrangement is quite reasonable since the identification of the weld with the welder who made it is most useful during construction. At that time if a faulty weld were detected by NDE or hydrostatic test, other welds made by that same welder could be investigated. However, once the boiler has passed all required NDE and the hydrostatic test, the welds are presumably sound, and there is no longer much purpose to be served by retaining the records of which welder made any particular weld. This was affirmed by Interpretation I-92-78.

WELDING NONDUCTILE MATERIALS

As explained in the introduction, the ultimate goal of welding procedure and performance qualification is to achieve a weldment whose properties are at least the equivalent of the base material being joined. Usually the WPS is proven to be capable of producing such a weldment by bend and tension tests of specimens cut from test coupons, which must show adequate strength and ductility. However, there are some nonpressure part materials, for example, tube lug materials intended for high-temperature service in the boiler, which lack the ductility to pass a bend test. The important property for such materials is not ductility, but rather strength at elevated temperature, 1150°F or higher. Typical alloys found suitable for such service often contain high percentages of chromium and nickel and lack the ductility required of pressure boundary materials.

In 1992 Subcommittee I addressed this problem by permitting a test called a macro-examination instead of the usual mechanical tests required for procedure and performance qualification of groove and fillet welds. As explained in PW-28.1.2(b), a weld test coupon representing a typical production configuration may be evaluated by using the macro-examination method. The weldable quality of the nonpressure part materials is verified by examining a single cross section of the weld. Visual examination of the weld metal and the heat-affected zone of both the pressure part and nonpressure material must show complete fusion and freedom from cracks.

WELDING VARIABLES

In the development of a WPS, it is necessary to establish which variables of the welding process are so-called **essential variables** in the production of qualifying welds and those which are not (**nonessential variables**). (Another category of variable is a **supplemental essential variable**, which becomes an essential variable only if notch toughness testing is required by the referencing book section. Section VIII sometimes requires such testing for vessels in low-temperature service, where the possibility of brittle fracture is a concern. Because Section I vessels are usually quite warm when pressurized, Section I does not require notch toughness, or impact testing, and supplemental essential variables are not used in welding for Section I construction.)

The categorization of welding variables differs somewhat, depending on the welding process used and whether the variable applies to procedure or performance qualification. An essential variable for a weld procedure is one which, if changed, would affect the properties of the weldment and thus require requalification of the procedure, i.e., additional testing and issuance of a new PQR to support the changed WPS. A nonessential variable for a weld procedure is one that can be changed without requiring requalification, although it would require a change in the WPS. Essential variables for weld procedures include such things as material thickness, P-Number (which refers to material category, see below), F-Number (which refers to filler metal alloy, see below), the use or omission of backing, and the use or omission of preheat or postweld heat treatment. Examples of nonessential variables for procedure qualification are groove design, root spacing, method of back gouging or cleaning, change in electrode size, and the addition of other welding positions than any already qualified.

An essential variable for welder performance is one that would affect the ability of a welder to deposit sound weld metal. Examples of such variables are deletion of backing in single welded groove welds (because it is more difficult to weld without backing), a change from one F-Number to another, a change in deposited weld metal thickness beyond the range qualified, and the addition of other welding positions than any already qualified. Note that this latter variable is an essential one for performance, since welding in a new position, such as vertical instead of horizontal, could certainly affect the ability of the welder to deposit a sound weld. However, it is a nonessential variable for procedure qualification, because, in the words of QW-100.1, "it is presupposed that the welder performing the welding procedure qualification tradition test is a skilled workman." Such a welder would presumably deposit sound metal in any position, and the properties of the resulting weldment would be unaffected.

Section IX has a great number of tables summarizing procedure and performance variables for all common welding processes.

P-NUMBERS, S-NUMBERS, F-NUMBERS, AND A-NUMBERS

P-Numbers

One important essential variable is, of course, the type of base material being welded. There are several hundred different materials permitted for Code construction by the various book sections. If every change in base material meant that weld procedure specifications and welder performance had to be requalified, the qualification of procedures and welders for all these materials would be an enormous task. To reduce this task to manageable proportions, the ASME has adopted a classification system in which material specifications are grouped into categories, based on similar characteristics such as composition, weldability, and mechanical properties. Each category is defined by a **P-Number**. Within the P-Number category are subcategories called **Group Numbers** for ferrous base metals that have specified impact requirements. Although Section I doesn't use Group Numbers in relation to impact testing, it does sometimes establish different postweld heat treatment requirements for the different Group Numbers within a given P-Number category.

A welding procedure that has been qualified for a particular base material may be used for any other base material with the same P-Number. (On the other hand, if a base material has not been assigned a P-Number, it requires an individual procedure qualification.) Similarly, a welder who has been qualified for a base metal of a particular P-Number may be qualified for quite a few other P-Number base materials (see QW-423).

A numerical listing of P-Numbers and Group Numbers can be found for all Section II materials, by specification number, in Table QW/QB-422 of Section IX. That table was reorganized and expanded into its present form in the 1994 Addenda. In addition to the P-Numbers, the new version lists certain useful information about each material: UNS number (a number that categorizes each material by its nominal composition), minimum specified tensile strength, chemical composition, and product form. The previous version of Table QW/QB-422 was organized differently; it listed all P-Numbers by category: first all the materials that were P-Number 1, then all the P-Number 3 materials, and so on. (There is no longer a P-Number 2 category; it used to cover wrought iron, such as was used in some staybolts.) Under the old system, unless a user knew the P-Number of a given material, he or she had to search all the categories to find it. However, many users did like certain aspects of the old system, and in keeping with a tradition of placing certain superseded material in an appendix where it would remain available, Subcommittee IX put a material listing in the old format in Nonmandatory Appendix D. That arrangement groups materials by P-Number and permits a user to locate material specifications with matching P-Numbers more easily.

Although a specific definition of each P-Number in terms of chemical composition is not given in the Code, the P-number categories for commonly used Section I materials are as follows:

- P-No. 1 covers plain carbon steels, C-Mn-Si, C-Mn, and C-Si steels.
- P-No. 3 covers low alloy steels obtained by additions of up to ¾% of Mn, Ni, Mo, or Cr, or combinations of these elements.
- P-No. 4 alloys have higher additions of Cr or Mo or some Ni, to a maximum of 2% Cr.
- P-No. 5 covers a broad range of alloys varying from 2.25 Cr-1 Mo to 9 Cr-1 Mo. This category is divided into three groupings, 5A, 5B, and 5C. The 5A grouping covers alloys up to 3% Cr, 5B covers alloys with from 3 to 9% Cr, and 5C covers alloys whose strength has been improved by heat treatment.
- P-No. 8 contains the austenitic stainless steels.

S-Numbers

There is another material categorizing system that Section IX uses for materials which are not included in the material specifications listed in Section II. As explained in QW-420.2 of Section IX, these are materials that are permitted for use by the ASME B31 Code for Pressure Piping and by certain Code Cases. These materials are grouped into **S-Number** groupings similar to the P-Number groupings and are listed with them in Table QW/QB-422 of Section IX. The use of S-Numbers is not mandatory. Qualification of a welding procedure with a base metal in one P-Number qualifies that procedure for all other base metals in the same S-Number grouping, but not vice versa. That is, qualification of a weld procedure for an S-Number material does not qualify that procedure for the corresponding P-number material. This is not the case for performance qualification, however. Qualification with P-Numbers qualifies a welder for the corresponding S-Numbers, and vice versa.

Base metals that have not been assigned either a P-Number or an S-Number require individual procedure qualification.

F-Numbers

To reduce the number of weld procedure and performance qualifications, Section IX also categorizes welding filler metals (electrode, bare wire, cored wire, etc.) based on their usability characteristics, which determine the ability of welders to make satisfactory welds. These filler metals are listed in Table QW-432 of Section IX, which groups the various ASME filler metal specifications (SFA specifications) into **F-Numbers**. Part C of Section II lists American Welding Society filler metal specifications that have been adopted by the ASME and designated as SFA specifications.

Table QW-433 shows the extent to which performance qualification using one F-number qualifies a welder for other F-Numbers. Generally speaking, qualification with a filler metal that is more difficult to use qualifies a welder to use a filler metal that is easier to use.

A-Numbers

A last set of numbers called **A-Numbers** is designated in Table QW-442 for the classification of ferrous weld metal for procedure qualification. This table gives the chemical composition of the weld metal for the various A-Number groupings. Weld metal composition, i.e., A-Number, can be an essential variable for the WPS, but not for performance qualification.

The choice of welding electrodes and filler material for any given application is left to the judgment of the designer. It is often advantageous to match the weld metal composition with the base metal. Some general advice to this effect is found in PW-5.4:

Welding electrodes and filler metal shall be selected to provide deposited weld metal of chemical composition and mechanical properties compatible with the materials to be joined and the service conditions anticipated.

There is a pitfall to be avoided when selecting welding electrodes and filler metal for pressure retaining welds in 2¼Cr-1Mo materials for service at design temperatures exceeding 850°F. In the late 1980s, after the occurrence of some piping failures, Subcommittee I was advised to add a new requirement (PW-5.6), namely, that the weld metal must have a carbon content greater than 0.05%, except for circumferential buttwelds no larger than 3½ inches in outside diameter. That minimum carbon content is thought to be needed to maintain the elevated temperature strength of the weld metal. Interpretation I-92-41 reiterates the details of this requirement.

WELDABILITY

Materials are said to have good weldability if they can be successfully joined, if cracking resulting from the welding process can be avoided, and if welds with approximately the same mechanical properties as the base material can be achieved.

PW-5.2 prohibits welding or thermal cutting of carbon or alloy steel with a carbon content of over 0.35%. Higher carbon can cause zones of high hardness that impart adverse properties to the weld and heat-affected zone (HAZ) and may increase susceptibility to cracking, because regions of high hardness often lack ductility. In fact, carbon strongly affects weldability due to its influence on hardenability during heat treatment. Heat treatment includes not only deliberate heat treatment such as solution annealing, it also includes the thermal cycle that the weld and HAZ undergo in the welding process.

Various empirical expressions, known as carbon equivalency (CE) formulas, have been developed to predict a steel's hardenability by heat treatment. Thus they have also proven to be rough indicators of the relative weldability of steels, their susceptibility to cracking during welding, and the need for preheat to prevent cracking during or after welding. This type of cracking is variously described as cold cracking, hydrogen cracking, and delayed cracking (because it may occur some hours after welding is completed). One common representation of a carbon equivalency formula used to determine whether a steel needs preheat or postweld heat treatment to avoid cracking is given in the *Welding Handbook, Volume 1*, Second Edition, American Welding Society:

$$CE = \%C + \%\frac{Mn}{4} + \%\frac{Ni}{20} + \%\frac{Cr}{10} + \%\frac{Cu}{40} - \%\frac{Mo}{50} - \%\frac{V}{10}$$

Thus carbon equivalency (*CE*) is the sum of weighted alloy contents, where the concentration of alloying elements is given in weight percent. From this equation, it is seen that carbon is indeed the most potent element affecting weldability. The equation also shows the relative influence of other elements that might be present in pressure vessel steels, compared to that of carbon. Note that some elements such as molybdenum and vanadium actually contribute to weldability, at least in this particular carbon equivalency formula.

Associated with the various carbon equivalency formulas are weldability guidelines. A certain value of carbon equivalency, perhaps 0.35% or less, might indicate good weldability; a higher value might indicate that preheat and/or postweld heat treatment are required to avoid excessive hardness and potential cracking in the welds; and a still higher value might mean that the material is very difficult to weld and both high preheat and postweld heat treatment will be needed to obtain satisfactory welds. Another, slightly different example of a carbon equivalency formula is given in the supplementary (nonmandatory) requirements of ASME Specifications SA-105 and SA-106. For SA-105 forgings, the carbon equivalent is limited to 0.47 or 0.48, depending on thickness; for SA-106 pipe, the limit is 0.50. There would be little likelihood of making successful welds in these materials with *CE* values higher than those limits, even with high preheat.

Although various carbon equivalency formulas have been developed over the years for specific alloys, no universal formula has been devised that works well for all the low alloy steels. In fact the more complex the alloy becomes, the more difficult it is to devise an equivalency formula. More recent formulas for low alloy steels involve more complex interactive terms rather than the simple additive terms of the formula cited above.

DESIGN OF WELDED JOINTS

The design of welded joints is covered briefly in PW-9, which says that they "shall preferably be of the double-welded butt type, but may also be of the single-welded butt type with the filler metal added from one side only when made to the equivalent of the double-welded butt joint by providing means for accomplishing complete penetration." PW-9 then gives requirements for a tapered transition in joints between plate materials of different thickness, as discussed in the next section. Figure PW-9.1 illustrates what is intended.

For longitudinal shell joints, which are the most important because they are the most highly stressed, it is necessary to line up the center lines of the adjoining plates within the alignment tolerances permitted in Table PW-33. If these plates are not in such alignment, the membrane pressure forces carried by each shell plate would be offset from each other, creating a circumferential bending moment at the weld joint. This would be a source of additional stress in the shell. For circumferential shell joints, the centerlines of the plates are not required to line up, because the membrane force in the axial direction is less than half of that in the circumferential direction, so the bending moment induced by the offset is much reduced, and there is also margin to sustain extra stress in the axial direction. Long experience has shown that this is a satisfactory design approach.

Then in the mid-1990s a boiler manufacturer built a boiler drum in which the inside surfaces of the shell plates, instead of their centerlines, were lined up at the longitudinal weld joint. The boiler was in service by the time the error was discovered. The drum clearly did not conform to the rules of Section I. The manufacturer undertook some careful finite element analysis of the drum and evaluated the results using the design-by-analysis methods of Section VIII, Division 2. The analysis included the thermal stresses caused by start-up and shutdown of the boiler and a fatigue analysis of the combined pressure and thermal stresses. The manufacturer presented the results of the analysis to Subcommittee I and asked for a Code Case permitting construction of a boiler drum without lining up the centerlines of the shell plate at the longitudinal weld joint. The manufacturer pointed out that Section VIII permits such construction and that the drum in question could have satisfied the design rules of Section VIII. After careful review and consideration, Subcommittee I approved the case, Case 2210. It permits the inside surfaces of the shell plates to line up, under fairly restrictive conditions with respect to configuration thickness and pressure allowed. Once the case was approved, the longitudinal weld configuration used by the manufacturer became permissible, and the jurisdiction where the boiler was installed granted the owner a continued operating permit for the boiler, which is now considered a so-called state special.

BENDING STRESS ON WELDS

As explained in the chapter on the determination of allowable stresses, most of the design formulas in both Section I and Section VIII are based on the use of membrane stress as the basis for design. In a boiler, the tubes, the piping, the headers, and the shell of the drum are all designed for the membrane stress in their cylindrical walls. Bending stress and the effects of stress concentration are not calculated, although they are known to be present to some extent. This is explained, not in Section I, but in UG-23(c) of Section VIII. In fact, there are bending stresses in such pressure vessel components as flat heads, stayed heads, dished heads, hemispherical heads, and cylinder-to-cylinder junctions.

In recognition of the existence of these largely unavoidable bending stresses, and perhaps still reflecting some lack of confidence in the soundness of welded construction, Section I contains a number of design rules intended to minimize bending stresses on welded joints. For instance, PW-9.13 requires a gradual taper, no greater than 1:3, between plates of unequal thickness. A 1:3 slope, or taper, for plate materials represents an angle of a little over 18°. PW-9.3 advises that its rules for transition slopes do not apply to joint designs specifically provided for elsewhere in Section I, and less restrictive rules are provided for the weld end transition where components such as valves and pipes or fittings are welded together. (See Fig. PG-42.1 of Section I.) For such applications, a maximum slope of 30° is permitted.

Long after these rules were formulated, finite element analyses have demonstrated their validity. That is, the discontinuity stresses developed at these gradual changes in thickness have generally been found to be unremarkable and well within the limits that would apply if the design-by-analysis methods of Section VIII, Division 2 were to be applied.

The tapered transition rules of PW-9.3 and the welding end transition rules illustrated in Fig. PG-42.1 still left some doubt about the treatment of certain other weld joints, and in 1977 Subcommittee I clarified that treatment by Interpretation I-77-22, Welded Joints Between Materials of Different Thickness, as follows:

> Question: *Do the requirements of PW-9.3, calling for a tapered transition section of at least three times the difference in section thickness, apply to circumferential butt welds between two sections of header, between a pipe and the end of a header, or between a header or a nozzle and some other pressure part component such as a valve or a fitting? In all these cases the longitudinal axis of the two components coincide.*
>
> Reply: *The last paragraph of PW-9.3 now includes the phrase "This paragraph is not intended to apply to joint designs specifically provided for elsewhere in the Code...." The type of circumferential butt joint described in the question is provided for in PG-42.4.2, for which the most recent revision has been issued in the Winter 1976 Addenda. The term "component bodies" as used in PG-42.4.2 and shown on Fig. PG-42.1 is intended to include the situations described. Since the rules in PG-42.4 apply, the requirements of PW-9.3 are not mandatory for the circumferential butt welds.*

This Interpretation clarified the intent of Section I rules: joints in plate materials should have a 1 on 3 tapered transition, but joints involving pipelike components and fittings may have a steeper welding end transition (a 30° slope) as shown in Fig. PG-42.1.

For decades paragraph PW-9.4 had prohibited construction in which bending stresses were "brought directly on a welded joint." This paragraph had also specifically prohibited a single-welded butt joint subject to bending stress at the root of the weld, as illustrated in Fig. PW-9.2. In 1996, Subcommittee I recognized that bending stresses routinely occur on welded joints in all boilers and pressure vessels with no untoward results and that this broad prohibition against bending stresses was unwarranted. Accordingly, the general prohibition against subjecting weld joints to bending stress was deleted, and PW-9.4 was revised so that it now prohibits only the single-welded corner joints illustrated in Fig. PW-9.4.

ACCEPTANCE CRITERIA FOR WELDS

Part PW includes fabrication rules covering welding processes, base metal preparation, assembly, alignment tolerances, and the amount of excess weld that may be left on the weld joint (so-called reinforcement). In general, butt welds in pressure-containing parts must have complete penetration, and the weld groove must be completely filled. To assure that it is and that the surface of the weld metal is not below the surface of the adjoining base metal, extra weld metal may be added as *reinforcement* on each face of the weld. A table in PW-35 gives the maximum amount of reinforcement permitted, as a function of nominal thickness and whether the weld is circumferential or longitudinal. Requirements for butt welds are found in PW-9, PW-35, and PW-41.2.2. PW-9 and PW-35 stipulate that butt welds must have full penetration. PW-41.2.2 adds that complete penetration at the root is required for single-welded butt joints and that this is to be demonstrated by the qualification of the weld procedure. Lack of penetration is a reason for rejection of welds requiring radiography. PW-35.1 limits undercuts to the lesser of 1/32 inches or 10% of the wall thickness, and they may not encroach on the required wall thickness. Some concavity in the weld metal is permitted at the root of a single-welded butt joint, if it does not exceed the lesser of 3/32 inches or 20% of the thinner of the two sections being joined. When concavity is permitted its contour must be smooth and the resulting thickness of the weld, including any reinforcement on the outside, must not be less than the required thickness of the thinner section being joined.

It is generally not possible to measure concavity directly, and in 1983 an inquiry was received by Subcommittee I (resulting in Interpretation I-83-36) asking the following question:

> Question: *When concavity at the root of a weld is shown by radiography, can the actual depth of 3/32 in. or 20% of the thinner section being joined be measured by radiographic density only?*
> Reply: *Section I does not address measurement of concavity.*
> Question: *Is concavity at the root of a weld as determined by radiography acceptable if the contour of the concavity is smooth and the density change is not abrupt?*
> Reply: *One of the criteria for acceptability of concavity is that the contour be smooth. PW-41.2.2 contains several additional criteria.*

A short while later another inquiry was received asking several questions about acceptance criteria for welds. They were answered by Interpretation I-83-59, as follows:

> Question: *What are the acceptance criteria if two indications of different types are present in the same radiograph?*
> Reply: *The acceptance criteria in PW-51.3 for any type of imperfection apply independently of any other type of imperfection in the same radiograph.*
> Question: *What are the acceptance criteria if an indication of a slag inclusion is noticed to be interposed between two or more isolated indications which are acceptable per Appendix A-250, Summer 1977 Addenda?* (Authors' note: Appendix A-250 is virtually the same in 1998 as it was in 1977.)
> Reply: *The slag inclusion is itself an indication; the acceptance criteria are given in PW-51.3.2 and A-250.*
> Question: *What is the acceptance standard for the following indications of imperfections in the radiograph?*
> *(a) Excess penetration at the root of the weld.*
> *(b) Mismatch at the root of the weld.*
> *(c) Concavity/setback/suck up at the root.*
> *(d) Tungsten inclusion in weld metal.*
> Reply: *Acceptance standards are given in the following paragraphs:*
> *(a) PW-41.2.1 and T-234.1 of Section V.*
> *(b) PW-33 and T-234.1 of Section V.*
> *(c) PW-41.2.2 and T-234.1 of Section V.*
> *(d) PW-51.3.4 and Appendix A-250.*
> Question: *Is the acceptance standard in PW-51 applicable regardless of whether defects exist at the root of the weld or are interposed at the fusion zone between two layers of weld metal or at the boundary of the fusion zone between weld metal and base metal?*
> Reply: *Yes.*

A 1992 Interpretation, I-92-75, dealt with the complete penetration of butt welds in boiler tubes and how it may be demonstrated:

> Question: *Is it a requirement of Section I that all boiler tube buttwelded joints be complete joint penetration welds?*
> Reply: *Yes.*
> Question: *When radiography is not required, and visual examination of the root is not practical, what demonstration in addition to the requirement of PW-41.2.2 and a successful hydrostatic test is necessary to accept circumferential butt welds covered by PW-41?*

Reply: *None.*

Question: *Is it a requirement of Section I that the manufacturer be responsible for meeting the applicable requirements of Section I?*

Reply: *Yes. The manufacturer is responsible for meeting the requirements of Section I and of his Quality Control Program.*

When Subcommittee I replied that Section I doesn't address the measurement of concavity, it was tacitly acknowledging the difficulty of making such measurements. This reply also reflects the industry's good experience with single-welded circumferential butt joints in boiler and superheater tubes using all of the above-described acceptance criteria. Some part of this success is also due to the fact that the axial stress in these tubes at the butt joints is generally less than half of the allowable design stress, as explained in Radiographic Examination of Welds at the end of this chapter.

TUBE-TO-HEADER ATTACHMENT USING PARTIAL PENETRATION WELDS

A review of the acceptable types of nozzles and other connections illustrated in Fig. 16.1 shows a number connected by partial penetration groove welds. One very widely used detail for tube-to-header connections is shown in Fig. PW-16.1(z). The application of this detail confuses some designers, and over the years, Subcommittee I has answered several inquiries about its use. As can be seen in the side view (section 1-1) of the connection, the following requirements are imposed (all nomenclature is defined in PW-16.2):

1. There must be a groove weld whose depth, t_w, is equal to the wall thickness, t_n, of the attached tube or nozzle, but not less than ¼ inches.
2. A fillet weld with a throat equal to t_c must be provided on top of the groove weld.
3. The tube must project below the bottom of the groove weld at least 1/16 of an inch. The sketch shows a recess of that depth, but by Interpretation I-92-95, Subcommittee I explained that the tube need not rest on such a shoulder and that a through-hole in the header matching the outside diameter of the tube was permissible so long as the tube extended into the hole a minimum of 1/16 of an inch beyond the machined weld groove. (The purpose of a recess or insertion into a hole matching the OD of the tube is to ensure that the tube is centered properly, so that adequate weld is provided all around the tube.)

From the two views of PW-16.1(z), it can be seen that adequate header thickness is needed to provide for the depth of the groove weld and that beyond some maximum ratio of tube diameter to header diameter, the above requirements cannot be met. This is because the groove weld would have little or no depth on the hillside portion of the header, as is shown in the left sketch of PW-16.1(z). Advice to this effect is given in two hard-to-find interpretations. Although they were issued in 1977 and 1978, one appeared on page 161 in Volume 5 of the interpretations as part of errata to Volume No. 3 of the Interpretations, with no title, but under the heading Section I, Weld Joint Acceptability Under Figure PW-16.1. The other appeared in 1984 as Interpretation I-83-08R, on page 61 of Volume 15 of the interpretations.

ALIGNMENT TOLERANCE

PW-33 addresses the alignment tolerances that apply when plates are butt welded together. These tolerances are expressed as the amount of offset permitted at the edges of the plates. Table PW-33 lists these tolerances for longitudinal and circumferential welds as a function of the thickness of the plates to be joined. Any offset within the allowable tolerance must be faired at a 1 on 3 taper across the width of the finished weld. The addition of weld metal beyond what otherwise might be the edge of the weld may be needed in order to achieve the 1 on 3 taper if a line connecting the edges of the weld preparation is steeper than 1 on 3.

In 1995 a dispute arose regarding the alignment tolerance required when tubes are butt welded end to end. As explained in Radiographic Examination of Welds at the end of this chapter, these circumferential tube joints do not require radiography, and there are accordingly no acceptance standards to guide the parties. The inquirer wanted to know whether the alignment tolerances of PW-33 for plates applied to circumferential joints in tubes. In Interpretation I-95-01, Subcommittee I replied that they do not. That left the question of just what tolerances do apply to such joints. The subcommittee considered the problem and decided to revise the title of PW-33 so that it would cover alignment tolerance for shells and vessels, including pipe or tube when used as a shell. At the same time a new paragraph, PW-34, was added to cover alignment when tubes or pipe are butt welded end to end. That paragraph specifies only that the alignment shall provide for complete weld penetration at the inside surfaces.

BACKING RINGS

PW-41.2.2, in discussing single-welded butt joints, stipulates that if complete penetration at the root cannot otherwise be assured, the procedure shall include a backing ring or equivalent. Occasionally, a dispute arises as to what material may be used for a backing ring. The answer to this question is found in PW-41.2.4, which says only that material for backing rings shall be compatible with the weld metal and the base material and shall not cause harmful alloying or contamination. Thus the manufacturer is free to use its best judgment in choosing any material that meets those criteria. Backing rings may be left in place or removed as desired.

PREHEATING

Deposited weld metal is cooled very quickly by its surroundings. Once it solidifies, it forms various microstructural phases—solid solutions of iron and carbon (and other alloying elements). These phases include ferrite, pearlite, bainite, martensite, and austenite (in the higher alloy materials). The formation of any particular phase depends on the cooling rate and alloying elements, such as chrome or nickel, that may be present. Certain phases, e.g., untempered martensite or bainite, can impart unfavorable properties to the resulting weldment (high hardness, lack of ductility, lack of fracture toughness). One way to address this problem is through the use of preheating, which by raising the temperature of the surrounding base metal, slows the cooling of the weld and may prevent the formation of undesirable phases.

Preheat serves another useful function. It assures that the weld joint is free of any moisture, which can otherwise serve as a source of hydrogen contamination in the weld. Hydrogen can dissolve in liquid weld metal and by various mechanisms cause cracking in the weld (known variously as cold cracking, hydrogen-assisted cracking, and delayed cracking). Maintaining the weldment at preheat temperature during and after welding promotes the evolution of dissolved hydrogen. Although preheating a weld joint can be beneficial, Section I does not mandate the use of preheat except as a condition for the omission of postweld heat treatment (as provided in Table PW-39). However, it is often specified for new construction welding and for repair welding, especially when the walls are relatively thick. A brief guide to preheating practices is included in Nonmandatory Appendix A-100.

It is important to remember that while Section I does not normally mandate preheat, B31.1, Power Piping, does. Consequently, all welds (including tack welds) in boiler external piping require preheat. The preheat temperatures required for the various P-No. materials are given in B31.1, paragraph 131, which also specifies the size of the region around the weld that must be maintained at the preheat temperature. These preheat requirements are similar to those in Nomandatory Appendix A-100 of Section I. It turns out that for many commonly used carbon and low alloy steels up to 2 inches thick, the preheat temperature is only 50°F. Note also that the B31.1 definition of thickness for consideration of preheat is the *greater* of the nominal thicknesses at the weld for the parts to be joined. This is different from the B31.1 definition (in 132.4) of nominal thickness to use when considering PWHT requirements. Unfortunately, the Section I definition of nominal thickness (in PW-39.3) for consideration of PWHT requirements differs from that of B31.1. This is discussed further in the next section.

POSTWELD HEAT TREATMENT

As explained under preheating, rapidly cooling weld metal and the adjacent heat-affected zone are subject to the formation of adverse phases, which can result in zones of high hardness, lack of ductility, and poor fracture toughness in the weldment. In addition, the differential cooling of the weld metal compared to the surrounding base metal can lead to the formation of very high residual stresses. Again, the extent to which any of these effects takes place is a function of the cooling rate, alloy composition, relative thickness of the parts being joined, and whether preheat is applied. Experience has shown that preheat and postweld heat treatment (PWHT) are often unnecessary for relatively thin weldments, especially those of plain carbon steel. However, the general rule of PW-39 is that all welded pressure parts of power boilers shall be given a postweld heat treatment unless otherwise specifically exempted (e.g., by PFT-29 or PMB-9) or exempted by the notes in Table PW-39. The materials in Table PW-39 are listed in accordance with the P-Number grouping of QW-422 of Section IX.

There are some materials for which PWHT is neither required nor prohibited. These are the P-No. 8 (austenitic stainless alloys), P-No. 31 (copper and copper-based alloys), and P-No. 45 (nickel and nickel-based alloys). There is also one material group, P-No. 10 Group No. 1 (ferritic stainless alloys) for which PWHT is always required.

Remember the rules for postweld heat treatment of boiler external piping are found in B31.1, paragraph 132, and those rules differ somewhat from those of Section I. This is another pitfall for the unwary, should an Authorized Inspector discover that the wrong rules were followed, particularly just as the job is nearing completion. One difference between the two codes is that Section I requires the PWHT of a small pipe connection to another pipe or header, while B31.1 often does not.

While preheat is aimed at preventing or limiting the formation of adverse phases, postweld heat treatment is used to ameliorate their effects. Postweld heat treatment is capable of tempering, or softening, hard phases, and it can restore a large measure of ductility and fracture toughness in the weld and heat-affected zone. At the same time, it relieves most of the residual stress caused by the cooling of the weld metal. The goal of PWHT, therefore, is to restore the properties of the weldment as nearly as possible to those of the original base metal.

Exemptions from PWHT

The exemptions from PWHT permitted by the twelve tables in PW-39 are listed for various types of welds: circumferential welds in pipes, tubes, or headers; fillet welds attaching pressure parts such as socket welding fittings; fillet welds attaching connections to vessels; fillet welds attaching nonpressure parts to pressure parts; and seal welds on tubes, closures, or fittings secured by mechanical means such as rolling or threading. For each type of weld, exemptions from PWHT may be available if certain limits are not exceeded. The limits are based on such parameters as material thickness, carbon content, chrome content, fillet weld throat size, component diameter, opening diameter, or a requirement for preheat at a particular temperature. One common misunderstanding regarding PWHT exemptions is what Section I means by the term **circumferential welds**. This term applies only when the two parts such as two tubes, or two pipes, or two sections of a cylindrical shell are butt welded end to end. The weld between a tube stub and a header, or the weld between a handhole or manhole frame and a drum, is not considered a circumferential weld and therefore must be given a PWHT unless a specific exemption can be found that covers the situation. (This was affirmed by Interpretations I-92-23 and I-92-77.) The rationale behind this definition of a circumferential weld is that the abutting parts are likely to have comparable thickness and one would not be a significantly greater heat sink than the other, so that there need be no concern about rapid cooling of the weld that would potentially leave it with undesirable characteristics for which PWHT would be advisable.

Sometimes manufacturers of firetube boilers stop construction just short of completion and place the boilers in stock, in anticipation of orders. Occasionally, a purchaser wants a nozzle added to a boiler as an auxilliary steam connection or to provide for a larger size safety valve, and the addition of such a nozzle would normally require PWHT. PWHT can damage a firetube boiler with rolled tubes, because rolled joints loosen at PWHT temperatures. (Rolled tubes are held in place by residual stresses deliberately induced in the tubes and the tubesheet by the rolling process, and at PWHT temperatures these stresses relax.) The problem of adding nozzles to a boiler without either repeating PWHT or doing a local PWHT was brought to Subcommittee I. After due consideration the committee developed Code Case 1918, Attachment of Nozzles or Couplings to a Boiler Vessel After PWHT. This case allows the attachment of not more than two nozzles or couplings subject to quite a number of restrictions intended to ensure that PWHT can be safely omitted. The nozzle must be made of P-No. 1 Group 1 or 2 material, its size is limited to NPS 4 with a maximum wall thickness of ½ inches, vessel thickness is limited to ¾ inches, only certain weld configurations of PW-16 are permitted, and a minimum preheat of 200°F must be applied. Despite these many restrictions, the case can be very useful if some small connection needs to be added to a boiler vessel after the PWHT has been completed. Moreover, the case applies to all boilers, not just firetube boilers. In the early 1990s, an attempt to incorporate Case 1918 into Section I was not approved by the Main Committee, for a variety of reasons, and it is expected to be among those few that remain in effect indefinitely.

In PWHT the components to be treated are placed in a heat treatment furnace and slowly heated to the temperature specified in Table PW-39. They are held at that temperature for the time specified, usually one hour per inch of thickness up to a certain thickness plus 15 minutes additional for every inch above that thickness. This rule is intended to ensure that the centermost portions of thick sections have sufficient time to reach the minimum holding temperature. The nominal thickness in Table PW-39 used to determine PWHT requirements is the thickness of the weld, the pressure retaining material, or the thinner of the sections being joined, whichever is the least. The time at temperature requirement can be satisfied by an accumulation of postweld heat treatment cycles. When it is impractical to heat treat at the specified temperature it is permissible to do the heat treatment at lower temperatures for longer periods of time, but only for P-Number 1, P-Number 3, P-Number 9A, and P-Number 9B materials. Table PW-39.1 lists these alternative longer times. The table shows that for a 50° decrease in the minimum specified holding temperature, the holding time doubles. This shows how strongly temperature-dependent the underlying creep relaxation process is.

Unlike some other book sections, Section I usually does not specify any particular heating or cooling rate to be used in PWHT. PW-39.3 stipulates only that the weldment shall be heated slowly to the required holding temperature. It goes on to say that after being held at that temperature for the specified time, the weldment shall be allowed to cool slowly in a still atmosphere to a temperature of 800°F or less. This is to avoid high thermal stresses and potential distortion caused by large differential temperatures throughout the component. Table PW-39 does have two groups of material for which it gives advice on the cooling rates to be used following PWHT. Those are the P-No. 7 and the P-No. 10E materials. Notes in the PW-39 tables applicable to those materials specify a maximum cooling rate of 100°F/hr in the range above 1200°F, after which the cooling rate "shall be sufficiently rapid to avoid embrittlement." In some cases this might be 400°F/hr. Presumably, the manufacturer has sufficient knowledge of metallurgy to determine how rapid that rate must be.

Local Postweld Heat Treatment

Local postweld heat treatment is permitted as an alternative to heating a complete assembly. It is also used following repair welding, when the complete assembly may have already received its PWHT and repeating the whole process would be unnecessary and uneconomical. PW-39 requires that local PWHT be accomplished by heating a circumferential band around the vessel to the prescribed holding temperature. The minimum width of the band is three times the plate thickness on either side of the welded joint. If a nozzle or other attachment is being given a local PWHT, the circumferential band must extend on either side of the nozzle by at least three times the wall thickness. PW-39.6 provides similar circumferential band rules for local PWHT of welded joints in pipes, tubes, and headers.

The basis of the heated circumferential band method is this: The welds and heat-affected zone must be brought up to the prescribed temperature and held there, and the band widths specified are simply a means of assuring that this happens. To avoid creating high thermal stresses that might undo some of the benefits of the PWHT, it is necessary that the entire circumference be brought up to a uniform temperature. This can be seen by considering the temperature distribution in the vessel. For example, the prescribed PWHT temperature for a P-No. 1 material is 1100°F. If the rules for local PWHT are followed, the resulting temperature distribution consists of a band at 1100°F, with the temperature decreasing along the vessel in either direction away from the band. The axial plot of temperature would look fairly symmetrical, with a flat top at the 1100°F band and axial temperature gradients sloping down to cooler portions of the vessel on either side. With such a temperature distribution, the heated band is permitted relatively free and uniform thermal expansion, thereby avoiding high thermal stresses.

By contrast, consider what might happen if a local PWHT were attempted on a nozzle at the middle of a dished head by heating only the nozzle and a narrow region of the head surrounding it. The heated portion is constrained from expanding by the much cooler surrounding portion of the head. Thermal stresses exceeding the yield strength of the material might occur, leaving the vessel with high residual stresses after it cools. This may be more readily appreciated by considering a simpler but analogous problem, that of a large flat plate with a small circular hot zone in the middle of the plate. The hot circular zone is constrained from expanding by the surrounding cooler material, and thermal stresses greater than yield may easily develop, leaving high residual stresses after the plate cools. Since one of the goals of PWHT is to mitigate residual stresses, Section I always requires heating a band all the way around the vessel. Although Section I doesn't offer any advice on the subject of local PWHT on a head, it is the authors' opinion that to do a proper local PWHT of a weld or a nozzle in a head without causing high residual stresses near the weld, the whole head should be heated to the prescribed temperature.

Minor Repairs After PWHT

Occasionaly, some minor local damage is done to completed boiler components after postweld heat treatment, and a little repair welding is needed. Unless an exemption from PWHT is available in PW-39, the PWHT would have to be repeated for the whole component. In the late 1970s, Subcommittee I granted some relief from this problem when the welding was considered so minor that any further postweld heat treatment could reasonably be waived. New provisions were added to PW-40.2 covering tube-to-header or tube-to-drum welded joints that have already received their required PWHT. In such cases minor local additional welding for certain listed materials without repeating the PWHT became permissible, subject to limitations on tube size, depth of reworked weld, use of preheat, and qualification of the welding procedure for the thickness of rework welding to be performed without PWHT. This approach is analogous to that of PW-54.3, which permits the welding of nonpressure parts to pressure parts after the hydrostatic test (see discussion under Hydrostatic Testing).

Interpretations on PWHT

There have been a number of interpretations on the subject of PWHT. Several are worthy of mention. One was Interpretation I-80-10, which dealt principally with *local* PWHT. Several questions were posed. One was whether a gas torch could be used to perform the PWHT. Subcommittee I replied that Section I did not restrict the method of heating; a gas torch was an acceptable method. Another question was whether temperature-indicating crayons could be used to indicate postweld heat treatment temperatures. The reply was that Section I did not restrict the method of temperature indication, provided the method selected permitted control of PWHT within Section I limits for the material involved.

Another inquiry on PWHT asked whether there was a *maximum holding temperature* for PWHT. Subcommittee I replied in Interpretation I-81-04 that Section I did not define such a limit. (This was slightly inaccurate, since Table PW-39 did at that time stipulate a maximum holding temperature of 1175°F for SA-203, a P-No. 9B, low nickel alloy sometimes used for pressure vessel plates.) There may be some negative effects of holding weldments at temperatures significantly above the minimums specified in Table PW-39. Some softening of the material may take place, evidenced by a slight reduction in the yield stength. However, the as-received yield strength of most materials is usually somewhat above the specification minimum, and would probably remain so, even after prolonged heat treatment. Another possible effect of such treatment on low alloy steels might be a slight lowering of mechanical properties at elevated temperature.

It happens that B31.1, Power Piping, doesn't list just a single minimum PWHT holding temperature. It lists a temperature range, and advises that the upper limit of the range may be exceeded, provided the actual temperature does not exceed the lower critical temperature of any of the materials (at which point some undesirable metallurgical change might occur). Power Piping also provides Table 129.3.2, which gives approximate lower critical temperatures for various commonly used piping materials. A note under the table indicates that the information is for guidance only and that the user may use other values obtained for specific materials.

TEST PLATES

One of the last subjects covered in Part PW is that of test plates for vessel materials. This subject is covered in considerable detail in PW-53. Most of this text has changed very little in the last 40 years, except that in the 1960s a clause was added to PW-53.1 that exempts cylindrical pressure parts constructed of P-No. 1 materials from these test requirements. Since most boiler drums are made of these plain carbon steel materials, the tests are not required. In fact, most of this testing duplicates the testing already covered by the material specifications and the weld procedure qualification. Moreover, 30 years of experience have shown that the testing called for by the material specifications is sufficient for the materials ordinarily used for pressure vessels.

INTERPRETATIONS ON WELDING

PW-28 makes it clear that welding of pressure parts to each other, and the welding of load-carrying nonpressure parts to pressure parts, requires the use of qualified welders using qualified procedures. However, when the welding process is automatic, welding procedure and performance qualification is not required for attachments such as fins that have essentially no load-carrying function. A number of questions about the welding of fins have been considered by Subcommitee I.

One fin question was considered under Interpretation I-81-18. The question was whether and under what conditions a boiler manufacturer could buy tubes with machine resistance welded fins from an outside organization (presumably one without a Code symbol stamp) and use the tubes for Section I construction. This inquiry was answered in 1981 as follows:

> Reply: *Tubing with machine resistance welded fins falls in the category of miscellaneous parts and the rules of PG-11 apply, including requirements for marking and inspection.*

Such fins are usually considered to "have essentially no load-carrying function" (see PW-28.1.2). In 1981 PW-28 permitted such fins to be joined to pressure parts by any machine welding process performed in accordance with a welding procedure specification, without requiring procedure and performance qualification testing. The rules were changed around 1990 so that procedure and performance qualification testing is required for all types of welding except automatic welding (in which the equipment performs the welding without adjustment of the controls by the welding operator). However, under either the old rules or the later rules, finned tubes could be furnished as standard welded pressure parts under the provisions of PG-11.3, which call for the parts manufacturer to do any welding in accordance with virtually all the requirements of Sections I and IX. As explained in Chapter 11 under Standard Pressure Parts, the parts manufacturer need not have a Code symbol stamp to provide welded parts.

The calibration of welding equipment is not mentioned in the Code. In 1983 an inquiry on this subject was received that applied to Sections I, IV, and VIII. It was answered in Interpretation I-83-45:

> Question: *Is it a requirement of Section I, IV, VIII, Divisions 1 and 2, that manual, semiautomatic, and automatic welding equipment be calibrated, and if so, that the devices used for calibration be calibrated?*
>
> Reply: *No.*

In considering the calibration question, the various committees apparently agreed that the industry has developed welding methods and equipment that provide satisfactory welds when the procedures and welders are qualified in accordance with Code rules. The Code also mandates NDE for the welds (visual, radiographic, or ultrasonic examination), which is intended to detect significant imperfections, whatever their cause.

PW-31 deals with the assembly of parts to be welded. It mentions *tack welding* as a means of holding the edges of parts to be welded in alignment. Tack welds must be made using a fillet weld or butt weld procedure qualified in accordance with Section IX. Once they have served their purpose, tack welds may either be removed completely or their starting and stopping ends have to be prepared by grinding so that they can be satisfactorily incorporated into the final weld. Tack welds to remain in place must be made by qualified welders and examined visually for defects and, if found defective, must be removed.

Often a manufacturer authorized to use a Code stamp subcontracts the rolling of a shell to an outside organization, one without an ASME Certificate of Authorization to use a Code stamp. The question then arises as to whether and under what circumstances the subcontractor may use tack welding. This question was answered in 1983 for Sections I, IV and VIII, Division 1, by Interpretation I-83-46, as follows:

> Question: *A manufacturer, holder of a valid U, S, H, or M Certificate of Authorization, subcontracts an outside organization to roll a shell, which is to be part of a stamped boiler or vessel, and to perform tack welding to secure the longitudinal seam alignment. The outside organization is not part of the Certificate holder's organization. Under what conditions may this work be performed?*
>
> Reply: *In accordance with paragraphs PW-31 of Section I, HW-810 of Section IV, and UW-31 of Section VIII, Division 1, tack welds, whether removed or left in place, shall be made using a fillet weld or butt weld procedure qualified to Section IX. Tack welds to be left in place shall be made by welders qualified to Section IX and shall be examined visually for defects, and if found to be defective, shall be removed. It is not necessary that a subcontractor performing such tack welds for the boiler or vessel manufacturer be a holder of an ASME Certificate of Authorization. The final boiler or vessel manufacturer shall maintain the controls to assure that the necessary welding procedure and performance qualifications are met in order to satisfy Code requirements.*

Questions sometimes arise as to what extent pipe and tube made by electric resistance welding (ERW) must meet any of the Section I rules for welding and welded products. ERW pipe and tube are made by an automatic welding process under what are supposed to be carefully controlled factory conditions. Included in the manufacturing process are various nondestructive examination methods, whose purpose is to assure the soundness of the welds. These NDE methods may include eddy-current testing, ultrasonic testing, and hydrostatic testing. When ERW products were introduced years ago, there were occasional problems with faulty welds that escaped detection, and the Code committees imposed an arbitrary 15% penalty on the allowable stress for such products. More recently, following years of generally satisfactory experience, the penalty has been dropped for some tubular products when extra NDE is provided for the full length of the ERW seam. Nevertheless, some committee members lack full confidence in ERW tubing because of the possibility that NDE may not always be effective in weeding out bad welds, for one reason or other. There have been several interpretations on the subject, which have established that the weld in ERW products does not require PWHT or radiography by the boiler manufacturer and that ERW pipe may be used for the shell of an electric boiler built to the rules of Part PEB without inspection and Manufacturer's Partial Data Reports. These 1986 Interpretations are I-86-40, I-86-55, and I-86-56, as follows:

> Question: *When ERW pipe or tube which complies with a material specification is used as a nozzle, does PW-11 require that the longitudinal seam be radiographed?*
> Reply: *No.*
> Question: *When ERW pipe or tube which complies with a material specification is used as a nozzle does Table PW-39 require that the longitudinal seam be subjected to postweld heat treatment?*
> Reply: *No.* (Note: Remainder of reply omitted because it was on another subject.)
> Question: *May SA-178 Grade A ERW tubing produced by a supplier who is not a Code Certificate Holder be used as the shell of a boiler constructed to the rules of Part PEB?*
> Reply: *Yes.*

Note that in asking this last question, the inquirer was laboring under a misapprehension. The tube manufacturer is never required to have a Certificate of Authorization to use a Code symbol stamp, since the manufacturer is not engaging in Code construction when it manufactures tubing, even though that tubing is used in Code construction. The underlying question is simply whether ERW tubing may be used for the shell of an electric boiler; the answer is that it can be so used. This interpretation clarifies the application of footnote 6 to PG-11 on standard pressure parts. That footnote says that fusion-welded pipe for use as the shell of a vessel is subject to the same requirements as a shell fabricated from plate, including inspection and Manufacturers' Partial Data Reports. However, ERW pipe has been welded without the addition of filler metal, by a fully automatic process, and the welding and other aspects of pipe quality assurance are considered adequately controlled by the governing ASME material specification. Consequently, the inspection referred to in footnote 6 (which would have consisted of assuring proper long seam welding, PWHT, and radiography) is not needed, and the piece of pipe can be treated as a standard pressure part that doesn't require inspection and a Manufacturers' Partial Data Report. (Of course, the rest of the boiler, including head-to-shell welds and nozzle-to-shell welds, require the full application of Section I rules; it is only the ERW seam that is exempt.)

In 1992 Subcommittee I issued Interpretation I-92-82, essentially covering the same ground as I-86-56 (above), advising that ERW pipe material listed in PG-9.1 may be used as the shell of an electric boiler, subject to the PEB-5.2 minimum shell thickness requirement of 3/16 of an inch.

PW-40 covers the repair of weld defects. It stipulates that imperfections detected visually, by leak tests or by the examinations mandated by PW-11 (RT or UT), must be removed. The weld joint must then be rewelded and reexamined. A 1986 Interpretation, I-86-66, relaxed the otherwise rather strict rules concerning who can make weld repairs on Section I components, as follows:

> Question: *May a non-Certificate Holder's welders be used in his shop, provided that they are under the control and supervision of a Certificate Holder under the conditions and requirements of PW-1.2 and PG-105.3, to make repairs either on pressure parts in a new vessel or on new vessel parts which are to be stamped with the ASME Section I symbol?*
> Reply: *Yes.*

The above interpretation is consistent with the rules for welded standard pressure parts in PG-11. Those rules permit a non-Certificate Holder to weld standard pressure parts provided the welding complies with PW-26 through PW-39. That proviso is essentially the equivalent of calling for welders and procedures to be qualified in accordance with Section IX. The two paragraphs referenced in the interpretation call for such welding and also for the oversight of the Authorized Inspector.

One design feature that is not explicitly covered by the rules of Section I is the attachment of nozzles by single fillet welds. There have been several interpretations on the subject: I-79-18, I-86-82, and I-92-

86. The reply in all these cases was that Section I does not permit the attachment of small-diameter nozzles, or internally threaded fittings, or tubes to headers, by single fillet welds only. Where Section I does permit single fillet welds, additional mechanical support must be provided by some other means, such as by expanding. In such situations, the weld is considered a seal weld, and the strength of the joint is attributed to the expansion of the tube. Several such combined attachment methods for tubes are described in PWT-11.1 and PFT-12.2.

In the early 1990s Subcommittee I attempted to include in PW-15 a specific prohibition against the attachment of nozzles and other connections by single fillet welds, but the action was blocked at the Main Committee by some members who considered properly designed fillet welds a satisfactory means of attachment and by others who noted that Section VIII permits the use of at least some single fillet welded connections. However the members of Subcommittee I continue to have a number of concerns about using such connections, among which are these: Section I is a system Code, covering the boiler and attached piping. The piping often imposes mechanical loading on the boiler connections, due to its thermal expansion. In a single fillet welded connection, the fillet welds would have to resist these external forces and moments in addition to their function of attaching the connection and transferring pressure loads. Moreover, it is not always possible to ensure full fusion at the root of a fillet weld, and that root can be a weak point with respect to stress concentration and susceptibility to corrosion. Also, when Section VIII permits the use of single fillet attachment welds [see Fig. UW-16.1, UW-16.2, and UW-16(e) and (f)], several restrictions are imposed: Fittings attached to the outside of a vessel by single fillet welds are limited to a maximum size of NPS 3. When a nozzle neck or tube is attached from the inside only and there are external loads on the nozzle, the clearance permitted between the outside of the nozzle and the vessel wall is restricted [see Fig. UW-16.1(v) and (w)]. Thus the vessel wall could help resist any significant bending moment on the nozzle, thereby limiting the loads transmitted to the single fillet weld.

The present status of single-welded connections on Section I boilers is that they are not explicitly prohibited, but three interpretations have been issued by Subcommittee I since 1978 saying they are not allowed. The Subgroup on Design of Subcommittee I is considering whether single fillet welded connections could perhaps be justified under certain conditions, such as for very small fittings and connections, or connections with little or no external loading, e.g., the connection for a pressure gage. PW-41.5 presently permits the use of socket-type joints to connect pipe or tubes with an outside diameter up to 3½ inches. In such joints the pipe or tube need only be inserted ¼ of an inch into the socket. A fillet weld is then used to attach the tube or pipe to the socket. The resulting socket welded connection bears a strong resemblance to a single fillet welded connection. It would seem that years of satisfactory experience with these relatively small socket welded connections may convince Subcommittee I to permit some limited use of tube or nozzle attachment by single fillet welds in the future. The Subgroup on Design is presently (1998) considering such a proposal.

RADIOGRAPHIC EXAMINATION OF WELDS

As a general rule, Section I requires a radiographic examination of all longitudinal and circumferential butt welded joints. However, radiography is not required for those types of weld joints for which it is not possible to do meaningful radiography. Examples of welds exempt from radiography on that basis include fillet welds, partial penetration welds, corner joints such as those illustrated in Fig. PW-9.2, and most of the nozzle attachment welds illustrated in Fig. PW-16.1. In order to obtain useful results from radiography, the thickness at the weld joint must not vary abruptly, because potential flaws, such as lack of penetration, are recognizable by variations in a relatively uniform pattern of radiographic density. The joints just enumerated cannot by their nature yield radiographs of uniform density.

The paragraphs which call out the requirements for radiography, PW-11 and PW-41, also grant a number of exceptions. Some of the most important exceptions are for circumferential butt welds in components that do not exceed certain sizes and wall thicknesses, depending on whether the contained fluid is water or steam and whether the welds are in contact with furnace gases or radiation from the furnace. Under these exemption rules, which are found in PW-41, most circumferential tube welds used in boiler and superheater tubes need not be radiographed. The same is true for circumferential joints in waterwall tubes of the furnace. In large boilers, the latter are usually fabricated into waterwall panels, which are joined end to end. Aligning these tubes when panel sections are welded together is somewhat difficult because the individual tubes in a given panel are not perfectly spaced, due to small variations in the welding process joining the fins to the tubes.

The fins (also called membrane) between the tubes hold them rigidly together and prevent any adjustment for weld fit-up, so a short length of fin is left off at each end of the tubes until the panels are welded together. It is then possible to make slight adjustments to improve the fit of the tube ends. After completion of the butt welds the omitted length of fin is installed. However, even after all this care is taken, the alignment of the tube ends is sometimes not perfect. It happens that Section I exempts most of these circumferential welds, which are in relatively small tubes, from radiography. (The basis of this exemption is explained below.) Occasionally, a boiler owner decides to have the welds radiographed anyway, as a check on the quality of the welding. Any misalignment shows up as an indication of a potential flaw on the radiograph, and a dispute can then arise between the manufacturer and the owner as to whether the welds comply with Section I, because in a weld that does require radiography, such an indication might be cause for rejection. The owners misinterpret PW-40 on repair of weld defects as applicable to welds that do not require radiography.

In 1983 such a situation led Subcommittee I to issue Interpretation I-83-28, entitled Acceptance Standards When Radiography Is Not Required:

> Question: *When radiography is not required by Section I, but is used by the fabricator, must the acceptance criteria meet the requirements of PW-51?*
> Reply: *No, it is the intent of Section I that radiographic standards for which Section I does not require radiography are outside the scope of the Code.*

That same year Subcommittee I provided further guidance about acceptance standards for butt welded joints that don't require radiography, in Interpretation I-83-79. The second question of that Interpretation was as follows:

> Question: *What are the acceptance requirements for butt welded joints that are not radiographed, particularly in regard to the full penetration requirement?*
> Reply: *PW-35 and PW-41.2.2 contain rules applicable to acceptance of butt welded joints that are not radiographed.*

The paragraphs referred to, PW-35 and PW-41.2.2, call for full penetration at the root of the weld, noting that such penetration shall be demonstrated by the qualification of the welding procedure.

After several such disputes were brought to Subcommittee I in the early 1990s, the committee revised the introduction to PW-11 to make it clear that this problem is not a Code question, it is rather a contractual matter between the manufacturer and the user. PW-11 now gives the following advice:

> *This Section exempts selected welds from radiographic and ultrasonic examination. Experience has demonstrated that such welds have given safe and reliable service even if they contain imperfections which may be disclosed upon further examination. Any examination and acceptance standards beyond the requirements of this Section are beyond the scope of the Code and shall be a matter of agreement between the Manufacturer and the user.*

Basis of RT Exemption for Circumferential Welds

Section I exempts these circumferential welds from radiography and ultrasonic examination because the longitudinal stress in the tube crossing the welds is usually quite low. The tube is essentially a long cylinder, and in the absence of any local bending loads or axial loads other than pressure (which is usually the case), the axial pressure stress in a long cylinder is something less than half of the hoop membrane stress. Since the tube is designed for the hoop stress, the axial stress is less than half of the allowable stress. Thus even if half the metal were missing at the weld (due, e.g., to lack of fusion or lack of penetration), there would still be enough metal remaining to carry the pressure stress in the axial direction without exceeding the allowable design stress. The tube on either side of the weld is capable of making up for some weakness in the weld with a slight increase in hoop stress. This is the equivalent of the ability of a cylinder to carry stress around a nozzle opening with a modest increase in local stress, as permitted by Section I's philosophy of designing for average hoop membrane stress.

There are occasions, however, when the axial stress can be significantly higher than the modest pressure stress mentioned above, due to superimposed mechanical loads. Sometimes tubes are used as vertical supports, called stringer tubes, to support other, horizontal tubes, when the span of those tubes would otherwise be too long. Stringer tubes might typically be used in the heat recovery area (HRA) of a boiler. The stringer tube is, in effect, a steam-cooled support out in the gas stream, where it is too hot to provide any other type of support. In that case, the designer must choose the weight to be carried so that the combined axial stress due to pressure and mechanical load doesn't exceed the allowable stress. The designer also has the option of using a thicker tube if needed to carry heavy loads.

Another instance of an axial stress adding to the stress caused by internal pressure can occur in waterwall tubes due to internal pressure in the furnace or external pressure on the outside of the furnace depending on whether the boiler is a forced draft or an induced draft unit. Some large utility boilers are designed with bag house filters as part of the flue gas cleanup system, and these filters can add a large pressure drop to the flow path through the boiler. As a result, the furnace walls could be subject to a pressure of as much as 36 inches of water. Even without bag house filters, a significant differential pressure can develop between the inside and outside of the furnace. Horizontal structural members, called buckstays, are provided on the outside of the boiler to carry this load. The tubes act as continuous vertical beams in carrying this differential pressure on the furnace wall to the buckstays. The tubes thus develop axial bending stresses that the designer must add to the axial pressure stress in the tube. Here, again, it is the designer's job to consider the differential pressure and the buckstay spacing and to select a spacing that will not overstress the tube.

Acceptance Standards for Radiography

Acceptance standards for radiography are given in PW-51. That paragraph invokes the radiographic examination procedures of Article 2 of Section V, Nondestructive Examination. However, acceptance criteria for any imperfections disclosed by the examinations are provided in PW-51 and Appendix A-250. There are four types of indications that are considered rejectable imperfections and must be removed after which the joint must be rewelded and reexamined. These four are:

1. Any indication characterized as a crack or zone of incomplete fusion or penetration.
2. Any other elongated indication on the radiograph greater than a certain length, which is a function of the thickness, t, of the weld.
3. Any group of aligned indications that have an aggregate length greater than t in a length of $12t$, except when the distance between the sucessive imperfections exceeds $6L$, where L is the length of the longest imperfection in the group.
4. Rounded indications larger than those shown in Appendix A-250.

Appendix A-250 provides detailed acceptance criteria for welds with rounded indications determined by radiography. The indications may be from any imperfection in the weld, such as porosity, slag, or tungsten. The terminology paragraph of the appendix defines a **rounded indication** as one whose length does not exceed three times its width. Such an indication may be circular, elliptical, conical, or irregular in shape and may have a tail, in which case the tail must be included in the size evaluation. An indication with a greater length to width ratio than 3 is considered to be an **elongated indication**, whose acceptance is governed by the the rules in PW-51.3.2. **Aligned indications** are defined in Appendix A-250 as a sequence of four or more rounded indications that touch a line parallel to the length of the weld drawn through the center of the two outer, rounded indications. Indications are further categorized as relevant and nonrelevant indications. A **nonrelevant indication** is one smaller than certain sizes given in A-250.3.2, which vary depending on the thickness, t. In this case, t is defined as the thickness of the weld, of the pressure retaining material, or of the thinner of the two sections being joined, whichever is least. Table 1 of A-250 lists examples of the maximum size of nonrelevant indications and also the maximum size of acceptable rounded indications, both as a function of thickness, t. A **relevant indication** is one larger than the limit for a nonrelevant indication. There are also **isolated indications**, which are separated from adjacent indications by an inch or more, **random rounded indications**, and **clustered indications**, which show up to four times as many indications in a local area as are shown in the illustrations of random rounded indications.

Appendix A-250 contains a series of charts in Figs. 3.1 through 3.6, representing full-scale, 6-inch radiographs. These charts illustrate isolated, random, and clustered indications for different weld thicknesses greater than $\frac{1}{8}$ inch. The charts represent the maximum acceptable concentration limits for rounded indications. Detailed acceptance criteria for the various types of indications are covered in the text of A-250. Although it is apparent that these rules are somewhat arbitrary, experience has shown that indications deemed acceptable do not harm the boiler.

ULTRASONIC EXAMINATION OF WELDS

Ultrasonic examination methods are not often used in Section I construction since PW-11 limits their use to electroslag welds in ferritic materials and in situations where the geometric unsharpness, as defined in that paragraph, would exceed 0.07. Again, Section I refers to Section V, Nondestuctive Examination, Article 5, for ultrasonic technique and standards, but provides its own acceptance criteria for any imperfections disclosed by the ultrasonic examination. Those criteria are given in PW-51.3.1 and PW-51.3.2 and are essentially the same as those for radiography.

As can be appreciated from the variety of topics discussed in this chapter, the subject of welding and postweld heat treatment is a complex one. There are many factors that must be addressed in order to achieve quality welds. Decisions in this area should be made by engineers with training and experience in welding and metallurgy.

NONDESTRUCTIVE EXAMINATION AND HYDROSTATIC TESTING

NONDESTRUCTIVE EXAMINATION

There are several related terms applied to Code construction that are somewhat imprecisely used. These are examination, inspection, and testing. Within the context of Section I, **examination** usually describes activities of the manufacturer; **inspection** refers to what the Authorized Inspector does; and **testing** refers to a variety of activities, usually of the manufacturer. Some confusion arises from the fact that nondestructive examination (NDE) includes activities defined as examinations, but which are often called tests. NDE is an indispensable means of assuring sound construction, since when properly used, these examinations are capable of discovering hidden flaws in material or welds.

The examinations referenced by Section I in the NDE category are the following: radiographic examination, ultrasonic examination (which is mandated as an alternative or supplement to radiographic examination), magnetic particle examination, and liquid penetrant examination (also called dye penetrant examination). Although not explicitly mentioned in Section I, the use of visual inspection is certainly implied and is called out in B31.1 for application to the boiler external piping. These examinations are often referred to in the industry by the shorthand terms RT (radiographic test), UT (ultrasonic test), MT (magnetic particle test), PT (penetrant test), and VT (visual inspection).

Section I generally follows the Code practice of referring to Section V, Nondestructive Examination, for the rules on how to conduct the various examinations, but provides its own acceptance standards. For example, PW-11 calls for certain welds to be examined in accordance with Article 2 of Section V, to the acceptance standard of PW-51. Similarly, PG-25, Quality Factors for Steel Castings, calls for magnetic particle or dye penetrant examination of all surfaces of castings in accordance with Section V, but provides acceptance criteria right there in PG-25.

QUALIFICATION AND CERTIFICATION OF NDE PERSONNEL

Personnel performing and evaluating radiographic, ultrasonic, and other nondestructive examinations are required to be qualified and certified as examiners in those disciplines, in accordance with a written practice of their employer (PW-51, PW-52). This written practice must be based on a document called *Recommended Practice for NDT Personnel Qualification and Certification, SNT-TC-1A*, published by the American Society for Nondestructive Testing. (Note the use of the word testing, rather than examination in this title, showing again that the industry tends to use these terms interchangeably.) Recommended Practice SNT-TC-1A is revised periodically. The edition current as of this writing is the 1992 edition, finally adopted in 1996 by the various book sections of the Code. SNT-TC-1A establishes three categories of examiners, depending on experience and training, designating them as Level I, II, or III, with Level III being the highest qualification. The lowest ranking examiner, Level I, is qualified to perform NDE following written procedures developed by Level II or Level III personnel. A Level I individual can also interpret and accept the results of nondestructive examinations in accordance with written criteria (except for RT and UT examinations). A Level II individual has sufficient additional training and experience so that under the direction of a Level III person, he or she can teach others how to conduct, interpret, and accept the results of nondestructive examinations. A Level III examiner is the most qualified and can develop and write procedures, as well as establish a written practice for his or her employer. A Level III examiner can also administer examinations to qualify Level I, II, and III examiners.

A significant difference between the 1992 SNT-TC-1A and the previous, 1984 edition, is in the requirements for the qualification of a Level III examiner. In the older edition, it was possible for examiners experienced in performing tests, and in developing and writing NDE procedures, to be qualified as Level III examiners on the basis of that experience, and many individuals were so qualified, without having to pass an examination. In the 1992 edition, the requirements were tightened, and the practice of grandfathering based on experience is no longer permitted. To become a Level III examiner, it is now necessary to pass an appropriate examination. This is causing a problem when a Level III examiner who has never had to take an examination is subject to periodic review (every five years) for renewal of certification and in some cases is told that he or she must now take an examination. A suggested solution to this problem is as follows: PW-51.5 does not say a certificate holder must follow SNT-TC-1A to the letter. It merely states that this document be used as a guideline by the employer for basing the written practice for certification and qualification of NDE personnel. Consequently, it is permissible for an employer to take exception in the employer's written practice to the requirement for qualification of Level III personnel by examination only and to substitute criteria from the 1984 edition of SNT-TC-1A, which allowed the employer to waive examination for Level III individuals, on the basis of ability, achievement, experience, and education, as defined in paragraphs 4.3 and 6.3 of that edition.

The American Society for Nondestructive Testing has another document, entitled *Standard for Qualification and Certification of Nondestructive Testing Personnel, CP-189*, which is an alternative to SNT-TC-1a (remember the latter is only a recommended practice). CP-189 requires Level III individuals to pass an examination given by the ASNT, and calls for the employer to give a further examination, since the first examination may not assure the full range of capability needed. In 1997 Subcommittee I voted to allow CP-189 to be used as an alternative to SNT-TC-1A, with the choice left to the certificate holder. This alternative standard appeared in the 1997 Addenda to Section I.

PW-51 requires a complete set of radiographs for each job to be retained and kept on file by the manufacturer for at least 5 years. This requirement is based on the reasonable idea that the radiographs might be of assistance in determining responsibility (or lack of responsibility) for any defects, alleged or actual, subsequently discovered, or other problems that might occur in service. Similarly, PW-52 requires a manufacturer who uses ultrasonic examination to retain a report of that examination for a minimum of 5 years. Again, these records might prove useful should problems occur in service.

HYDROSTATIC TESTING

The hydrostatic test is one of the last steps in the construction of the boiler. Hydrostatic test requirements are given in PG-99 and PW-54. These tests may be made either in the manufacturer's shop or in the field, using water. Unlike Section VIII, Section I does not permit the use of other fluids or pneumatic testing.

The hydrostatic test serves a number of purposes. Many members of the Code committees believe that its major purpose is to establish that the boiler (or pressure vessel) has been properly constructed and that it has a significant design margin, or safety margin, above and beyond its nominal maximum allowable working pressure. (The hydrostatic test pressure is normally 1.5 times MAWP.) In this sense, the hydrostatic test is seen to demonstrate the validity of the design as a pressure container. Another important aspect of the hydrostatic test is that it serves as a leak test. Any leaks revealed by the test must be repaired, and the boiler must be retested (see PW-54).

The first edition of Section I in 1915 called for a hydrostatic test at 1.5 times the maximum allowable working pressure, a basic rule essentially unchanged to this day. We may surmise that the choice of this pressure followed the practice of the time and was seen to provide evidence that the boiler had been constructed with a significant design margin. If a higher test pressure had been chosen, the stress in the components would have begun to approach their yield strength, a situation to be avoided.

Prevention of Brittle Fracture

PG-99 requires the hydrostatic test to be conducted using water ''at no less than ambient temperature, but in no case less than 70°F.'' This stipulation about the water temperature is intended to minimize the possibility of catastrophic brittle fracture of heavy-walled pressure parts during the test. Brittle fracture is a type of behavior that can occur when metal is under tensile stress when its temperature is at or below its so-called **nil ductility transition temperature** (NDTT). Above this temperature, the metal behaves in a ductile manner; below this temperature its behavior is brittle. Contrasting examples of these two types of behavior are the bending of a wire hanger compared with the bending of a glass rod. The wire, which is ductile, bends easily when its yield strength is exceeded and can be restored to its original shape. The glass rod can carry a certain amount of bending load, but then suddenly fractures in a brittle manner when its yield strength is reached.

Any flaw, notch, or other discontinuity can raise stress at some local area to the yield point. If the material is ductile, it can yield locally, with little harm done. If the material is below its NDTT, it behaves in a brittle manner; when stress reaches the yield point, the material may tear or form a crack, which can then grow suddenly through the thickness causing a catastrophic failure. The ability of a material to resist tearing or cracking is a measure of its fracture toughness.

Brittle fracture is generally not a concern for relatively thin materials. The manufacturers of steels used in boilers have developed melting practices that usually result in nil ductility transition temperatures well below 70°F, assuring adequate fracture toughness during hydrostatic tests. The component of greatest concern in a large utility boiler (so far as brittle fracture is concerned) is the drum, since it has very thick walls. (It happens that thick vessels and headers are more susceptible to brittle fracture than thinner components.) After a number of brittle failure accidents in the 1970s, the manufacturers of large boilers took steps to lower the NDTT of heavy-walled parts such as drums (by slightly modifying specifications for material ordered to achieve a finer grained material with greater fracture toughness). Also, in some instances the manufacturers of large boilers recommended to their customers that any future hydrostatic tests of existing boilers be conducted using warm or even hot water, to assure ductile behavior at the time of the test. This became potentially hazardous for the Authorized Inspector, and Subcommittee I placed a 120°F limit on metal temperature during the hydrostatic test at the request of the National Board. Also around this time, some boiler manufacturers switched from the use of SA-515 to SA-516 plate for heavy-walled boiler drums, because the latter has greater fracture toughness. Even so, heavy SA-516 plate is often examined metallurgically by boiler manufacturers to determine its NDTT, and if that temperature is not well below 70°F, a hydrostatic test using water warmer than 70°F is recommended.

The concerns described above can be illustrated by considering the design of a thick-walled drum for a large high-pressure boiler. Such a drum would typically be made of SA-516 grade 70 plate. The allowable stress used for designing the drum is about 17 ksi. Accordingly, during the hydrostatic test, the actual average membrane stress in the drum wall is approximately 1.5 times 17 ksi, or 25.5 ksi. At first glance, this seems to be a reasonable margin below the cold yield strength of this material, which is about 38 ksi. Unfortunately, this is not the case, because in a real vessel, there are always irregularities of various kinds present that act as stress raisers. Examples include welds with undercuts, grooves, or ridges. Ligaments between openings can also act as stress raisers, as can changes in vessel geometry at transitions between materials of different thickness and at nozzle-to-shell junctions. These stress-raising irregularities, sometimes loosely called notches, may cause a stress concentration of two or more. Thus in a drum undergoing hydrostatic testing, it is very likely that local surface stress at some of these notches may approach or exceed the yield stress. At such a time, it is important that the boiler plate material be warm enough for its behavior to be ductile, so that local yielding can occur without significant danger of initiating a brittle failure. When the boiler is in normal service, it is, of course, at a temperature that ensures ductile behavior. It is also at a pressure much lower than that during the hydrostatic test, so that the stress even at notches is likely to be well below the yield stress.

Test Pressure for Drum-Type Boilers

For most boilers, the hydrostatic test is conducted by slowly raising the pressure to 1.5 times the MAWP. Close visual inspection is not required during this stage, in the interest of safety of the Inspector. The pressure is then reduced to the MAWP, and the boiler is carefully examined for leaks or other signs of distress.

In Chapter 15 on Determination of Allowable Stresses, it is mentioned that both Sections VIII and I are presently (1997) considering the possibility of increasing certain allowable design stresses below the creep range by changing the design factor on tensile strength from about 4 to about 3.5. There has been some concern that to avoid yielding in any of the pressure parts, this might require a reduction in the hydrostatic test pressure. Subcommittee I assigned a task group to investigate the ramifications of what would be a modest increase in some design stresses. The hydrostatic test pressure is one of the issues considered. On completion of the investigation, Subcommittee I concluded that a reduction in hydrostatic test pressure was unnecessary and that it would suffice to delete existing paragraph PG-99.3.3. The words of that paragraph were then modified and incorporated into the introduction to PG-99, so that they would apply to hydrostatic tests of any kind of boiler. Those words, as voted by the subcommittee in September 1997 and later modified slightly by the Main Committee, are as follows:

> *At no time during the hydrostatic test shall any part of the boiler be subjected to a general primary membrane stress greater than 90% of its yield strength (0.2% offset) at test temperature.*

The new words characterizing the limit as applicable to general primary membrane stress are a first use of such terminology by Section I. They were chosen to make clear that the goal is to limit average membrane stress through the wall of the components, and not the bending stress or peak stress at the surface.

(Note: In a July 1998 development, final approval of the above change was delayed by a request to clarify whether the 90% yield strength limit applied at the nominal test pressure of 1.5 times MAWP or at the maximum permitted test pressure, which is 6% higher. Subcommittee I had intended the former. The issue was expected to be resolved in time to permit inclusion of new language in the 1999 Addenda, to be published in July 1999.)

Forced-Flow Steam Generators with No Fixed Steam and Waterline

The usual hydrostatic testing procedure is modified for forced-flow steam generators with no fixed steam and waterline. These boilers are designed for different pressure levels along the path of water-steam flow, with a significant difference between economizer inlet and superheater outlet. In one such boiler, for example, those design pressures were 4350 psi and 3740 psi, respectively. The design pressure or MAWP at the superheater outlet is the design pressure stamped on this type of boiler and is called the **master stamping pressure**. In the first stage of the hydrostatic test, a pressure equal to 1.5 times the master stamping pressure (but no less than 1.25 times the MAWP of any other part) must be applied. In the example given, the 1.5 factor controls, at 5610 psi. For the second, or close examination stage, the pressure may be reduced to the MAWP at the superheater outlet.

The somewhat different treatment for hydrostatic tests of forced-flow steam generators with no fixed steam and waterline comes about from several factors peculiar to this type of boiler as compared to the natural circulation type. If the former type were tested at 1.5 times the MAWP of its component with the highest design pressure, the result would be excessive and might cause yielding of some other components. (In fact, PG-99.3.3 warns that no part of the boiler may be subjected to a stress greater than 90% of its yield strength at test temperature.) In the example cited, the test pressure would have been 1.5 times 4350 psi, or 6525 psi rather than 5610 psi. The subcommittee compromised by requiring a test at 1.5 times the master stamping pressure, with the proviso that no part should be tested at less than 1.25 times its MAWP. This provides a leak test and demonstrates an adequate design margin for this type of boiler, which has no drum, and whose pressure is controlled by a pump, not by firing rate.

Heat Recovery Steam Generators

In recent years, gas turbines with heat recovery steam generators (HRSG's) have assumed increasing importance in energy conservation programs. As a consequence, HRSG's with new configurations and design arrangements have evolved. Some of these boilers have multiple circuits with different design pressures, with low, intermediate, and high pressure. For such boilers, each independent design pressure circuit is given a separate hydrostatic test at 1.5 times its design pressure (MAWP). There are, however, other types of HRSG in which the various components along the flow path from economizer inlet to superheater outlet are designed for different pressures even though there are no intervening valves between the different pressure zones. In 1991 Subcommittee I was asked two questions about such a boiler: what maximum allowable working pressure should be stamped on the boiler, and what hydrostatic test pressure should be used? The committee advised in Interpretation I-89-68 that the MAWP to be stamped was the lowest of the several design pressures used, and the hydrostatic test pressure was 1.5 times that pressure. It is quite clear that with no intervening valves, there is no way to conduct tests at different pressure levels, and the lowest design pressure zone establishes and limits the hydrostatic test pressure to be applied.

WELDING AFTER HYDROSTATIC TEST

Formerly the hydrostatic test completed the construction of the boiler, and no further work (i.e. welding) was permitted. However, around 1980 the Committee added provisions to PW-54 permitting nonpressure parts to be welded to the pressure parts after the hydrostatic test, if certain conditions were met. (Welding is limited to P-No. 1 materials; attachment is done by stud welds or small fillet welds; 200°F preheat is applied when the thickness of the pressure part exceeds 3/4 of an inch; and the completed weld is inspected by the Authorized Inspector before he or she signs the Manufacturer's Data Report Form for the completed boiler.) These provisions granted relief from delays that arose when miscellaneous structural steel parts to be welded to the pressure parts were not available when the pressure parts were ready for the hydrostatic test.

INTERPRETATIONS ON HYDROSTATIC TESTING

Over the years there have been quite a few other inquiries about various aspects of hydrostatic testing not explicitly covered by PG-99. Subcommittee I's replies to these inquiries now form a useful body of information. Some of the more important of these are reviewed below.

Same Hydrostatic Test Pressure Used for All Parts of the Boiler

Occasionally someone raises a concern as to whether the usual hydrostatic test pressure of 1.5 times MAWP is appropriate for the feedwater, blowoff, and main steam piping, since that piping does not have the same design pressure as the boiler. As explained in Chapter 5, the feedwater and blowoff piping design pressure is as much as 225 psi higher than the MAWP of the boiler, and the main steam piping may be designed for a pressure lower than the MAWP of the boiler. Thus the normal hydrostatic test is less than 1.5 times design pressure for the feedwater and blowoff piping and may be more than 1.5 times design pressure for the main steam piping.

A little thought shows this concern is groundless. Remember first that the feedwater and blowoff piping have been deliberately made stronger than the rest of the boiler to account for their so-called shock service, by arbitrarily increasing their design pressure. Then consider that the 1.5 times MAWP hydrostatic test pressure is also an arbitrary value, which has been in the Code since 1915, when today's nondestructive examinations and quality control systems were not available. From the preceding discussion of hydrostatic test rules for forced-flow steam generators with no fixed steam and waterline (which date from the mid-1960s), it is clear that Subcommittee I accepts a hydrostatic test of 1.25 times MAWP in certain circumstances. There is also a precedent for tests at lower pressure, in Section VIII, which permits pneumatic testing at 1.25 times MAWP as an alternative to a hydrostatic test at 1.5 MAWP.

Consider now the main steam piping, for which the standard hydrostatic test might exceed 1.5 times the piping design pressure. Is this a real concern? Would the stresses during the test be too high? A little thought shows they would not, because the main steam piping is usually designed for high-temperature service, using allowable stresses that are typically only about half of those permitted at hydrostatic test temperature. Thus during the hydrostatic test, the stress in the main steam piping would probably not even reach the cold allowable stress.

Note also that PG-99 makes no mention of different hydrostatic test pressures for different parts of the boiler. It simply instructs that a single test pressure is to be applied to the completed boiler (which of course includes the boiler external piping). Thus in 1981 when an inquirer asked if it were permissible to test a main steam line in just such a situation, i.e., the boiler MAWP was higher than the design pressure of the main steam piping, Subcommittee I replied yes, the main steam line may be hydrostatically tested with the boiler as outlined in PG-99. (See I-82-11, Maximum Hydrotest Pressure for Drum-Type Boilers in Interpretations No. 11.) PG-99.1 offers guidance on conducting the hydrostatic test. For most boilers it specifies that the pressure should be raised gradually to 1.5 times the MAWP and be under such control that the required test pressure is never exceeded by more than 6%. It also notes that close visual inspection is not required when the boiler is at full test pressure. (This last provision was added in the 1960s. There is an apocryphal story that it was added after Subcommittee I learned to its dismay that an Authorized Inspector lost the sight of his one good eye due to a leak from a boiler he was inspecting while it was at a pressure of 1.5 times MAWP. A recent attempt to confirm this story proved unavailing. It appears that an Inspector did suffer an eye injury while observing such a test, but the one-good-eye story was someone's unfortunate attempt to poke fun at the Authorized Inspection Agencies for making life easy for the manufacturers by assigning one-eyed Inspectors to verify compliance with the Code. In any case, failure to wear appropriate safety glasses when making a close inspection during a hydrostatic test is an unsafe practice, even at design pressure.)

Exceeding the Hydrostatic Test Pressure

The rule that the test pressure must not be exceeded by more than 6% [i.e., not more than $(1.06)(1.50)(\text{MAWP})$ = $(1.59)(\text{MAWP})$], while arbitrary, is not unreasonable. As explained in Chapter 15, one of the criteria for setting allowable stress is two-thirds of the yield strength of the material. Therefore, a hydrostatic test at 1.5 times MAWP could theoretically utilize 100% of the yield strength of those components whose allowable stress is based on the yield strength at room temperature. There would then be little margin for overshooting the prescribed test pressure. However, this is not the case, for several reasons. First of all, for most Section I materials, the allowable stress at room temperature is actually based on one-fourth of the ultimate tensile strength, which is quite a bit lower than two-thirds of the yield strength. Moreover, the yield strength (and ultimate tensile strength) of most materials as received from the mill is somewhat stronger, often by 10% or more, than the minimum called for by the material specification, although no credit may be taken for this extra strength. Also, any components intended for service at temperatures higher than 650°F are designed using allowable stresses lower than those at room temperature (allowable stress is essentially constant from room temperature to 650°F and then begins to fall off as the temperature increases). Thus at test temperature, these components have extra strength available to provide a margin against yielding.

During a typical hydrostatic test, the boiler is first filled with water using a pump capable of providing a large volume of water at relatively low pressure. Various vents are left open to permit all the air to escape, and when the boiler is full of water, all valves are closed and a different kind of pump is used, one that can achieve the high pressure required. Often the pump used for the hydrostatic test is a piston-type pump that slowly builds pressure as any remaining air pockets are filled with water. When the water-holding volume of the boiler reaches the stage where it is essentially solid water, each piston stroke raises the pressure substantially, due to the essentially incompressible nature of water. It is thus difficult to achieve the desired test pressure without overshooting the mark, and PG-99.1 permits this pressure to be exceeded by 6%. Test personnel must be attentive at this time to avoid exceeding this 6% test pressure tolerance.

As might be expected, the maximum permitted hydrostatic test pressure has occasionally been exceeded, and in a few cases, the Authorized Inspector had refused to accept the boiler, because PG-99.1 says that the desired test pressure is *never* to be exceeded by more than 6%. What to do? Inquiries came to Subcommittee I, where the sentiment was on the side of common sense, namely, that if the manufacturer could demonstrate to the AI, by some means, perhaps calculations, that stresses during the test had not exceeded the yield strength of the parts, and no damage had been done to the boiler, the AI could accept the boiler. In 1981 Interpretation I-81-27, Maximum Hydrostatic Test Pressure, was issued, as follows:

> Question: *Under what circumstances may the hydrostatic test pressure stated in PG-99 be exceeded?*
>
> Reply: *The test pressure may be exceeded when the manufacturer demonstrates to the satisfaction of the Authorized Inspector that no component has been overstressed.*

This Interpretation is consistent with Section VIII's design philosophy, as expressed in UG-99, the counterpart to Section I's PG-99. Section VIII does not establish an upper limit for hydrostatic test pressure. Section VIII, UG-99(d) says that if the test pressure exceeds the prescribed value, either intentionally or accidentally, to the degree that the vessel is subjected to visible permanent distortion, the Inspector shall reserve the right to reject the vessel.

The authors consider such an approach to be reasonable, and at one point in the 1980s, Subcommittee I actually voted to add a similar provision to Section I to settle future inquiries on this topic. However, the Main Committee would not approve the change. Among the objections offered was the assertion that it was too difficult for an Inspector to see the slight deformation that would indicate yielding had taken place. The fact that Section VIII had for years granted the Inspector the discretion to make such a judgment was insufficient to convince the negative voters, and Subcommittee I abandoned the effort. The interpretations cited here would suffice to resolve any problems about overshooting the prescribed hydrostatic test pressure.

Not long after Interpretation I-81-27 was issued, a manufacturer wanted to use certain standard components for a series of boilers with a range of design pressures. (These particular parts were designed for the highest MAWP.) The manufacturer wanted to avoid testing at several different pressures by simply testing all of the parts at 1.5 times the MAWP of the boiler with the highest design pressure. He submitted an inquiry to Subcommittee I in 1983, which was formulated by the committee into Interpretation I-83-75, comprising the following two questions and replies:

> Question: *Is the pressure specified in PG-99.1 plus the 6% deviation the maximum hydrostatic test permitted?*
>
> Reply: *Yes.*
>
> Question: *May a manufacturer intentionally exceed this maximum hydrostatic test pressure if it is demonstrated to the satisfaction of the Authorized Inspector that no component has been overstressed?*
>
> Reply: *No.*

Subcommittee I answered in this fashion because apparently some members had reservations about the wisdom of permitting hydrostatic tests with no specified limit on the test pressure. Thus the committee thought it better to reiterate the maximum pressure permitted and to prohibit exceeding that pressure deliberately. However, the earlier interpretation, I-81-27, is still valid, even if it is not intended for routine use. Thus Subcommittee I established the policy that while it may be acceptable to exceed the specified maximum hydrostatic test pressure inadvertently, this may not be done deliberately. In the opinion of the authors, what is important is that a way has been provided for the AI to accept a boiler accidently subjected to a pressure greater than the maximum normally permitted. There is certainly no good reason to reject an undamaged boiler. For that matter, there is no good reason to reject the undamaged parts of a boiler that has been damaged, if any parts that were damaged can be repaired or replaced. The manufacturer would have to persuade the AI that the boiler as repaired was the equivalent of a newly constructed, undamaged boiler. Of course, in that case, the AI would require a new hydrostatic test.

Hold Time for Hydrostatic Test

In 1980 an inquirer asked how long the hydrostatic test pressure should be held at 1.5 times the MAWP. Subcommittee I responded in Interpretation I-80-15 that the pressure was to be maintained an amount of time satisfactory to the Authorized Inspector and then reduced to the MAWP at the Inspector's discretion for close examination of the boiler.

Pressure Drop During Hydrostatic Test

In the same Interpretation (I-80-15), the inquirer asked whether a pressure drop was permitted during the hydrostatic test. Subcommittee I replied that the pressure was to be maintained at the MAWP while the boiler is examined. In effect, the committee was advising that a slight pressure drop was permissible so long as the pressure didn't fall below the MAWP. The pressure drop presumably would be caused by some slight leakage in the water supply system, not through any welds or pressure part material, which would represent failure of the test. This point was explained further in the following Interpretation.

Slight Leakage Through Valve Seats During Hydrostatic Test

In 1992 an inquirer asked whether slight leakage through the seat of a closed valve was permissible during the hydrostatic test. Subcommittee I responded in Interpretation I-92-56 that such leakage was permissible. This was a typical case of the committee's practical approach based on the field experience of its members, who were satisfied that such leakage was not uncommon and did not invalidate the test. The inquirer had a logical follow-up question, as explained next.

Use of Pump to Maintain Hydrostatic Test Pressure

The inquirer of Interpretation I-92-56 asked also whether a pump may be used to maintain pressure during the hydrostatic test (during which some slight leakage through valve seats was taking place), so long as the test pressure was maintained within the limits prescribed by PG-99. Again, on the basis of its practical experience, Subcommittee I saw nothing wrong with such a practice.

Minor Leakage from Mechanical Joints During Hydrostatic Test

In September 1997, Subcommittee I answered an interesting inquiry about the purpose of the hydrostatic test and whether some minor leakage through mechanical joints (such as a rolled tube joint) could be permitted. (At this writing the Interpretation hasn't yet been given a number; the item number was BC97-206.)

Question: *Is the purpose of the hydrostatic test required by PG-99 of Section I to demonstrate the strength and integrity of the boiler and for detecting leakage?*
Reply: *Yes.*
Question: *Is it acceptable to have minor leakage from mechanical joints during the hydrostatic test?*
Reply: *Yes, if acceptable to the Authorized Inspector.*

After careful consideration and discussion, the committee was persuaded that the answer to the second question was in accord with industry practice and would provide guidance to the Authorized Inspector in exercising his or her judgment, while still leaving the Inspector ultimate responsibility for acceptance of the test.

Location and Type of Test Gages

There have been a number of interpretations clarifying the application of test gages used during the hydrostatic test.

Interpretation I-78-27 explained that the test gage does not have to be mounted directly on the pressure parts. This interpretation also permits the use of a single test gage installed in the system near the pump to be used for the hydrostatic test of a number of components connected in series.

In a similar situation, Interpretation I-89-58 explains that the required test gage may be mounted anywhere on the line between the pump and the pressure parts or on a manifold connecting several pressure parts undergoing hydrostatic testing. The same inquirer also asked whether it was necessary for the test gage to be mounted at the highest point on the pressure parts being tested. The committee responded that this was not necessary, but cautioned that the hydrostatic head on the gage must be taken into account when conducting the test.

Reflecting the steady improvements in modern technology, an inquiry was submitted to the committee in 1992 asking if it was permissible to use a pressure transmitter with a remote readout in lieu of a digital pressure gage, provided the readings gave the same or greater degree of accuracy as obtained with a dial pressure gage. Subcommittee I replied in Interpretation I-92-52 that the use of such devices was permissible, provided the remote pressure readout was visible to the operator controlling the applied pressure. Implicit in this interpretation is the acceptance of a digital gage whose accuracy is equal to that of a conventional dial gage.

A last inquiry on test pressure gages in 1993 asked whether PG-99.4.2 restricted the range of the gage to double the intended maximum test pressure. Since the paragraph in question uses nonmandatory language, the committee could only reply in the same vein, that the maximum range of the gage should preferably be about double the maximum test pressure.

Subsequent Hydrostatic Tests

A 1980 inquiry asked whether any subsequent hydrostatic tests were required after the initial hydrostatic test performed by the boiler manufacturer. Subcommittee I in Interpretation I-80-15 replied with this advice:

> *Subsequent inservice inspection hydrostatic test pressures are not addressed in Section I, which is for new construction only. Such subsequent tests must follow the requirements of the jurisdiction in which the boiler is operating.*

Another interpretation on the subject of subsequent hydrostatic testing deals with a boiler that was completely assembled and hydrostatically tested in the shop. The Authorized Inspector then signed the Master Data Report. The boiler external piping was subsequently disassembled for shipment. The question was whether Section I required a further hydrostatic test after reassembly of the same components. Subcommittee I replied in Interpretation I-89-50 with a fundamental tenet of the Code: After the Authorized

Inspector has signed the Master Data Report for a complete boiler, the boiler is no longer under the jurisdiction of the Code. Accordingly, no further testing is required.

Hydrostatic Testing of Boiler External Piping

An inquiry on hydrostatic testing of boiler external piping asked whether the rules for such testing, including test temperatures, were those found in PG-99 of Section I. In Interpretation I-83-64, the committee replied in the affirmative. As explained in Chapter 1, the boiler external piping is part of the complete boiler unit and is hydrostatically tested with the complete boiler, even though most design and construction rules for this piping are found in the B31.1, Power Piping Code. However, B31.1, paragraph 137.3 on pressure testing simply refers to PG-99 and notes that the testing must be conducted in the presence of the Authorized Inspector. This reminder was included because B31.1 normally does not require third-party inspection of the piping it covers. Further discussion of potential problems associated with hydrostatic testing of boiler external piping and valves can be found in the Special Considerations section of Chapter 5.

Temperature Ratio Applied to Hydrostatic Test Pressure

An unusual inquiry was received in 1984, asking whether the temperature ratio of Appendix A-22.8 had to be applied to the hydrostatic test pressure. It should be explained that this temperature ratio approach is used by Section VIII for establishing hydrostatic test pressure. Section VIII requires the basic 1.5 times MAWP test pressure to be further increased (multiplied) by the lowest ratio (for the vessel materials) of the allowable stress value S at test temperature to the allowable stress value at design temperature. This can be a significant multiplier if the vessel is designed for high-temperature service. However, Subcommittee I has never considered that such a multiplier was necessary for Section I construction. (The authors concur with this view, which has been borne out by many years of satisfactory experience.) Section I requires the use of the temperature ratio only for the determination of maximum allowable working pressure by proof test under the proof testing provisions of Appendix A-22. That was the substance of the reply published in 1984 as Interpretation I-83-92.

Applying Insulation Before Hydrostatic Test

There have been several interpretations dealing with the question of applying insulation to the boiler before the hydrostatic test. The concern in this situation was the possibility that the insulation or refractory might cover a leak that the Inspector otherwise might find during the test. The first of these inquiries came in 1985 and asked if it was permissible to test sections of a boiler in the shop at MAWP (as a leak test), then to apply insulation and ship the insulated sections to the field, where they would be assembled and given a hydrostatic test in accordance with PG-99. In Interpretation I-86-01 (later reversed by I-89-21), the committee said this was not permissible because all surfaces must be available for inspection during the hydrostatic test. Four years later, Subcommittee I received a closely related and somewhat more narrowly stated inquiry that was answered in Interpretation I-89-21:

> Question: *May portions of welded power boilers that do not contain longitudinal welded joints made with the addition of filler metal be covered with insulation or refractory prior to the hydrostatic test required by PG-99?*
> Reply: *Yes; however, the Authorized Inspector may require the hydrostatic test pressure to be maintained at the maximum allowable working pressure for an extended period of time, sufficient to assure there is no indication of leakage. The Authorized Inspector may also require removal of insulation for cause.*

The committee discussion that finally led to an agreement on this reply covered a number of issues. Several members of the committee (including the authors) who had disagreed with the assertion in Interpretation I-86-01 that all surfaces must be available for inspection during the hydrostatic test pointed out that in large utility boilers, it is not only unnecessary but well nigh impossible for the Inspector to see all surfaces during the hydrostatic test. It is the authors' opinion that a leak will soon make itself known, whether it is behind refractory and insulation, or at a stub-to-header weld, or wherever it might be. Moreover, such a leak is hardly a major safety hazard, it is simply a minor problem requiring repair, and it is the boiler manufacturer's responsibility to see to it that any leaks are repaired. Thus the manufacturer who installs insulation and refractory before the hydrostatic test does so for its own convenience and at its own risk. The committee was finally able to agree on the above reply, once language was included in it to caution that the Inspector could call for the pressure to be maintained for an extended time to assure there was no indication of leakage and could also require the removal of insulation if there was such an indication.

Thus Interpretation I-89-21 effectively reverses the assertion in Interpretation I-86-01 that all surfaces must be available for inspection during the hydrostatic test. Note also that longitudinal welds, which are usually the most important and highly stressed welds, may not be covered during the hydrostatic test.

Subcommittee I's deliberations leading to the insulation-before-hydrostatic-test ruling just described illustrate the workings of the committee when, as is often the case, an inquirer asks about some aspect of Section I construction that is not explicitly covered in the rules. The committee must then rely on the common sense, experience, and memory of its members to formulate a reply that the majority can support and that does not conflict with previous interpretations or set a bad precedent for the future.

Painting Before Hydrostatic Test

In a somewhat related inquiry about painting before the hydrostatic test, the committee had no trouble in agreeing that this was permissible and said so, in Interpretation I-86-69 in January 1988. The members knew that painting pressure parts before testing was a routine practice and that no significant leak could be prevented or hidden by painting.

Hydrostatic Test Procedures

There have been a few inquiries about procedures to be used during the hydrostatic test. In 1986 an inquirer asked whether Section I prohibited a pneumatic pressurization of the boiler system before performing the required hydrostatic test. The committee replied in Interpretation I-86-42 that such a test was not prohibited, but added a caution that pneumatic tests may be hazardous and recommended that special precautions be taken.

Another procedural question was whether it was permissible to conduct a hydrostatic test of a boiler by filling it completely with water, connecting it to a vessel that is partially filled with water, and then pressurizing that vessel with air or another gas. The committee replied in Interpretation I-89-19 that the means by which hydrostatic pressure is applied are beyond the scope of Section I. In other words, Section I does not prohibit such a procedure.

Hydrostatic Test of Mechanically Assembled Boiler External Piping

For quite some time Subcommittee I has been struggling with a number of issues having to do with boiler external piping assembled by mechanical means, i.e., without welding. Two issues pertinent to the present discussion are whether this type of piping requires the usual hydrostatic test as part of the complete boiler unit and whether it is necessary to have an Authorized Inspector witness and sign for a hydrostatic test of such piping. The committee has been somewhat ambivalent in its replies, because of a tradition that the installation of threaded piping could be safely left to a plumber or others not authorized to engage in

Section I construction. Support for this tradition can be found in Note 2 of PG-104.1, which apparently exempts organizations that install boiler external piping by mechanical means (as opposed to welding) from the requirement to provide Code certification and stamping for it. However, the committee has issued a number of interpretations since about 1990 saying that an Authorized Inspector has to witness the hydrostatic test of the completed boiler even if the boiler external piping is assembled by mechanical means. In late 1997, after long and difficult deliberations, a series of changes (to PG-99, PG-104.1, PG-112.2.2.5, and A-357), a new PG-109.4, and a new alternative form P-4B were approved by the Main Committee. These changes should resolve the confusion about such piping. The new rules provide that mechanically assembled piping need not be assembled by a Code symbol holder, but that responsibility for documentation and hydrostatic testing of such piping must be assumed by the holder of an S, A, or PP stamp. This would include a hydrostatic test, witnessed by an Authorized Inspector. The new provisions and the new form P-4B appeared in the 1998 Addenda included in the 1998 edition of Section I.

A number of rulings have been issued related to the hydrostatic testing of mechanically assembled piping. In one such situation, a boiler and flanged boiler external piping for it were manufactured and hydrostatically tested separately by two different manufacturers in the presence of Authorized Inspectors. The question was whether those two separate tests sufficed or whether a hydrostatic test of the complete field-assembled boiler was required. The committee answered in Interpretation I-92-73 that an additional hydrostatic test was required on the completely assembled boiler. Similar rulings in the past have been made in the case of any kind of components (not just mechanically assembled components) that had been individually tested and were then assembled into a complete unit. The committee's reasoning was that if the complete unit is not tested, there is no assurance of the leak tightness of those joints not previously connected.

THIRD PARTY INSPECTION

Third party inspection refers to the system evolved by the Code committee to assure that manufacturers or other Code symbol stamp holders will actually follow the Code rules. In the simplest situation, a boiler manufacturer and the boiler purchaser are the first two parties, and an Authorized Inspector (AI) is the independent third party. It is the function of the AI to assure and verify that the manufacturer complies with the Code.

An Authorized Inspector is defined by Section I in PG-91 as an inspector employed by a state or municipality of the United States, a Canadian province, or an insurance company authorized to write boiler and pressure vessel insurance. The employer of an Authorized Inspector is called an Authorized Inspection Agency (AIA). Thus an AIA can be either the inspection agency of an insurance company or of a jurisdiction that has adopted at least one section of the Code. In the United States, it has been traditional that the Authorized Inspection Agencies providing authorized inspection for ASME Code symbol stamp holders are private insurance companies such as the Hartford Steam Boiler Inspection & Insurance Company, Factory Mutual, or Kemper National, among others. In Canada, until 1996, the provincial governments had provided authorized inspection services through offices such as the department of labor. However, in an effort to cut the cost of government, two provinces (Alberta and Ontario) have recently privatized their authorized inspection activities, by spinning them off into self-sustaining private companies. The new Alberta organization is called the Alberta Boilers Safety Association (ABSA), and the new Ontario organization is called the Technical Standards and Safety Authority (TSSA). The Canadian government passed a bill allowing the delegation of authority over this public safety program, formerly vested in the provincial governments, to not-for-profit nongovernment organizations. The new organizations are Crown Corporations, the administrators of which are dual employees of the jurisdiction and the newly privatized corporations. The National Board of Boiler and Pressure Vessel Inspectors has accepted the new corporations as representing the jurisdiction. The chief inspectors of the new corporations are National Board members, as they were before the change. Other provinces may follow the example set by Alberta and Ontario. As Code construction becomes increasingly an international activity, the National Board intends to recognize and accept other foreign jurisdictions and their inspection agencies, as suitable arrangements can be devised.

The Authorized Inspector must be qualified by written examination under the rules of any state of the United States or province of Canada that has adopted the Code. When an Inspector is so qualified, he or she may obtain a commission, or certificate of competency, from the jurisdiction where the exam was given. The National Board of Boiler and Pressure Vessel Inspectors (the National Board) also grants commissions to those who meet certain qualifications and pass a National Board examination. The qualifications are these: The applicant must be employed by an Authorized Inspection Agency and must have a high school diploma plus 3 years of appropriate experience, of which college may count for 2 years. The applicant must also pass the National Board examination and be granted a Certificate of Competency from the jurisdiction where the exam was taken. Certificates of Competency can also be issued by the National Board itself when the examination is given outside the United States and Canada, where the jurisdiction is not recognized by the National Board. Finally, the AIA that employs the Inspector must apply to the National Board on the Inspector's behalf for issuance of a National Board commission.

As a condition of obtaining from ASME a Certificate of Authorization to use an ASME Code symbol stamp, each Section I manufacturer or assembler must have in force a contract with an Authorized Inspection Agency spelling out the mutual responsibilities of the manufacturer or assembler and the Authorized Inspector. The manufacturer or assembler is required to arrange for the Authorized Inspector to perform the inspections called for by Section I. Paraphrasing the words of A-300 (the Quality Control System): the manufacturer shall provide the Authorized Inspector access to all drawings, calculations, specifications, process sheets, repair procedures, records, test results, and any other documents necessary for the Inspector to perform his or her duties in accordance with Section I. Section I lists many duties of the Authorized Inspector, all of which are intended to assure Code compliance. In 1997 Subcommittee I approved the expansion of PG-90 on Inspection and Tests to include a comprehensive list of the AI's duties and references to the paragraphs where those duties are further described. The revised PG-90 appeared in the 1998 Addenda. However, even when Section I does not specifically describe the duties of an AI with respect to some provision of the Code, it is understood that the AI has very broad latitude in carrying out the mandate to verify compliance by the Code symbol stamp holder with all applicable Code rules.

During committee deliberations a number of members who happen to be Authorized Inspectors have explained that once they have established that a manufacturer is following its approved quality control system, they merely spot-check the manufacturer's activities. As a practical matter, an AI cannot be expected to check every detail of a manufacturer's Code construction operation.

It may be recalled from the discussion of the Preamble that "the Code does not contain rules to cover all details of design and construction," and that when complete details are not given, the manufacturer must provide details of design and construction as safe as those the Code does provide in its rules, subject to the acceptance of the Authorized Inspector. An AI may have sufficient experience to make a decision in such a situation, in which case approval may be a simple matter. At other times, because of new or unusual construction, the AI may seek guidance from higher authority within his or her organization (many Authorized Inspection Agencies maintain an engineering staff) or via an inquiry to the Code committee, to determine whether the proposed construction is acceptable.

If a symbol stamp holder violates the Code, the AI has several powerful remedies. The Authorized Inspector can require rework, additional NDE, or simply refuse to accept the boiler (or other component). Without the AI's signature on the Data Report Form, the boiler is not complete and cannot be sold or used where the Code is enforced. For repeated violations, the AI might also recommend that the contract for Authorized Inspection not be renewed. Finally, if the violations were flagrant, the matter could be brought to the attention of the Subcommittee on Boiler and Pressure Vessel Accreditation. That subcommittee would then conduct a hearing that could lead to the revocation of the offender's Certificate of Authorization to use a Code symbol stamp. Since the manufacturer needs its stamp to stay in business, third-party inspection provides a very effective means of assuring Code compliance.

CERTIFICATION BY DATA REPORTS AND STAMPING

CERTIFICATION AND ITS SIGNIFICANCE

In order for the Code to be effective, there must be some means of assuring that it has been followed. Each phase of the work of designing, manufacturing, and assembling the boiler must be done in accordance with the Code. Assurance that a boiler is constructed to the Code is provided in part by a process called Code certification. In the simplest case of a complete boiler unit made by a single manufacturer, the manufacturer must certify on a form called a Manufacturers' Data Report Form that all work done by the manufacturer or others responsible to the manufacturer complies with all requirements of the Code. When some portion is performed by others not responsible to the manufacturer, the manufacturer must obtain the other organization's Code certification. In addition, the manufacturer must stamp the boiler with a Code symbol, which signifies that it has been constructed in accordance with the Code. Certification thus is an integral part of a quality assurance program which establishes that:

1. The organization that did the work held an appropriate ASME Certificate of Authorization to use a Code symbol;
2. The organization has certified compliance with the Code rules by signing and furnishing the appropriate Manufacturers' Data Report Form;
3. The organization has applied the Code symbol stamp to identify the work covered by its Data Report Form;
4. A qualified Inspector has confirmed by signing the Data Report Form that the work complied with the applicable Code rules.

In the case of a complete boiler unit that is not manufactured and assembled by a single manufacturer, the same principles are followed. There is always one manufacturer who must take the overall responsibility for assuring through proper Code certification that all the work complies with the requirements of the Code. If, for example, the manufacturer of a boiler buys the drum (or any other part) of the boiler from another manufacturer, that other manufacturer must follow similar certification and stamping procedures. The part manufacturer must have a Certificate of Authorization to use the appropriate Code symbol, must certify compliance with all the Code rules on a Manufacturers' Data Report Form, must stamp the part, and an Authorized Inspector must sign the form as confirmation that the work complied with the applicable Code rules. That form becomes part of the complete documentation assembled by the manufacturer of record. In general, the manufacturer of record has the duty of obtaining from all organizations that have done any Code work on the boiler their proper Code certification.

The same approach is used to certify that portion of the work called field assembly, for those boilers that are too large to be completely assembled in the shop. The Manufacturers' Data Report form typically has a certification box for use by the assembler of a boiler.

MANUFACTURERS' DATA REPORT FORMS

Section I has ten different Manufacturers' Data Report Forms [MDRF's] that have evolved over the years to cover various types of boilers and related components. These forms serve several purposes. First, they provide a documented summary of certain important information about the boiler: its manufacturer, purchaser, location, and identification numbers. Most forms also provide a concise summary of the construction details used: a list of the various components (drum, heads, headers, tubes, nozzles, openings), their material, size, thickness, type, etc., and other information such as the design and hydrostatic test pressures and maximum designed steaming capacity.

The forms and their use are described in PG-112 and PG-113, and in Appendix A-350, which contains examples of all the forms and a guide for completing each one. Also in the Appendix (page A-357) is a so-called Guide to Data Report Forms Distribution. This guide explains which forms should be used in nine different circumstances involving several types of boilers that may be designed, manufactured, and assembled by different stamp holders. There are so many forms and combinations of forms that can be used that the subject can often be quite confusing. What is important to remember is the purpose of the forms: to provide a summary of essential information about the parts comprising the boiler and to provide for certification as appropriate by all those involved (the various manufacturers and inspectors) that all components of the boiler were constructed to the rules of Section I. At the same time, it is also important to understand that the MDRF's were never intended to be an all-inclusive catalog or parts list for the boiler or a substitute for the manufacturers' drawings. If more information is needed than is found on the forms, it can be found on those drawings.

Certain of the forms are sometimes called Master Data Report Forms. A Master Data Report Form, as the name implies, is the lead document when several different forms are used in combination to document the boiler. The P-2, P-2A, P-3, P-3A, and P-5 forms can be used as Master Data Report Forms. The other forms (the P-4, P-4A, P-4B, P-6, and P-7) usually supplement the information on the Master Data Report Forms and are attached to them.

Before explaining how the ten Section I forms are completed, it is helpful to list and describe them.

THE TEN SECTION I MANUFACTURERS' DATA REPORT FORMS

1. Form P-2, Manufacturers' Data Report Form for All Types of Boilers Except Watertube and Electric

This form was designed primarily for firetube boilers.

2. Form P-2A, Manufacturers' Data Report Form for All Types of Electric Boilers

This form was specifically designed for electric boilers, when Part PEB, dealing with such boilers, was added to Section I in 1976.

3. Form P-3, Manufacturers' Data Report Form for Watertube Boilers, Superheaters, Waterwalls, and Economizers

This form can be used to document a complete boiler or major subcomponents. Both the P-3 and P-4 forms contain a table identified as line 12, which has a column for listing heating surface of various components. Some advice is given next to the table, noting that total boiler and waterwall heating surface is to be stamped on the drum heads and that any other heating surface is not to be used for determining minimum safety valve relieving capacity. The details of this heating surface calculation are explained in PG-101. That paragraph is a vestige of former rules for calculating the required safety valve relieving capacity for certain types of boilers from the type and area of heating surface in the boiler. However, since a significant revision to the safety valve requirements appeared in the 1993 Addenda, heating surface is seldom used to determine minimum required safety valve capacity. See further discussion under safety valves.

4. Form P-3A, Engineering-Contractor Data Report Form for a Complete Boiler Unit

This form differs significantly from those listed above. It is used when an engineering-contractor organization assumes the Code responsibility for the boiler normally taken by a boiler manufacturer. Since the engineering-contractor has no manufacturing facilities, all parts comprising the boiler are purchased from other Code symbol stamp holders, who are manufacturers. They furnish the necessary documentation for those parts, using other MDRF's such as the P-4 or P-4A, which the engineering-contractor attaches to the P-3A. The P-3A form requires the engineering-contractor to provide a design specification for the boiler, that is, a list of components giving the design pressure and temperature of each.

5. Form P-4, Manufacturers' Partial Data Report Form

This form gets its name from the fact that it covers only an individual component part of the boiler such as a drum or header. In general, this is not a stand-alone form, it merely provides supplementary data for the information shown on another form, the so-called Master Data Report Form, to which it is attached. However, it is a stand-alone form when used to document replacement parts for an existing boiler, as explained in PG-112.2.4.

One of the weaknesses of the P-4 form used to be that it never made clear who was taking design responsibility for the parts being documented. The answer to this question is not so simple as it might first appear, because sometimes a parts manufacturer may design the part, while at other times the manufacturer may just fabricate the part in accordance with a drawing he was given (a common practice called "build to print"). After considerable discussion and study by the committee, the problem of design responsibility for parts was addressed by the addition of a new paragraph, PG-112.2.4(3), in the 1996 Addenda. That paragraph calls for the parts manufacturer to indicate on the "remarks" line of the form the extent to which it has performed the design function.

6. Form P-4A, Manufacturers' Data Report for Fabricated Piping

PG-112.2.5 explains that this form is used to record all shop or field-welded piping that is within the scope of Section I but that is not furnished by the boiler manufacturer. (When the boiler manufacturer furnishes such piping it can be included on the manufacturer's own Data Report Form without the P-4A.) Note that PG-109.2 relaxes the stamping and marking requirements for welded piping that is NPS 2 and smaller. It need not be stamped, nor listed on the data report. However, it must be traceable to the data report.

Like the P-4 form just described, the P-4A form is also not definitive regarding who has design responsibility for the piping, mostly because of the nature of the PP stamp holder's business, which often involves fabrication of piping specified or designed by someone else (build to print). The P-4A form has three entries that concern design on its line 5. The entries are:

1. Design conditions (pressure and temperature);
2. The name of the organization that specified the design conditions; and
3. The name of the organization that did what is described as Code Design.

There is no certification box or any other place on the form where any organization actually assumes Code design responsibility by explicitly certifying that the design of the piping meets the rules of Section I. Unless the organization whose name appears in the space marked "Code Design by" has an appropriate Code symbol stamp (S or PP), it seems the responsibility would fall by default on the PP stamp holder who fabricates the piping. Although the system appears to be working satisfactorily as is, the committee is considering whether further guidance is needed in this area.

7. Form P-4B, Manufacturers' Data Report for Field Installed Mechanically Assembled Piping.

As explained at the end of Chapter 7 under Hydrostatic Test of Mechanically Assembled Piping, the 1998 Addenda to Section I established new rules for this type of piping in PG-99, PG-104.1, PG-109.4, and PG-112.2.5. The new rules provide that mechanically assembled piping need not be assembled by a

Code symbol holder, but that responsibility for its documentation and hydrostatic testing must be assumed by the holder of an S, A, or PP stamp. This would include a hydrostatic test witnessed by an Authorized Inspector. Documentation of the piping requires the use of a new form, the P-4B form. This form resembles the P-4A form, but differs from the latter in the certification required. The stamp holder must certify only that the field assembly of the described piping conforms to the requirements of Section I. There is no certification of shop or field fabrication compliance.

8. Form P-5, Summary Data Report for Process Steam Generators

This form was devised to cover process steam generators of the waste heat or heat recovery type. These are usually field-assembled arrangements of one or more drums, arrays of heat exchange surface, and associated piping.

All components must be certified by their manufacturers on individual Data Report Forms. The organization having Code responsibility for the complete system (the manufacturer or engineering-contractor of record) must collect and list on the summary form P-5 all Data Report Forms for components comprising the boiler. This approach to documenting a boiler that is field assembled from a group of parts furnished by different manufacturers is similar to the use of the P-3A form by an engineering-contractor organization.

As the use of waste heat and heat recovery boilers has become widespread, certain documentation problems became apparent. Most of these waste heat boilers are designed as multiple pressure units. Since the existing data report forms don't lend themselves to reporting different maximum allowable working pressures, the committee devised a Code Case that permitted such multiple pressure boilers to be documented as a single boiler. That case, Code Case 2173, stipulated how this could be done, with separate documentation for the parts designed for each different pressure and a summary form. A separate hydrostatic test pressure is required for the components at each design pressure level. All the MAWP's must be stamped on the nameplate. In 1996 the provisions of the case were approved for incorporation into Section I and appeared in the 1997 Addenda, A97.

9. Form P-6, Manufacturers' Data Report Supplementary Sheet

This form is used to record additional data when space is insufficient on any other Data Report Form. Many boilers, large utility boilers especially, cannot be documented on a single page; some take 50 pages or more. It is common to use supplementary sheets with whatever format or arrangement best suits the circumstances.

10. Form P-7, Manufacturers' Data Report for Safety Valves

This form was added to Section I by the 1996 Addenda, in response to concerns that changes to safety valve set pressure on existing boilers might subject piping to pressure for which it was not designed. Although this is a postconstruction issue usually considered beyond Section I's concerns, the committee felt that a summary listing of all safety valves, their location, set pressure, and relieving capacity would be useful to the Authorized Inspector during construction of the boiler and perhaps subsequently. Since the information called for on the P-7 form is in most cases already available from the boiler manufacturers' drawings, the new form is not much of an administrative burden. As for the problem that gave rise to the P-7, the National Board is considering changes to its Inspection Code that would treat any significant in-service changes to safety valve set pressures as an alteration to the boiler requiring a review of the potential consequences of the changed pressure.

COMPLETING THE DATA REPORT FORMS

These forms typically provide a place for the manufacturer to certify that the material used to construct the boiler meets the requirements of the Code (no specific edition) and that the boiler design, construction, and workmanship conform to Section I, with the particular edition and Addenda specified. In so certifying, the manufacturer must also identify which Code symbol it is using, and the number and expiration date of its Certificate of Authorization to use it.

Also provided on these forms are other certification boxes, or certificates of compliance, covering shop inspection, field assembly, and field assembly inspection. Some of these boxes, such as field assembly, may not be applicable and would be left blank.

The certificate of shop inspection is completed and signed by the Authorized Inspector who has inspected the shop fabrication of the boiler or other components. The AI provides information about the AI's employer, commission held, parts of the boiler inspected, and documentation examined for parts not inspected. The AI then, by his or her signature, certifies that, to the best of his or her knowledge and belief, the manufacturer of the boiler has constructed it in accordance with Section I.

Many boilers are not completed in the shop because of their size or for other reasons. In such cases, the various components are assembled in the field. The field assembly is in most cases a Code activity and must be done by an organization holding an appropriate Code symbol (either an A or an S). This organization must complete a box certifying that the field assembly of the boiler conforms with the requirements of Section I and provide particulars about its Certificate of Authorization.

The field assembly of the boiler is subject to inspection by the assembler's AI, who also completes a certification box giving information about his or her employer, the commission the AI holds, and which parts of the boiler the AI has inspected. The AI must also have witnessed the hydrostatic test of the boiler. The Inspector then signs and dates the form, stating that to the best of his or her knowledge and belief, the manufacturer and/or the assembler has constructed and assembled the boiler in accordance with Section I.

COMMENTS ON THE DATA REPORT FORMS

A review of the various Data Report Forms shows considerable variation and certain inconsistencies among them. This is chiefly due to their development at different times, to meet changing circumstances. In recent years, Subcommittee I has been reviewing all the forms with a view to clarifying and modernizing them. One difficult aspect of this effort is the clear assignment of design responsibility for different components of the boiler when more than one manufacturer is involved.

Such a situation used to arise, for example, if an engineering-contractor organization purchased all components of a boiler from individual manufacturers who provided them with Partial Data Report Forms P-4. Until recently, when a P-3A Engineering-Contractor Data Report served as the Master Data Report Form, no Code stamp holder was explicitly assuming design responsibility, since neither the P-3A nor the P-4 form addressed this responsibility. (Although, as a practical matter, the S stamp holder of record was considered to have design responsibility for the complete boiler.) Accordingly, the P-3A form was changed to require the engineering-contractor to certify that the design of the boiler complies with Section I. The addition of a new paragraph, PG-112.2.4(3), in the 1996 Addenda, was another change intended to address the problem. That paragraph calls for the parts manufacturer to indicate on the P-4 form the extent to which the manufacturer is assuming design responsibility for the part.

Users of Section I should be aware that disagreements sometimes arise as to which data report form should be used in a particular case. Subcommittee I members do not consider this choice to be of overwhelming importance so long as it is clear who furnished and installed the various components and who is taking design responsibility. Several interpretations have affirmed this position, as shown below.

DISTRIBUTION OF THE DATA REPORT FORMS

When the Manufacturers' Data Report Forms are complete, Section I requires copies to be furnished to the purchaser, the inspection agency, and the municipal, state, or provincial authority at the place of installation. In addition, many jurisdictions require registration of the boiler with the National Board, which would then be sent copies of the Manufacturers' Data Report. Many boiler manufacturers routinely register all their boilers with the National Board. Thus the concerned organizations have available a valuable summary of the design data and the technical details of the boiler. This can be very helpful many years later for investigations, repairs, or alterations, when original plans may have disappeared or are not readily available. Surprisingly, this is often the case. Having such information as the applicable Code edition and the details of original components can facilitate any subsequent work on the boiler. The designer then knows the design pressure, the size, thickness, material, allowable stress, and design formulas originally used and can make appropriate choices for repair, replacement, or alteration. Copies of Data Report Forms on file with the National Board can be readily obtained from that organization, and this is a major benefit of National Board registration (a service available for both boilers and pressure vessels).

INTERPRETATIONS ON USE OF MANUFACTURERS' DATA REPORT FORMS

Although the use of computer-generated data report forms is becoming more and more common, there was a time when only the official forms obtained from the ASME or the National Board were deemed acceptable. However, in 1989, Interpretation I-89-27, Use of Computer-Generated Manufacturers' Data Report Forms, established that other forms were acceptable:

> Question: *Does Section I allow creation of Manufacturers' Data Report Forms by a Code symbol stamp holder using any printed or computer-generated forms, provided the size, arrangement, and content are identical, without addition or deletion, to the Data Report Forms in the latest Edition and Addenda of Section I?*
> Reply: *Yes.*

As mentioned earlier, there are so many data report forms, and so many different situations in which they might be applied, that occasionally disputes arise as to how the parts are to be documented. Subcommittee I has shown some flexibility in the use of two different forms for pipe and pipelike components. Either the P-4 Partial Data Report Form (intended for parts) or the P-4A (intended for piping) may be used. This was affirmed in Interpretation I-92-30, Manufacturers' Data Reports for Boiler Proper Piping and Boiler External Piping:

> Question: *May fabricated boiler proper piping constructed only by welding as covered by PW-41 and fabricated by a manufacturer or a contractor in possession of the pressure piping symbol stamp be reported on a Manufacturers' Data Report Form P-4A as called for in PG-112.2.5?*
> Reply: *Yes; alternatively, a Form P-4 may be used. See PG-109.3.*

A similar Interpretation, I-92-46, Stamping of Boiler Proper Piping, affirmed that certain components can be stamped either as parts or as piping. This interpretation may be found at the end of the following section on Code Symbol Stamps.

Other inquiries have dealt with the question of which piping must be documented on the data report forms, because Section I is not explicit on the subject. One of these was Interpretation I-80-18, External Piping:

> Question: *Do the rules of Section I, PG-104, require that boiler external piping within the scope of Section I and supplied by the boiler manufacturer, be specifically documented on the P-3 Data Report?*
> Reply: *No. The ASME P-3 Data Report Form contains (in item 11) spaces to record the number, size, and type of nozzle outlet for four categories of connection to the boiler. These are:*
>> *(1) steam*
>> *(2) safety valve*
>> *(3) blowoff*
>> *(4) feed*
> *Guidelines No. 21, 22, and 23 in A-352 amplify this to explain what information is to be recorded. The explanation includes a statement that small openings for water columns, controls,*

vents, drains, and instrumentation, or openings for connections within the boiler proper, such as risers and downcomers, are not intended to be included in this listing.

The P-3 Form does not contain any spaces or requirements for recording any other information, such as piping arrangements, length, thickness, or material specifications for boiler external piping in any category. The manufacturer who supplies the boiler external piping has the responsibility to comply with applicable rules (including both ASME Section I for certification and ANSI B31.1 for technical requirements). The Authorized Inspector has the duty to verify the manufacturers' compliance. Neither the manufacturer nor the Authorized Inspector has an obligation to use the P-3 Form as a record of Code compliance of their respective work on boiler external piping.

What the above interpretation affirms is that when the manufacturer of the boiler also furnishes the boiler external piping, the manufacturer need not document it in detail on the P-3 form. The reason is believed to be that in such cases the boiler manufacturer has the full responsibility for every part of the boiler, and that manufacturer's drawings will provide for all piping. However, when another manufacturer furnishes the external piping, it is necessary to have that piping documented in greater detail, on the P-4A form.

Another Interpretation advising about what piping had to be listed on a data report form was I-83-16, Form P-2A, Listing of Miscellaneous Piping:

Question: *Is it a requirement of Section I that piping items other than steam, feedwater, and blowoff listed on Form P-2A, Item 16, be included when completing Forms P-2A or P-6?*
Reply: *No.*

CODE SYMBOL STAMPS AND HOW THEY ARE OBTAINED

The ASME committee that formulated the first edition of the Code in 1915 recognized a need to identify in a unique way any boiler constructed to meet the Code. They decided to do this by having the manufacturer stamp the boiler with a Code symbol. Paragraph 332 of that original edition stipulates, in part, "Each boiler shall conform in every detail to these Rules, and shall be distinctly stamped with the symbol shown in Fig. 19, denoting that the boiler was constructed in accordance therewith. Each boiler shall also be stamped by the builder with a serial number and with the builder's name either in full or abbreviated, ..." The official symbol for such a boiler was then, and is now, an S enclosed in a clover leaf. That symbol is evidence that the boiler complies with the Code and represents an assurance of safe design and construction. A great many provisions of Section I are directed to making sure that this will be so.

Over the years, Section I has added other symbol stamps, and there are now a total of six covering different aspects of boiler construction. These symbol stamps are:

S	power boiler	A	boiler assembly
E	electric boiler	PP	pressure piping
M	miniature boiler	V	safety valve

Until the late 1970s, there was also an L stamp, used for locomotive boilers, but by that time there were only two manufacturers who were still authorized to use the L stamp. The committee decided to abolish what had become an obsolete stamp, and those last two manufacturers were given S stamps instead.

Except under certain carefully controlled circumstances, no organization may do Code work (e.g., fabricate or assemble any Section I boiler components) without having first received from the ASME a Certificate of Authorization to use one of the Code symbol stamps. Before an organization can obtain such a certificate, it must meet certain qualifications, among which are the following.

The organization (manufacturer, assembler, or engineering-contractor) must have in force at all times an agreement with an inspection agency (the Authorized Inspection Agency, or AIA), usually an insurance company authorized to write boiler and pressure vessel insurance, or with a government agency or government authorized agency (see Chapter 8) that administers a boiler law in that jurisdiction. This agreement covers the terms and conditions of the authorized Code inspection to be provided and stipulates the mutual responsibilities of the manufacturer or assembler and an inspector provided by the inspection agency. These inspectors, called Authorized Inspectors, must be qualified by written examination under the rules of any state of the United States or province of Canada whose boiler laws have adopted Section I. Authorized Inspectors provide the inspection required by Section I during construction or assembly and, after completion, of any Section I component.

A further requirement for obtaining a Certificate of Authorization to use one of the ASME Code symbol stamps is that the manufacturer or assembler must have, and demonstrate, a quality control system to establish that all Code requirements will be met. These pertain to material, design, fabrication, examination by the manufacturer or assembler, and inspection by the Authorized Inspector. An outline of the features required in the quality control system is provided in paragraph A-300 of the Appendix and is discussed in Chapter 10.

Finally, before the ASME issues (or renews) a Certificate of Authorization, the manufacturer's or assembler's facilities and organization are reviewed by a team consisting of a representative of the inspection agency (AIA) and an individual certified as an ASME Designee who is selected by the legal jurisdiction involved. The manufacturer or assembler must make available to the review team a written description of the quality control system that explains what documents and procedures will be used to produce or assemble a Code item. On recommendation of the review team, the ASME issues (or renews) a Certificate of Authorization to use a particular Code symbol stamp, normally for a period of 3 years.

The requirements for the makeup of the ASME review teams who survey manufacturers and the qualifications of its members have undergone a number of changes in recent years as the ASME has tried to adapt to international competition and the effects of trade agreements such as GATT and NAFTA. This evolution is expected to continue.

Applicants for Certificates of Authorization may not have a clear understanding of just what capabilities they must have to obtain any particular stamp, since nowhere does Section I provide a complete and specific list of such capabilities. The prospective certificate holder is ordinarily guided in this regard by their Authorized Inspection Agency.

Around 1980, Subcommittee I established a task group to develop what might be called generic requirements for applicants who wished to obtain the various Section I Code symbol stamps. It was thought that these might eventually be published as an aid to users of the Code, but the idea came to nought, probably because the AIA's were providing whatever information the applicants needed. Table 9.1 presents a simplified summary of the capabilities usually necessary to obtain the various stamps. Note that the Code has traditionally allowed certificate holders to subcontract almost every aspect of construction (such as nondestructive examination, postweld heat treatment, and even design) so long as the certificate holder retains and accepts responsibility for Code compliance of work done on their behalf. Further, the certificate holder's quality control system would have to describe this option and the monitoring and controls used to assure compliance with Section I.

WHAT THE CODE SYMBOL STAMPS COVER

The S Stamp

The S stamp covers the design, manufacture, and assembly of power boilers. This stamp is applied to the completed boiler unit. It may also be applied to parts of a boiler, such as a drum, superheater, waterwall, economizer, header, or boiler external piping. When some parts of a boiler are furnished by a certificate holder other than the manufacturer of the completed boiler, the manufacturer of such parts may use its own S symbol stamp plus the word ''part'' to identify them. Other parts of a boiler such as valves, fittings, and circulating pumps do not require inspection, stamping, and certification when furnished as standard pressure parts under the provisions of PG-11, as explained in Chapter 11.

TABLE 9.1
CAPABILITIES NEEDED TO OBTAIN SECTION I CODE SYMBOL STAMPS
(It is permissible to subcontract for certain of these capabilities. See text.)

CODE SYMBOL STAMP	DESIGN CAPABILITY	QUALIFIED WELDERS & PROCEDURES	NDE AVAILABLE	HEAT TREATMENT FACILITY AVAILABLE	Q.C. SYSTEM
S POWER BOILER	R	R Note 1	R	R Note 1	R
A BOILER ASSEMBLY	NR	R	R	R	R
PP PRESSURE PIPING	R	R	R	R	R
M MINIATURE BOILER	R	R	NR	NR	R
E ELECTRIC BOILER	R	NR	NR	NR	R
V SAFETY VALVE	R	R Note 2	R	R Note 2	R

LEGEND: R = REQUIRED
NR = NOT REQUIRED

Note 1: It is possible to manufacture certain small boilers without welded joints. In such cases, qualified welders, weld procedures, heat treatment facilities, and nondestructive examination would not be required.

Note 2: Not required for valves whose manufacture does not involve welding.

A manufacturer may assume Code responsibility for a miniature or electric boiler using the S stamp as a substitute for the M or E stamps. (Stated another way, an S stamp may be used for miniature or electric boilers.)

The S stamp may also be used to cover field assembly of a boiler and/or boiler external piping. However, if the boiler or piping was furnished by others and the S stamp holder was merely acting as an assembler of the boiler and piping, the assembly work would be more likely to be done under the authority of an A certificate (see the A stamp, below), so as to avoid confusion as to which organization was the manufacturer of record and which was the assembler.

The S stamp can also be issued to entities called engineering-contractor organizations, mentioned in PG-104, Note 1. These are organizations that may not have manufacturing facilities but do have boiler design capability and sometimes field-assembly capability. The work this type of organization may do is discussed below with interpretations on Code symbol stamping, at Interpretation I-92-12.

Subcommittee I has always considered the S stamp the senior stamp among all the rest, which in theory at least, allows the holder to engage in any type of Section I construction: design, manufacture, or assembly, so long as the holder's quality control system covers that activity.

The M Stamp

The M stamp is applied only to miniature boilers, that is, to boilers that do not exceed the limits specified in PMB-2 of Section I. The miniature classification is thus limited to boilers that are small, have maximum allowable working pressure that does not exceed 100 psi, and do not require radiography or postweld heat treatment.

The E Stamp

The E stamp is applied to certain electric boilers when the manufacturer of the boiler is not authorized to apply the S or M stamps, principally because the manufacturer lacks the capability to do Code welding. In such cases, that manufacturer connects the necessary trim, fixtures and fittings to a boiler pressure vessel furnished and Code stamped by others with the S, M, or U stamp (in the latter case, subject to the vessel requirements of PEB-3). The manufacturer then applies the E stamp to the completed assembly and becomes the manufacturer responsible for the completed boiler.

The A Stamp

The A stamp covers the field assembly of a boiler when done by someone other than the manufacturer of the boiler. In such a case, the assembler shares Code responsibility for the completed boiler jointly with the boiler manufacturer, and the assembler's A stamp must supplement the manufacturer's stamp. (See PG-107.2 and PG-108.2.)

The A stamp may also be applied for the assembly of boiler external piping.

Note that the A stamp does not cover manufacture of a boiler or parts of a boiler; it covers only assembly of Code-stamped parts or Code material manufactured by others. The A stamp holder is neither authorized nor required to take Code design responsibility.

The PP Stamp

The primary application of the pressure piping stamp PP is to the fabrication of boiler external piping and to its installation by welding when the piping is not furnished by the boiler manufacturer.

The PP symbol may also be applied to fabrication and shop or field assembly of boiler parts such as superheater, waterwall, or economizer headers, or any construction involving only circumferential welds as covered by PW-41.

There is a task group of the Subgroup on Piping presently reviewing the issue of who is taking design responsibility for the piping furnished by PP stamp holders and documented on the P-4A form normally used for fabricated piping. That form is not definitive regarding design responsibility, and the committee wants to be certain that some organization (stamp holder) is assuming it. Some further guidance may be added to the form or to the guide for filling it out.

The V Stamp

The safety valve stamp V may be applied by manufacturers or assemblers of safety valves for power boilers to show that the parts and assembly of the valves comply with Section I requirements, including capacity certification. For further discussion, see Chapter 12: Safety Valves.

INTERPRETATIONS ON CODE SYMBOL STAMPING

There have been many Interpretations on the use of Code symbol stamps that clarify what the rules permit and prohibit. Since the stamping rules cover so many situations it is helpful to review some of the important interpretations on the subject. One of these, I-78-24, Stamping of Boiler External Piping for Boiler Installations, explains some basic principles and applications of the stamping and certification process:

> Question: *What is the definition of boiler unit as used in PG-104?*
> Reply: *Boiler unit refers to everything within the scope of the jurisdiction of Section I; this comprises the boiler proper plus the boiler external piping (Preamble, third paragraph).*
> Question: *Does the boiler manufacturer use his S symbol stamp to ''cover'' the entire boiler unit?*

Reply: *Yes, but only in the sense explained in PG-104. If some portions are supplied by, or Code work is done by, others not responsible to the boiler manufacturer, these other organizations must hold proper Certificates of Authorization and furnish their certification and stamping to supplement the boiler manufacturer's certification.*

For example, if some or all of the boiler external piping is supplied by a fabricator other than the boiler manufacturer, such piping should be "covered" by that fabricator's S or PP stamp and documented with the P-4A Form. (See PG-109.)

If the field assembly is by other than the boiler manufacturer, provisions of PG-107.2 require use of the assembler's A symbol to supplement the manufacturer's S symbol, and the assembler is identified on the field assembly block of the data form.

Question: *Does the stamping referred to in ANSI B31.1 for boiler external piping differ from that of ASME Section I?*

Reply: *No, B31.1 refers to Section I for all certification (data reports, inspection, and stamping) to cover boiler external piping.*

Question: *In the context of the Preamble of Section I, is boiler external piping considered a "pressure part"?*

Reply: *The fourth paragraph of the Section I Preamble begins, "Superheaters, economizers, and other pressure parts connected directly to the boiler without intervening valves shall be considered part of the boiler proper...." Boiler external piping is not an "other pressure part" as used in that phrase, even though any piping is considered a pressure part in Code usage. Figures PG-58.3.1 and PG-58.3.2 attempt to illustrate the commonly encountered division points between "boiler proper" and "boiler external piping."*

Another lengthy Interpretation that illustrates which stamps may be used by different organizations involved in manufacture and assembly of boiler components and also affirms the S stamp as the senior stamp is I-81-22, Clarification of Limits of Certificates of Authorization:

Question: *In PG-109.1, installation of boiler external piping fabricated by anyone other than the boiler manufacturer is allowed by a manufacturer or contractor in possession of a Certificate of Authorization for an S stamp, A stamp, or PP stamp.*

(a) *Under what conditions does PG-109.1 permit the use of the A stamp?*

(b) *When the boiler external piping is to be installed, but not fabricated by the contractor, may the S stamp still be used or must the contractor have a PP stamp?*

(c) *May a contractor who did not design the boiler unit use his S stamp for assembly? Can he also use his S stamp for assembly of boiler external piping?*

Reply for (a): *PG-109.1 permits a manufacturer in possession of a Certificate of Authorization for use of the A stamp to install boiler external piping, by welding, in accordance with the applicable rules of ANSI/ASME B31.1, Power Piping.*

Reply for (b): *The holder of an S stamp may use that stamp to install boiler external piping by welding.*

Reply for (c): *Yes to both questions.*

Question: *Are the terms "manufacturer and fabricator" as used in PG-104 and PG-109.1 synonymous with respect to the S stamp?*

Reply: *The term "fabricator" is not used in PG-104 and PG-109.1. Note 1 of PG-104 mentions "fabricating facilities," and "fabricated by anyone other than the manufacturer of the boiler."*

(Note: This answer dodged the question. For practical purposes the two terms are synonymous in this context.)

Question: *May a holder of an S stamp, who is not the manufacturer (PG-104) of the boiler, use that S stamp to cover assembly work during construction of a complete boiler unit?*

Reply: *Yes.*

Question: *Does Note 2 of PG-104.1 pertain only to shop-assembled boilers as indicated in PG-107?*

Reply: *No. Note 2 of PG-104.1 applies to all boiler external piping within the scope of Section I, whether the boiler is field assembled or shop assembled, when the boiler manufacturer does not furnish that piping.*

Question: *What is the definition of a shop-assembled boiler?*

Reply: *A shop-assembled boiler is one for which the manufacture of the boiler, with or without the boiler external piping, is completed in the boiler manufacturer's shop.*

Question: *Is it a mandatory requirement that any contractor who may have occasion to work on a Code boiler be required to have a stamp in order to perform the necessary work even though the welders are qualified to Section IX? If this is not a Code requirement, would it be a jurisdiction or National Board requirement?*

Reply: *If the inquiry pertains to new construction, any work that falls within the scope of the Code must be performed by an organization holding an appropriate Certificate of Authorization. Rules governing work on boilers after they have been placed in service vary, depending on the requirements, if any, adopted by the administrative authorities of the state, municipality, or province where the boiler is installed.*

It happens that the reply to the above question of what is a shop-assembled boiler raises a question of just what completion of the boiler can mean, depending on whether the boiler is field assembled or shop assembled. A field-assembled boiler is considered complete after the boiler external piping is installed and the boiler has passed the hydrostatic test. However, for various reasons, smaller, shop-assembled boilers, especially those with threaded piping, are often not provided with boiler external piping installed in the shop. This situation sometimes arises because with the piping attached, the boiler would be too large to ship, or because purchasers have traditionally furnished their own piping, or perhaps because the boiler is being built for stock, in anticipation of future sale to a customer whose external piping requirements are not yet known.

When threaded boiler external piping is involved, it is often the custom to subject the boiler proper, either with no external piping attached or perhaps with some minimal piping and valves installed to a hydrostatic test. The data report forms are then completed and signed, and the boiler is considered complete. Any further completion and installation of the boiler external piping is done by others, perhaps a plumber, in the field. This practice has been followed for a long time in many jurisdictions and is supported by the words of PG-104.1, Note 2, which in effect waive Code certification and data report forms covering piping that is not welded. However, at present, Indiana and Iowa insist on a more formal and comprehensive completion of a boiler with threaded piping. They insist on a hydrostatic test in the field in the presence of an Authorized Inspector and certification of the piping. As explained in Chapter 7, under Hydrostatic Test of Mechanically Assembled Boiler External Piping, Subcommittee I in 1997 resolved a number of conflicting issues associated with what is called "piping installed by mechanical means" (threaded or bolted piping).

The situation differs when the boiler external piping is welded. When welded external piping is furnished by a stamp holder other than the boiler manufacturer, Note 2 stipulates that the organization that furnishes and installs it is responsible for proper Code certification, stamping, and Data Report P-4A. That same organization is also responsible for distributing copies of Form P-4A to the inspection agency and the proper authorities, as noted in PG-113.4

Interpretation I-83-56, Use of S Symbol for Fabrication of BEP, affirms that an S stamp holder who is not the manufacturer of the boiler may fabricate boiler external piping in his shop and stamp it with the S symbol.

It is sometimes difficult to draw a clear distinction between the activities that constitute manufacturing piping and those that constitute assembly of the piping. Both activities could involve cutting, bending, and preparing pipe for welding, and then welding lengths of pipe together. This has led to some confusion as to whether the A stamp holder could manufacture, or fabricate, welded piping. The key difference is that an assembler may not take design responsibility, one of the capabilities a fabricator must have if he is to obtain an S or PP stamp, since the fabricator is often expected to take design responsibility for that piping.

The committee addressed this misapprehension twice, in Interpretation I-80-17, Certificates of Authorization, and in I-83-68, Fabrication of Pressure Piping Using an A Symbol Stamp. Here are those Interpretations:

Question: *May an A stamp holder, whose quality control system covers all Code mandated requirements for shop fabrication of pressure piping, fabricate such piping and provide Code certification of it, using his A stamp authorization?*
Reply: *No. If an applicant has a quality control system which covers all Code mandated requirements for shop fabrication of pressure piping, and can show implementation of this system, the applicant is eligible for the PP stamp authorization, and should obtain it if he wishes to manufacture such piping components. The same organization may or may not be an A stamp holder.*
Question: *May an organization in possession of an A Code Symbol Stamp fabricate welded pressure piping described in PG-109.1 or parts of boilers described in PG-109.3?*
Reply: *No. An A stamp holder may only assemble such piping or parts by welding; an S or a PP stamp is required for their fabrication.*

As mentioned above in the discussion of Certificates of Authorization, a stamp holder may subcontract many of the activities for which he is ultimately responsible. These could include such activities as nondestructive examination, postweld heat treatment, bending operations, and cutting and threading of pipe. The key requirements for subcontracting such activities are control by the stamp holder's quality system and the avoidance of welding, which is an activity normally requiring the use of qualified welders using qualified welding procedures under the direct supervision of the stamp holder. One Interpretation dealing with such subcontracting is I-86-38, somewhat misnamed as Definition of Joined by Mechanical Means:

Question: *When piping within the scope of Section I is being installed by a Certificate holder, is it permissible for a subcontractor who is not in possession of a Code symbol stamp to perform such functions as cutting and threading of pipe?*
Reply: *Yes; however, responsibility must be accepted by the Certificate Holder.*

Another Interpretation dealing with such subcontracting is I-95-17, Pipe Bending by Organization Not Authorized to Use a Code Symbol Stamp:

Question: *May an organization holding an S or a PP stamp subcontract the bending of pipe to an organization without a Code symbol stamp, provided the bending operations involve no welding and the stamp holder's quality control system covers the subcontracted activity?*
Reply: *Yes.*

A similar Interpretation dealt with threading and machining of piping, which may also be done by an organization without a Code symbol stamp. This was Interpretation I-92-04, Code Stamping of Pressure Piping:

Question: *Is Code stamping required prior to shipment on piping which has undergone bending, threading, or machining, (but not welding)?*
Reply: *No.*

Another example extends the principle of subcontracting to include even design calculations. This was affirmed by Interpretation I-92-48, Code Responsibility for Design Calculations:

Question: *May a manufacturer certify that engineering and design calculations performed by others, not holding a Certificate of Authorization, but under contract to the manufacturer, comply with all the requirements of the Code?*
Reply: *Yes.*

Interpretation I-86-13 stated this principle in another way:

> Question: *Boiler external piping as defined in PG-58 is fabricated and installed by separate organizations holding appropriate Certificates of Authorization in accordance with paragraphs PG-104, PG-109, and PG-112. Is the organization that prepares the purchase specifications and/or performs the Code design of the piping as identified on line 4 of the Form P-4A required to be the holder of a valid Certificate of Authorization?*
> Reply: *No.*

Note that under the circumstances just described, one, or the other, or both the PP stamp holders who fabricate the piping would have Code design responsibility for the piping.

In Note 1 of PG-101 it is explained that the term "boiler manufacturer" can also mean an S stamp holder that is a so-called engineering-contractor organization, with or without fabricating facilities, that has the capabilities of designing a boiler and of assembling the fabricated parts of a boiler in the field. Such an organization must provide for the design of the unit, that is, it must provide a design specification that establishes the design pressure and temperature conditions for each component of the boiler. The activities of an engineering-contractor organization were described in Interpretation I-92-12, under the uninformative title Certificate of Authorization:

> Question: *May an organization which provides the specifications, the design, the calculations, and the drawings for the components of a boiler, but which has no manufacturing facilities, obtain a Certificate of Authorization to use the S symbol stamp?*
> Reply: *Yes.*
> Question: *May the holder of the Certificate of Authorization to use the S symbol stamp subcontract any or all aspects of boiler construction so long as he retains Code responsibility for the completed boiler and his quality control system provides for those portions of the work which may be subcontracted?*
> Reply: *Yes.*

Although Note 1 of PG-104 appears to limit the issuance of an S stamp to engineering-contractor organizations that have the capability of field assembling parts fabricated by other stamp holders, in practice, this is not necessarily so. This can be seen from the above interpretation, as well as the guide to information appearing on the Certificates of Authorization in Appendix A-370.

Several other Interpretations have dealt with the seemingly endless uncertainties that arise as various stamp holders and their subcontractors collaborate in the construction of a boiler. One of these has to do with the furnishing of nameplates. There is no doubt that certain large components, such as a drum, are expected to have stampings or nameplates, as required by the various provisions of PG-106. However, what is less clear is what stamping or nameplates are required for smaller components such as headers or individual superheater elements. Interpretation I-86-32, Nameplates for Boiler Parts, gave some guidance:

> Question: *Does PG-104 of Section I require manufacturers to provide nameplates in addition to P-4 Partial Data Reports for individual boiler parts such as superheater elements, economizer platens, or waterwall panels?*
> Reply: *No; however, parts must be stamped by the manufacturer in accordance with PG-106.8. Additional guidance is provided in Interpretations I-83-101 and I-83-113.*

Interpretation I-83-101 is entitled Marking of Duplicate Boiler Parts:

> Question: *Where it is impossible to meet the stamping requirements of PG-111, is it permissible to use a nameplate and attach it to the part?*
> Reply: *It is intended that the manufacturer, subject to the approval of the Authorized Inspector, shall provide a stamping which will perform the function otherwise intended by the rules of the Code. A nameplate with the required stamping, securely attached by welding to the part, may be one acceptable solution.*
> Question: *What stamping is required when multiple duplicate parts are supplied for a field-assembled boiler by a manufacturer not taking responsibility for the completed boiler unit?*

Reply: *At least one part shall bear the full Code stamping and duplicate parts shall be identified by comparison of markings on the parts and the description of the part required to complete Form P-4.*

Question: *What stamping is required when multiple duplicate parts are supplied for a field-assembled boiler by a manufacturer not taking responsibility for the completed boiler unit, where it is impossible to meet the stamping requirements of PG-111?*

Reply: *It is intended that the manufacturer, subject to the approval of the Authorized Inspector, shall provide a stamping which will perform the function otherwise intended by the rules of the Code. A nameplate with the required stamping, securely attached by welding to at least one part with its location listed on Form P-4, may be one acceptable solution.*

Note that a nameplate or stamping on a part such as a superheater tube in a high-temperature zone will not last long. What is usually sufficient is some marking by which the manufacturer and Inspector can assure that the right parts have been installed in the right place.

Interpretation I-83-11 dealt with the Stamping of Piping Subassemblies:

Question: *Is it a requirement of Section I that piping subassemblies of boiler external piping (as defined in PG-58), which are shop fabricated by an organization in possession of a PP symbol stamp, be certified by that organization by completing the front of the P-4A Data Report and applying their PP symbol to each piping subassembly, if the piping is to be installed by welding by another organization?*

Reply: *Yes, provided that the subassemblies are larger than NPS 2. Where the subassemblies are NPS 2 or smaller, they shall be marked with an identification acceptable to the Inspector and traceable to the required Data Report (see PG-109).*

Question: *For the piping subassemblies defined in the preceding question, does Section I require the word "part" to be stamped below the PP symbol?*

Reply: *No.*

There is some flexibility in how certain piping components may be supplied and stamped either as parts or as piping. This was addressed in Interpretation I-92-46, Stamping of Boiler Proper Piping:

Question: *May fabricated boiler proper piping constructed only by welding as covered by PW-41, and fabricated by a manufacturer or contractor in possession of the pressure piping symbol stamp, be stamped in accordance with either PG-106.8 or PG-109.3?*

Reply: *Yes.*

The second question in this Interpretation dealt with some practical details of how multiple duplicate parts can be covered by stamping one part and providing identifying marks on the others (see precedent in PG-111.10).

Question: *A manufacturer or contractor who is not the manufacturer of the boiler fabricates a run of boiler proper piping consisting of multiple spool pieces constructed only by welding as covered by PW-41. May only one spool piece be stamped, the remaining spool pieces be identification stamped, and all spool pieces for this run of piping be reported on one P-4A Data Report Form?*

Reply: *Yes.*

A last Interpretation applies to an increasingly common situation in which purchasers of boilers to be installed outside of the U.S.A. and Canada want the boilers to be designed and manufactured to Section I of the ASME Code, but not field assembled by an ASME Code stamp holder under the supervision of an Authorized Inspector. Purchasers do this to lower costs, since it generally costs more to provide all of the quality control and oversight required for ASME Code construction. The boiler that results may be a perfectly satisfactory boiler, but it may not be stamped with a Code symbol or called a boiler that meets the ASME Code. This subject was addressed in Interpretation I-83-106, Use of ASME Marking:

Question: *May a power boiler which was designed in accordance with the rules of ASME Section I but which was not subject to authorized inspection or Code symbol stamping be marked on the nameplate with such information as "Design Code: ASME Section I"?*

Reply: *No. Such use of the term "ASME" on the nameplate is not in conformance with the "Statement of Policy on the Use of ASME Marking to Identify Manufactured Items," which appears in the front of Section I. That policy includes this sentence: "Markings such as 'ASME,' 'ASME Standard,' or any other marking including 'ASME' or the various Code symbols shall not be used on any item which is not constructed in accordance with all of the applicable requirements of the Code."*

QUALITY CONTROL SYSTEM

The 1960s was a time of rapid growth for the nuclear power industry. In 1963 a new ASME Code section, Section III, was published to cover construction of nuclear power plant components. One of the features of construction under Section III was a requirement that the manufacturer implement a quality assurance system, a formal system of oversight and record keeping to assure that construction takes place as intended, that the right material, parts, drawings, and procedures are used. The benefits of such a system soon became apparent. In 1973 the ASME Boiler and Pressure Vessel Committee adopted similar but less comprehensive and detailed rules for what is called a **quality control (QC) system** for the boiler and pressure vessel sections of the Code, Sections I, IV, and VIII. In Section I, these rules are found in PG-105.4 and Appendix A-300, both entitled Quality Control System. Each manufacturer is required to have a documented quality control system that is fully implemented into its manufacturing operations. It may be of interest to note that the B31.1 Power Piping Code has no provisions for a quality control system, although the governing committee has been considering the idea for many years.

The QC system is intended to control the entire manufacturing process from design to final testing and certification. The scope of such systems may vary significantly among manufacturers, since the complexity of the work determines the program required; however, the essential features of the QC system are the same for everyone. Each manufacturer is free to determine the length and complexity of its Quality Control Manual, making it as general or detailed as desired. It is not advisable to include detailed practices in the manual that are not actually followed in the shop. Also, when shop practices are revised, the manual must be updated to reflect the revisions. Otherwise the Authorized Inspector is likely to find nonconformities, which will cause considerable extra work before full compliance with the Code can be demonstrated. Implicit in the Code QC requirements is recognition that each manufacturer has its own established practices and these are acceptable if a documented system can be developed showing control of the operation in conformance with Code requirements. It is also recognized that some of the information included in a written description of a quality control system is proprietary, and no distribution is required other than to the Authorized Inspector.

An outline of features to be incorporated into a Section I quality control system is found in the A-302 paragraphs. These paragraphs are a list of requirements that have proven relatively easy to understand and have worked well. To begin, the authority and responsibility of those performing quality control functions must be established. That is, the manufacturer must assign an individual with the authority and organizational freedom to identify quality problems and to recommend and implement corrective actions should they be needed. In addition, the written system must include:

- The manufacturer's organization chart, showing the relationship between management, engineering, and all other groups involved in the production of Code components, and what the primary responsibility of each group is. The Code does not specify how to set up the organization, nor does it prevent the manufacturer from making changes, so long as the resultant organization is appropriate for doing Code work.

- Procedures to control drawings, calculations, and specifications. That is, the manufacturer must have a system to assure that current drawings, design calculations, specifications, and instructions are used for the manufacture of Code components.
- A system for controlling material used in Code fabrication to assure that only properly identified and documented material is used.
- An examination and inspection program that provides references to Nondestructive Examination Procedures, and personnel qualifications and records needed to comply with the Code. Also required is a system agreed upon with the Authorized Inspector for correcting nonconformities. A nonconformity is any condition that does not comply with the rules of the Code, whether in material, manufacture, or the provision of all required appurtenances and arrangements. Nonconformities must be corrected before the component can be considered to comply with the Code.
- A program to assure that only weld procedures, welding operators, and welders that meet Code requirements are used to produce Code components. This is normally one of the most detailed and important sections of the QC manual, which describes how qualified weld procedures are maintained and used and who within the organization is responsible for them. It also includes details on how welder qualifications are established and maintained.
- A system to control postweld heat treatment of welded parts and any other heat treatment, such as might be required following tube bending or swaging. This system must also provide means by which the Authorized Inspector can verify that the heat treatment was applied.
- A system for calibration of examination, measurement, and test equipment used in Code construction to assure its accuracy. The Code does not require such calibrations to be traceable to a national standard such as those maintained at the National Institute of Science and Technology (NIST). It merely states that the equipment must be calibrated.
- A system of record retention for such items as radiographs and Manufacturers' Data Reports.
- Procedures covering certain other activities of the manufacturer, such as hydrostatic testing.
- If a manufacturer intends to subcontract any aspect of Code construction, such as design, radiography, or heat treatment, the procedures and controls for the subcontracting must also be included in the manufacturer's quality control system.

Appendix A-300 concludes with the admonition that the quality control system shall provide for the Authorized Inspector at the manufacturer's plant to have access to all drawings, calculations, records, procedures, test results, and any other documents necessary for him or her to perform the inspections mandated by Section I. The objective is to assure the quality of the construction and compliance with the Code. Note also that the quality control system may not be changed without obtaining the concurrence of the Authorized Inspector.

As explained in Chapter 9, under Code Symbol Stamps, one of the conditions for the issuance or renewal of a Certificate of Authorization to use one of the ASME Code symbol stamps is that the manufacturer or assembler must have, and demonstrate to a review team, a quality control system intended to assure that all Code requirements will be met.

The issuance by ASME of a new Certificate of Authorization or the required triennial renewal of an existing Certificate of Authorization is based on a favorable recommendation after a joint review of the written quality control system by a representative of the manufacturer's Authorized Inspection Agency and a representative of the legal jurisdiction involved. In those areas where there are no jurisdictional authorities or when the jurisdiction declines to participate, the second member of the review team is a qualified ASME review team leader.

Standard Pressure Parts, Valves, and Valve Ratings

STANDARD PRESSURE PARTS

The rules dealing with standard pressure parts and how they may be provided are found in PG-11, Miscellaneous Pressure Parts. Ordinarily, Section I requires all pressure parts to meet applicable requirements for material, design, fabrication, welding, inspection, testing, and certification by stamping and data reports. These comprehensive Code construction requirements are considered necessary to ensure safe construction. Considerable relief from this burden is granted in the case of what are called standard pressure parts, provided these parts are designed and manufactured in such a way as to provide an equivalent assurance of quality and safety. Under these circumstances, mill test reports, inspection by an Authorized Inspector, and Partial Data Reports are not required.

Among the parts included in the category of standard pressure parts are such components as: pipe fittings, valves, flanges, welding necks, welding caps, manhole frames and covers, and the casings of certain pumps. Most of these parts are readily available as stock items in a variety of sizes. They don't have to be designed for each specific application, because the manufacturer has already established pressure and temperature ratings for them.

Section I establishes several categories of these parts. One category covers pressure parts which are cast, forged, rolled, or die-formed (i.e., not welded). Within this category, Section I designates two types: parts that comply with an American National Standard product standard (usually called an ASME/ANSI product standard) accepted by reference in PG-42 and parts that comply with a parts manufacturer's standard. There are some differences between the two in material and design requirements. Parts complying with a parts manufacturer's standard must be made of material explicitly permitted by Section I, but parts complying with an ANSI product standard may, in addition, be made of materials listed in that standard, provided Section I does not prohibit or otherwise limit the use of such materials. The ANSI product standard also establishes the basis for their pressure and temperature rating and specifies the marking on them; these ratings and markings are accepted by Section I. Some difficulties have arisen in recent years if the standard pressure part is not made of a material explicitly approved for Section I construction in Section II, Part D. This situation is explained in Chapter 3: Materials.

In the case of a parts manufacturer's standard, the manufacturer must establish the pressure and temperature rating of the part and the appropriate marking on it. Section I gives no further guidance on how this is to be done, and the manufacturers presumably use generally accepted engineering principles in developing their ratings. Perhaps because of this lack of specific guidance, Section I cautions in footnote 4 to PG-11: "The manufacturer of the completed vessel shall satisfy himself that the part is suitable for the design conditions of the completed vessel."

Another category of standard pressure parts established by Section I is designated by the somewhat awkward title ''welded standard pressure parts for use other than the shell of a vessel.'' This category includes the same kind of parts just described, except that welding is now involved. Since welding is a process involving many variables that when not properly controlled can result in unsatisfactory welds, Section I mandates that the welding must comply with the requirements of PW-26 through PW-29. The substance of these requirements is that the welders, welding operators, and welding procedures must be qualified under the provisions of Section IX of the Code. If a standard welded pressure part were used for the shell of a vessel, full Code requirements would be invoked, including inspection and data reports, since the pressure-containing function of a vessel shell is considered of prime importance to safety, and at least the longitudinal seam in the shell has to be radiographed.

PG-11 permits any required radiographic examination or heat treatment of welded standard pressure parts to be performed either in the plant of the parts manufacturer or in the plant of the manufacturer of the completed vessel. If radiography is done at the parts manufacturer's plant, the radiographs and inspection report must be sent to the vessel manufacturer and made available to its Authorized Inspector. If the parts manufacturer performs any required heat treatment, it must so certify and that certification must also be made available to the AI. As in the case of manufacturer's standard pressure parts, Section I again cautions that the manufacturer of the completed vessel must be satisfied that a welded standard pressure part is suitable for the design conditions specified for the completed vessel.

In addition to standard pressure parts, PG-11 permits the use of cast, forged, rolled, or die-formed nonstandard pressure parts (again note that welded parts are excluded) for components such as shells, heads, and cover plates. These parts are supplied and treated as though they were materials and must have appropriate mill test reports and marking. Once again the vessel manufacturer is charged to satisfy himself of the suitability of the parts for the intended design conditions.

In general, the marking required by PG-11 on the various kinds of pressure parts it covers serves to identify the parts manufacturer and the pressure and temperature rating of the part. It is also construed as the parts manufacturer's certification that the part complies with an appropriate material specification and is suitable for the rating indicated.

Note that these parts manufacturers need not be holders of certificates of authorization to use Code symbol stamps. Thus there is no Authorized Inspector monitoring the part manufacturer's operations. However, these standard parts are made under controlled shop conditions by manufacturers who know they are responsible for the satisfactory performance of their products, and the system has worked well without Code inspection and certification.

VALVES AND VALVE RATINGS

This discussion pertains to the pressure-containing, as opposed to the fluid-flow, aspects of valve selection, the latter being beyond the scope of Section I. Many valves are used on a boiler and its associated piping. The valves often establish the boundary of the boiler, that is, the jurisdictional limits of Section I, since boiler external piping is defined as that piping that begins where the boiler proper terminates and ''extends up to and including the valve or valves required by this Code.'' Thus most valves are associated with boiler external piping. A concise illustration of some of the valves typically required on a boiler is given in Figs. PG-58.3.1 and PG-58.3.2. Rules covering valves and valve ratings are found in PG-11, PG-42, PG-58, PG-59, PG-60, PW-42, and A-361. Specific valve application requirements are given in PG-58, PG-59, and PG-60. These requirements are stated in somewhat greater detail in paragraph 122.1 of ANSI/ASME B31.1, Power Piping, which since 1972 has had jurisdiction over boiler external piping for all aspects of construction except Authorized Inspection and Code Certification.

Usually only one valve is required on piping connected to the boiler. There are several exceptions to this rule, all safety related. A blowoff usually requires two valves, one slow opening, to provide for close control when water is blown out when the boiler is under pressure. Two main steam stop valves with an ample freeblow drain between them are required if boilers with manways are connected to a common steam header. When such boilers are out of service, an inspector or other worker could be working inside, and the object is to provide two layers of protection against inadvertent opening or leakage of the valves on the pipe leading to the common header. The Main Committee recently (1997) voted to approve the extension of this double-barrier requirement to any steam line that could supply steam to a boiler with a manway. This new provision appeared in the 1997 Addenda to Section I.

Often there is confusion about valves used on a boiler. They are mistakenly described as "Section I valves." They should rather be described as valves that meet Section I (or B31.1) requirements. The exception is safety valves, which do require stamping as Section I valves. The other valves do not; they are supplied under the category of standard pressure parts (PG-11), either under an ANSI product standard or a manufacturer's standard. The Power Piping Code, B31.1, accepts and rates valves on virtually the same basis as Section I (per B31.1, paragraph 107).

Most boiler valves comply with the product standard for valves, ASME/ANSI B16.34, Valves—Flanged, Threaded, and Welding End. This standard provides a series of tables, for various material groups, giving pressure–temperature ratings according to what are called pressure classes. The classes typically vary from the 150 class (the weakest) to the 4500 class. For each class, the allowable pressure is tabulated as a function of temperature. The allowable pressure at any temperature is the pressure rating of the valve at that temperature. Temperatures range from the lowest zone, $-20°F$ to $100°F$, to $850°F$ and higher, depending on the material group. The allowable pressure for the valve falls with increasing temperature, since it is based on the allowable stress for the material.

Flanged valves are available only as so-called standard class valves. Welding end valves are available in two classes—standard and special. The difference is that the special class valves are subject to more rigorous rules for nondestructive examination and defect removal and repair, and in consequence are permitted higher pressure ratings.

For Section I service, the designer must select a valve whose pressure rating at design temperature is adequate for the maximum allowable pressure (MAWP) of the boiler. However, this choice is complicated somewhat by the fact that both feedwater and blowoff piping must be designed for a higher pressure than the MAWP of the boiler, in recognition that this service is potentially more severe. (See Shock Service in Chapter 5.) This higher pressure, sometimes called the adjusted maximum allowable working pressure, is defined in 122.1.3 of ANSI/ASME B31.1, Power Piping, as the lower of (MAWP + 225 psi) or (1.25 times MAWP).

Because of this complication, Section I provides Appendix Table A-361 (previously Table PG-42), which gives the maximum allowable working pressure of the boiler for which various classes of valves and fittings are adequate. This table was provided as a convenience to the users, especially Authorized Inspectors, who could then quickly determine whether valves of a given class were adequate for their intended service. Many years ago, the underlying ANSI valve and flange ratings were expanded to cover more materials, and Table A-361 now is valid for components of only one material group. Quite a few inquiries were received on how the pressures in Table A-361 were determined, and Subcommittee I responded several times, with a general step-by-step procedure. The most recent of these replies was published in Volume 16 of the Interpretations, as Interpretation I-83-91, in July 1984. The multistep procedure for determining the maximum boiler pressure for which certain ANSI standard components may be used is explained in the fourth reply of this interpretation. The whole Interpretation is reprinted here, because it clarifies the application of valves in Section I service:

> Question: *Does PG-42 apply to butt welding end valves?*
> Reply: *Yes.*
> Question: *Does PG-42 permit the use of both Standard and Special Class valves?*
> Reply: *PG-42 permits the use of valves meeting ANSI B16.34, Steel Valves, Flanged and Butt Welding End. As explained in paragraph 2 of that standard, flanged valves are designated as Standard Class only. However, butt welding end valves may be designated either Standard Class or Special Class, and may also be assigned Intermediate Ratings. Note that a valve may*

not be designated Special Class unless it meets all requirements for that class given in paragraph 8 of ANSI B16.34.

Question: *Please explain the deflection limits and factor of safety mentioned in PG-42.4.5 (Summer 1983 Addenda).*

Reply: *For butt welding end valves, paragraph 6.1.4 of ANSI B16.34 provides a method for establishing the required wall thickness for valves with intermediate pressure ratings. When this method is used, the factor of safety will be maintained as required. However, it remains the responsibility of the designer to assure that valve deflection under load will not affect its proper operation.*

Question: *How are the pressures shown in Table PG-42 (now Table A-361) determined?*

Reply: *Section I rules (PG-11) permit the use of standard pressure parts such as valves and flanged fittings which comply with an ANSI product standard listed in PG-42. The ANSI standards (ANSI B16.34, Steel Valves, Flanged and Butt Welding End; and ANSI B16.5, Steel Pipe Flanges and Flanged Fittings) provide pressure–temperature ratings for these components. The actual choice of fitting is complicated somewhat by the fact that Section I and ANSI B31.1 mandate certain overpressure design requirements for boiler feed and blowoff line service (see ANSI B31.1, 122.1), and also because the rating of a fitting is a function of the material as well as of the pressure class. The user must therefore follow a multistep procedure which varies depending on whether saturated steam service or boiler feed and blowoff service is involved:*

    (1) Establish boiler maximum allowable working pressure P.

    (2) Establish feedwater and blowoff line pressures P_F or P_B as lesser of $P_F = (P + 225$ psi) or $P_F = 1.25P$.

    (3) Select fitting material.

    (4) Use either the ANSI B16.34 or ANSI B16.5 pressure–temperature ratings for that material group. Check successively higher pressure classes until one is found whose pressure rating, at the saturation temperature corresponding to P, equals or exceeds the required pressure rating (P_F or P_B for feedwater or blowoff lines, P for saturated steam lines). That is the pressure rating, or class, required for the fitting.

Table PG-42 (now Table A-361) is provided merely as a convenience to the user, but as explained in the table, the pressures listed are applicable only to Standard Class components of Material Group 1.1. For other materials and classes, the maximum allowable pressures must be determined using the procedure given above.

A design exercise involving the selection of a feedwater stop valve following the above-described procedure is provided in Appendix IV.

SAFETY VALVE REQUIREMENTS

INTRODUCTION

The ASME Boiler and Pressure Vessel Code was developed between 1911 and 1914 as a set of safety rules to address the serious problem of boiler explosions in the United States, which at that time were almost a daily occurrence. Average steam pressure in boilers in those days had reached about 300 psi, and the explosions took a heavy toll of lives and property. The principal design basis for boilers and pressure vessels is the safe containment of design pressure. Protection against overpressure was therefore a very important aspect of pressure vessel design. Accordingly, that first edition of what is now Power Boilers, Section I of the Code, included rules for the overpressure protection of boilers, based on the best industry practice at the time. The principles of today's Code rules for overpressure protection are little changed from those of the first Code.

That first ASME Code of 1914 has now grown into eleven so-called book sections, covering a wide variety of subjects. Five of these book sections deal specifically with the design of pressure vessels and include requirements for overpressure protection. The various pressure-relief devices used for overpressure protection may be categorized as pressure-relief valves (commonly called safety valves), and nonreclosing devices, such as rupture disks.

The function of a pressure-relief device is to open at a specified pressure and pass a sufficient amount of fluid to prevent pressure in the vessel or system being protected from exceeding an **allowable overpressure** above the design pressure. In some book sections such as Section I and Section VIII, the design pressure is referred to as the **maximum allowable working pressure (MAWP)**. Each book section provides rules for determining the required relieving capacity of pressure-relief devices, for establishing allowable overpressure, for determining the maximum set pressure of the pressure-relief devices, and for ensuring that pressure-relief devices will function as required should an overpressure condition occur. Although fundamentally similar, the rules vary somewhat in detail and complexity, reflecting the nature of the application and the somewhat different design philosophies of the committees governing the various book sections. It happens that the Code offers only limited explanation of its rules on overpressure protection, and aside from certain functional requirements, very few rules to guide the designers of pressure-relieving devices. The objective of this chapter is to summarize the Section I rules and to provide some explanation and discussion of them and how they are applied. Since similar rules apply to all the non-nuclear book sections, a brief explanation of those other rules is included, to point out some differences. Thus the discussion which follows covers all the non-nuclear book sections, with the major emphasis on the rules in Section I.

ASME has a policy that rules on a given subject that appear in more than one book section should be similar and consistent, insofar as possible. However, due to the diversity of application, there is some variation among the pressure-relief device rules that have evolved in five book sections of the Code. A governing subcommittee of the Boiler and Pressure Vessel Committee has the fundamental responsibility for writing, revising, adding to, and interpreting the rules of each book section of the Code. Assisting the book section subcommittees with technical advice are so-called service committees such as the Subcommittee on Design (SC-D) and the Subcommittee on Safety Valve Requirements (SC-SVR). The purviews of the book and of the service committees overlap somewhat, and the demarcation of their respective responsibilities is sometimes unclear. Traditionally, with regard to pressure-relief devices, matters of application and installation, such as which devices are permitted and where, fall within the domain of the book section committees. Technical matters dealing with design, testing, and capacity certification are covered by the SC-SVR. When inquiries dealing with these devices are received by the ASME, they are sent to the concerned book section committees. Only if those committees think it necessary or advisable is the inquiry forwarded to SC-SVR for a recommended reply or rule change; and even in such cases, the book section committee is free to accept, modify, or reject that recommendation. Thus, despite ASME policy on consistency of rules, the independence of all the committees involved has resulted in some slight gradual divergence among the book sections with respect to their rules for overpressure protection.

To provide overpressure protection, the designer must first determine potential sources of overpressure, calculate the rate at which fluid must be removed under emergency conditions to maintain pressure within maximum allowable limits, and select appropriate relieving devices to provide this required capacity. The Code provides a variety of rules for the design, manufacture, capacity certification, selection, and application of pressure-relief devices used to protect a broad variety of fired and unfired vessels. Early applications of the Code dealt with steam boilers, where the source of pressure is the application of heat to water in an enclosed vessel, producing steam. If the quantity of steam allowed to leave the vessel is less than that being generated, the pressure in the vessel rises. The same principle applies to other fluids heated in closed vessels. Providing sufficient overpressure protection is not always straightforward, however, and the Code rules differ depending on the type of vessel to be protected, as described next.

OVERPRESSURE PROTECTION

Relieving Capacity Required

The first task in selecting appropriate pressure-relief devices for a vessel or system is to determine the required relieving capacity (the rate at which fluid must be removed from the system during an emergency condition to maintain a safe level of pressure). This requires that all sources of fluid flow or energy into the system be considered and evaluated. Sources of energy might be heat, fire, chemical reaction, or mechanical work. Sources of fluid flow could include compressors, pumps, pressure-reducing valves stuck in the open position, malfunctioning control valves, inadvertent opening or closure of a valve connected to the vessel, failure of an internal tube in a heat exchanger, or some combination of these events. Each possible source of excess energy or fluid flow into the system must be taken into account to determine the worst case overpressure condition. The required relieving capacity is determined from this worst case; this determination can often require careful engineering study. The various book sections of the Code in some cases provide directly applicable rules and in other cases only limited guidance for determining the worst case overpressure condition. The following is a brief summary of those rules.

Summary of Rules for Determining Required Relieving Capacity, by Book Section

Section I still uses the basic rule from the first ASME Code in 1914 which, in brief, requires safety valve capacity sufficient to relieve all the steam a drum-type boiler can generate without allowing the pressure to rise more than 6% above the pressure at which any valve is set, and in no case more than 6% above the MAWP. This rule is now supplemented by others, for certain types of boilers and for various other components covered by Section I. For example, the required relieving capacity in lb/hr for an isolable economizer (one with valves on either side, which could be inadvertently left closed) is equal to the maximum heat absorbed in Btu/hr divided by 1000 (PG-67). This rule was based on the assumption that it takes 1000 Btu to boil a pound of water—a fair assumption at low pressures, somewhat unconservative at higher pressures. For high-temperature water boilers, this rule is reversed: the designer is told to provide relieving capacity equal to the boiler output in Btu/hr divided by 1000. If a superheater can be valved off from a boiler, its safety valves are potentially unavailable, and full relieving capacity must be provided on the boiler itself. Such a superheater must also be equipped with one or more safety valves having a discharge capacity based on the heating surface of the superheater (see PG-68.3). In the 1997 Addenda, new provisions were added to PG-68 to permit the boiler manufacturer to calculate the minimum safety valve discharge capacity for independently fired superheaters based on the maximum expected heat absorption (as determined by the manufacturer), in Btu/hr, divided by 1000. At the same time, the marking requirements of PG-106.4.2(5) were changed. Instead of the formerly required design temperature for an independently fired superheater, that paragraph now requires either the heating surface in square feet or the minimum safety valve discharge capacity calculated as just described. These changes came about with the incorporation of Code Case 2147. That case had been approved some years earlier to grant relief from the excess safety valve relieving capacity required when the 6 lbs of steam per hour per square foot of heating surface rule in PG-68.3 was applied to independently fired superheaters with extended surface, as are sometimes used in certain waste heat boilers.

Until 1994, when its rules were revised, Section I had provided a table (Table PG-70) for certain types of boilers. Table PG-70 stipulated the minimum safety valve relieving capacity required per square foot of heating surface, with the admonition that this was only a minimum and that more might be required. This minimum capacity varied with boiler type, firing method, fuel, and type of heating surface. The table was originally intended as a guide for Inspectors, who could then verify the boiler manufacturer's choice of safety valve capacity. In recent years, the capacities listed in Table PG-70 were found to be inappropriate for certain types of boilers, e.g., fluidized bed boilers and stoker fired boilers firing refuse. To correct this problem, and also to eliminate some confusion between the maximum designed steaming capacity mentioned in PG-67 and the minimum required safety valve relieving capacity specified in PG-70, these two paragraphs were revised in the 1993 Addenda to Section I. The new wording clearly placed on the boiler manufacturer the full responsibility for determining the maximum designed steaming capacity of the boiler and for providing at least that much safety valve relieving capacity. At the same time, a new rule was added, requiring the boiler manufacturer to stamp the maximum designed steaming capacity on the boiler nameplate. With these changes, the old Table PG-70 was no longer needed. However, to keep it available to Inspectors accustomed to using it, it was moved to a nonmandatory appendix, Appendix A-44, where it is now called Table A-44.

The term "all the steam the boiler can generate" has traditionally been interpreted to mean the maximum continuous rating (MCR) of the boiler at specified design conditions. Section I assigns the responsibility for the determination of this steaming rate to the boiler manufacturer, who calculates it based on experience, taking into account the various boiler systems: pumps, fans, and fuel. It is necessary in all such systems to provide some design margin to assure achieving design steam-generating capacity (MCR) and to permit the overfiring necessary when increasing load. Consequently, if all systems were pushed to their limits, it is possible to exceed the nominal design steam-generating capacity. However, the customary practice is to use the nominal capacity at MCR as the minimum safety valve relieving capacity required. This can be justified by the safety margins implicit in the certified capacities, in the design of the pressure parts, and in the overpressure limits of Section I, which are typically lower than those of Section VIII.

In 1965, Section I incorporated design rules for a then new type of boiler called a "forced-flow steam generator with no fixed steam and waterline, equipped with automatic controls and protective interlocks responsive to steam pressure." Such boilers have no steam drums; the fluid passes through the boiler only once, entering as water and leaving as steam. Unlike drum-type boilers, which are designed for a single design pressure and which achieve their pressure by firing, forced-flow steam generators with no fixed steam and waterline achieve their pressure from the boiler feed pump. These boilers are designed for different pressure levels along the path of water-steam flow, with a significant difference in pressure between the economizer inlet and superheater outlet, for example, 4350 psi versus 3740 psi.

Recognizing the special nature of these boilers, Section I provided alternative rules for their overpressure protection, in PG-67.4, which are quite different from the usual Section I rules. For drum-type boilers, Section I permits credit for use of spring-loaded safety valves only. However, for these once-through boilers, Section I permits a combination of safety valves and power-actuated pressure-relieving valves, the latter to provide a minimum of 10%, but to be credited with no more than 30%, of the required relieving capacity. Moreover, PG-67.4.1 permits the installation of an isolating stop valve between the power-actuated pressure-relieving valve and the boiler to permit repairs, provided certain conditions are met. In another departure from usual practice, any or all of the spring-loaded safety valves may be set up to 17% above the MAWP at the superheater outlet (the component with the lowest MAWP). Further, the power-actuated relieving valves must receive a signal to open when the MAWP at the superheater outlet is exceeded. When all pressure-relief valves are in operation, the pressure is permitted to rise to a maximum of 20% above the master stamping pressure, the design pressure at the superheater outlet. When still further special conditions are met, Section I permits certain forced-flow steam generators to be furnished with as few as two spring-loaded safety valves whose total relieving capacity is not less than 10% of the maximum design steaming capacity. These special conditions include the use of automatic controls and direct acting trip mechanisms for the fuel and feedwater supplies, actuated by the lifting of the valve stem of at least two spring-loaded safety valves. Further, the fuel and feedwater control circuitry must be fail-safe. These rules contrast with current practice in Germany and some other European countries, where overpressure protection for once-through units can now be provided by a power-actuated turbine bypass system alone, without spring-loaded pressure relief valves.

Section IV, Heating Boilers, covers three types of relatively low pressure equipment: steam heating boilers, hot water heating or supply boilers, and potable water heaters. The pressure-relieving capacity, in pounds of steam per hour, for both steam and water boilers may be determined in one of two ways: the maximum Btu output at the boiler nozzle divided by 1000 (again, the rule of thumb that 1000 Btu will boil a pound of water) or by use of a table of minimum steam-relieving capacity per hour per square foot of heating surface, similar to the method formerly used in Section I. The relieving capacity of safety relief valves on potable water heaters is specified in Btu/hr and, in general, must equal the maximum input to the heater. HG-402.7(a) advises that the relieving capacity in terms of Btu/hr may be determined by multiplying the capacity in pounds of steam per hour by 1000.

Of all the book sections, **Section VIII, Pressure Vessels**, covers the largest variety of vessels. These vessels are also potentially subject to overpressure conditions from many possible sources. When designers of Section I and Section IV equipment determine required relieving capacity, they often start with guidelines from the Code and knowledge of a boiler's maximum steam-generating capacity or heat input; by contrast, Section VIII designers are simply charged by UG-125 with preventing the pressure from rising more than a certain amount above the MAWP of the vessel, with few additional guidelines. Accordingly, the designer must often use the engineering principles outlined in the section on Required Relieving Capacity above and have a thorough knowledge of the system being protected to determine the appropriate relieving capacity. Section VIII comprises 3 volumes, Divisions 1, 2, and 3. Division 2 has alternative rules for the construction of pressure vessels, based on the design-by-analysis methods of Section III. However, the overpressure protection requirements of Divisions 1 and 2 are virtually identical. Division 3, Alternative Rules for Construction of High Pressure Vessels, was published in 1997. It contains special rules for overpressure protection.

Section X, Fiber-Reinforced Plastic Vessels, adopts the same basic overpressure rules as Section VIII, Division 1, but is somewhat more restrictive as to the types of nonreclosing pressure-relief devices permitted (see Table 12.1). As with Section VIII, the task of determining required relieving capacity is left to the designer.

Allowable Overpressure

Most pressure-relief devices, particularly pressure-relief valves, require a further increase in pressure above set pressure to reach a full open condition. Vessels with multiple valves may require an even greater increase

Pressure-Relief Devices Allowed by the ASME Code	BOOK SECTION			
	I	IV	VIII	X
Direct Spring Loaded Pressure-Relief Valve	•	•	•	•
Pilot Operated Pressure-Relief Valve			•	•
Power Actuated Pressure-Relief Valve	•			
Rupture Disk Device	◆		•	•
Breaking Pin Device with Pressure-Relief Valve			•	
Spring Loaded Non-Reclosing Pressure-Relief Device			•	
Buckling Pin Device			✚	

• Permitted
◆ In combination with pressure relief valve on organic fluid vaporizer only
✚ Permitted by Code Cases 2091, 2169, and 2267

TABLE 12.1
PRESSURE RELIEF DEVICES ALLOWED BY THE ASME CODE

over MAWP for all the relieving valves to open fully because some are allowed to be set above MAWP. Accordingly, each book section provides its own rules as to how much the pressure is allowed to increase above MAWP or design pressure during an overpressure event.

Section I provides for overpressure within very narrow limits. With some exceptions (see discussion of forced-flow steam generators, above), pressure may not rise more than 6% above the set pressure of any pressure-relief valve, and in no case more than 6% above the MAWP.

Section IV overpressure limits vary somewhat, depending on the type of equipment and its design pressure. For steam boilers, the design pressure is a minimum of 30 psi, but the operating limit set by Section IV is 15 psi, so the safety valves must be set at 15 psi, with an overpressure of 5 psi allowed. For hot water boilers with only a single pressure-relief valve, an overpressure that is 10% above MAWP is allowed. Where multiple valves are provided, the allowable overpressure is 10% above the set pressure of the highest set valve.

Section VIII limits on overpressure depend on the type of installation. In general, Section VIII mandates that pressure not be allowed to rise more than 10% or 3 psi, whichever is greater, above MAWP. When multiple devices are used, or additional devices are provided to protect against exposure to fire, overpressures of 16% or 21%, respectively, are allowed. For valves that protect liquified compressed gas storage vessels against exposure to fire, 20% overpressure is allowed.

Section X has the same overpressure requirements as Section VIII, except that special rules for protection of liquified compressed gas storage vessels exposed to fire are not considered.

Set Pressure of Pressure-Relief Devices

Once the required relieving capacity and allowable overpressure have been established, the next step is to determine the set pressure of the pressure-relief devices. The general rule for setting pressure-relief devices is that at least one device must be set to open at or below maximum allowable working pressure (the design pressure) of the system or vessel being protected. On many small boilers or pressure vessels, only one relief device is required, and this device is typically set at MAWP. When more than one device is used, the additional devices may be set at slightly higher pressures. This allowable increase in set pressure over

MAWP varies considerably throughout the Code (in general, from 3% to 10% above MAWP), depending on the application and the circumstances, as shown in Table 12.2.

Section I requires that at least one pressure-relief valve be set at or below MAWP. Additional valves may be set up to 3% above MAWP. In no case, however, may the complete range of settings of all the saturated steam safety valves on a drum-type boiler exceed 10% of the highest pressure at which any such valve is set. These rules apply primarily to safety valves on the drum. Section I does not advise a specific setting for safety valves on superheaters. Industry practice is to set these valves sufficiently below MAWP so that they will lift before the valves on the drum, thus helping to maintain a cooling flow of steam through the superheater.

Section IV steam boilers have an operating limit of 15 psi, so the safety valves must be set at 15 psi. Hot water heating boilers and hot water supply boilers are designed for pressures between 30 and 160 psi and/or temperatures not exceeding 250°F. (Above these limits, the boiler would, by definition, be a power boiler falling within the scope of Section I.) The safety relief valves used on hot water heating boilers must be set at or below the MAWP, but typically these boilers are supplied with valves set at 30 psi unless higher pressure is needed, as might be the case for tall apartment buildings.

Potable water heaters are designed for pressures up to 160 psi (with a minimum of 100 psi) and for temperatures not exceeding 210°F. Although Section IV calls for each such heater to have at least one safety relief valve or pressure–temperature relief valve, they are routinely furnished with the latter type, which is activated either by pressure or temperature. The purpose of the 210°F limit is to prevent any water discharged from relief valves from flashing to steam. The pressure setting of a single relief valve, or the low set valve when multiple valves are used, must be less than or equal to the MAWP of the heater, but is further limited to the lowest pressure for which any other component of the hot water system is designed.

With the exception of steam boilers, additional relief valves used on Section IV equipment may be set at pressures higher than MAWP (see Table 12.2).

Sections VIII and X require one valve to be set at or below MAWP. If additional valves are utilized, they may be set up to 5% above MAWP. Section VIII further allows a supplemental valve added to a vessel to protect against hazard due to fire to be set up to 10% above MAWP.

PRESSURE RELIEF DEVICE OPERATING REQUIREMENTS				
APPLICATION	ALLOWABLE VESSEL OVERPRESSURE (ABOVE MAWP OR VESSEL DESIGN PRESSURE)	SPECIFIED PRESSURE SETTINGS	SET PRESSURE TOLERANCE WITH RESPECT TO SET PRESSURE	REQUIRED BLOWDOWN
SECTION I Boilers	6% (PG-67.2)*	One valve ≤ MAWP Others up to 3% above MAWP (PG-67.3)	± 2 psi up to and including 70 psi ± 3% for pressures above 70 psi up to and including 300 psi ± 10 psi for pressures above 300 psi up to and including 1000 psi ± 1% for pressures above 1000 psi (PG-72.2)	Minimum: 2% of set pressure or 2 psi whichever is greater Maximum: 4% of set pressure or 4 psi whichever is greater, with some exceptions (See PG-72)
Forced-Flow Steam Generators	20% (PG-67.4.2)	May be set above MAWP, Valves must meet overpressure requirements (PG-67.4.2)	Same as above (PG-72.2)	Maximum: 10% of set pressure (PG-72)

*Note: References in parentheses are to Code paragraphs

1 psi = 6.89 kPa

TABLE 12.2(a)
SECTION I REQUIREMENTS

PRESSURE RELIEF DEVICE OPERATING REQUIREMENTS				
APPLICATION	ALLOWABLE VESSEL OVERPRESSURE (ABOVE MAWP OR VESSEL DESIGN PRESSURE)	SPECIFIED PRESSURE SETTINGS	SET PRESSURE TOLERANCE WITH RESPECT TO SET PRESSURE	REQUIRED BLOWDOWN
SECTION IV Steam Boilers	5 psi (HG-400.1e)	≤ 15 psi (HG-401.1a)	± 2 psi (HG-401.1k)	2 - 4 psi (HG-401.1e)
Hot Water Boilers	10% for a single valve	One valve at or below MAWP	± 3 psi up to and including 60 psi	None specified
	10% above highest set valve for multiple valves	Additional valves up to 6 psi above MAWP for pressures to and including 60 psi and up to 5% for pressures exceeding 60 psi (HG-400.2a)	± 5% for pressures above 60 psi (HG-401.1k)	

1 psi = 6.89 kPa

TABLE 12.2(b)
SECTION IV REQUIREMENTS

PRESSURE RELIEF DEVICE OPERATING REQUIREMENTS				
APPLICATION	ALLOWABLE VESSEL OVERPRESSURE (ABOVE MAWP OR VESSEL DESIGN PRESSURE)	SPECIFIED PRESSURE SETTINGS	SET PRESSURE TOLERANCE WITH RESPECT TO SET PRESSURE	REQUIRED BLOWDOWN
SECTION VIII Division 1 & 2				
All vessels unless an exception specified	10% or 3 psi—whichever is greater (UG-125c) (AR-150a)	≤ MAWP of vessel (UG-134a) (AR-141)	± 2 psi up to and including 70 psi	None specified
Exceptions: When multiple devices are used	16% or 4 psi—whichever is greater (UG-125 c1) (AR-150b)	One valve ≤ MAWP, additional valves up to 105% of MAWP (UG-134a) (AR-142a)	± 3% over 70 psi (UG-134 d1) (AR-120d)	
Supplemental device to protect against hazard due to fire	21% (UG-125 c2) (AR-150c)	Up to 110% MAWP (UG-134b) (AR-142b)		**Note:** Pressure relief valves for compressible fluids having an adjustable blowdown construction must be adjusted prior to initial capacity certification testing so that blowdown does not exceed 5% of set pressure or 3 psi, whichever is greater (UG-131 c3a) (AR-512)
Device to protect liquified compressed gas storage vessel in fire (Div. 1 only)	20% (UC-125c3a)	≤ MAWP (UG-125c3b)	−0% +10% (UG-134 d2)	
Bursting disk	Same as above	Stamped burst pressure to meet requirements noted above (UG-134 - Note 56) (AR-140 - Note 15)	± 2 psi up to and including 40 psi ± 5% of stamped burst pressure at specified coincident disk temperature (UG-127 a1a) (AR-131.1)	

1 psi = 6.89 kPa

TABLE 12.2(c)
SECTION VIII REQUIREMENTS

Set Pressure Tolerance

Each book section provides set pressure tolerances for a variety of applications. These are summarized in Table 12.2.

Blowdown

A pressure-relief valve will generally not reseat until the inlet pressure is reduced below set pressure. The difference between actual opening pressure and reseating pressure is called blowdown. For Section I, blowdown is generally specified as the greater of 2% of set pressure or 2 psi (minimum) to 4% of set pressure or 4 psi (maximum). Code Case 2071 permitted a 6% blowdown for safety valves whose set pressure doesn't exceed 250 psi. This case was incorporated into PG-72 in the 1998 Addenda to Section I. Section IV specifies 2–4 psi blowdown for steam boilers, but is silent on other types of equipment. Sections VIII and X have no specified blowdown requirements. Section VIII does, however, require that pressure-relief valves for compressible fluids having an adjustable blowdown construction be adjusted prior to initial certification testing, so that blowdown does not exceed 5% of set pressure or 3 psi, whichever is greater, during the test. This requirement is often misconstrued to be an operating requirement, which it is not. Blowdown requirements are summarized in Table 12.2.

SELECTION OF PRESSURE-RELIEVING DEVICES

When the required relieving capacity has been established and the allowable set pressure and overpressure determined, the pressure-relieving devices can then be selected. The choice of device is limited to those permitted by the particular book section covering the service in question. The first broad category of allowable overpressure protection devices comprises the reclosing pressure-relief devices.

The reclosing pressure-relief device most widely accepted by the various book sections of the Code is the spring-loaded pressure relief valve. This valve consists of a nozzle attached to an opening in the protected system with a disk held against the open end of the nozzle by a spring. When the system pressure acting against the disk overcomes the spring force, the disk opens and fluid flows from the system. One particular type of spring-loaded valve is the balanced valve, sometimes called a balanced bellows valve, wherein a bellows having an effective area equivalent to the valve seat area is attached to the disk to prevent backpressure in the valve's discharge system from acting on the disk and affecting its opening pressure.

PRESSURE RELIEF DEVICE OPERATING REQUIREMENTS				
APPLICATION	ALLOWABLE VESSEL OVERPRESSURE (ABOVE MAWP OR VESSEL DESIGN PRESSURE)	SPECIFIED PRESSURE SETTINGS	SET PRESSURE TOLERANCE WITH RESPECT TO SET PRESSURE	REQUIRED BLOWDOWN
SECTION X				
All vessels unless an exception specified	10% or 3 psi—whichever is greater (RR-130a)	≤ Design pressure (RR-120)	Ref. Section VIII (RR-111)	None specified
Exceptions: When multiple devices are used	16% or 4 psi—whichever is greater (RR-130b)	One valve ≤ design pressure. Others up to 105% of design pressure		
Additional hazard due to fire	21% (RR-130c)			

1 psi = 6.89 kPa

TABLE 12.2(d)
SECTION X REQUIREMENTS

Traditionally, spring-loaded valves for boiler or steam application have been called **safety valves**; spring-loaded valves for liquid application have been called **relief valves**; and multipurpose spring-loaded valves that might be used for steam, other compressible fluids, or liquid, have been called **safety relief valves**. None of these terms fully describes the design or function of a pressure-relief valve; and in many cases, the terms are used interchangeably. The more generic term, **pressure-relief valve**, has replaced the three definitions in some portions of the Code.

A spring-loaded pressure-relief valve is permitted by all sections of the Code as an overpressure protection device. Operational and construction requirements, however, vary among the book sections. The popularity of this design is due to the reliability of having few moving parts, the repeatability of an opening point controlled by a spring, and the ability of the valve to close when an overpressure condition is reduced to a safe level.

Another type of pressure-relief valve is the **pilot-operated pressure-relief valve**. This is a valve in which the major relieving device or main valve is combined with and controlled by a self-actuated auxiliary pressure-relief valve or pilot. The main valve consists of a nozzle and a disk similar to the spring-loaded valve, except that the disk is held in place by system pressure. When system pressure rises above the set pressure of the valve, the pilot senses that pressure and vents the pressure above the disk, allowing the main valve to open. As with spring-loaded valves, the pilot-operated pressure-relief valve is self-operated, using only system fluid as the operating medium. Pilot-operated pressure-relief valves have the advantage that their operation is less influenced by fluid conditions at the valve inlet; they generally can be tested in situ, and a high seating load is maintained up to the opening point of the valve. The valve has more operating parts, however, and because of typically small passages in the pilot, the cleanliness of the fluids on which the pilot-operated pressure-relief valve operates could become a concern. Pilot-operated pressure-relief valves are permitted by Sections VIII, and X, but not by Section I.

A third type of pressure relief valve is the **power-actuated pressure-relief valve**. These valves differ from pilot-operated pressure-relief valves in that they require an external energy source provided by electrical, pneumatic, or hydraulic systems and generally operate in response to signals from pressure- or temperature-sensing devices. These types of pressure-relief valves offer the benefits of a wide variety of control systems, but have the disadvantage of relying on an external source of power that may fail under emergency conditions. Power-actuated relieving valves are often furnished on drum-type boilers as a convenience to the operators, even though for that type of boiler Section I permits no credit for their relieving capacity. Power-actuated pressure-relief valves are permitted by Section I for use on forced-flow steam generators with no fixed steam and waterline, provided they are used in addition to self-actuated pressure-relief valves (see PG-67.4).

The second broad category of allowable overpressure protection devices comprises the nonreclosing devices. The most common nonreclosing device is a **rupture disk** device, which is designed to function by the bursting of a pressure-containing disk. The rupture disk is the pressure-containing and pressure-sensitive element of the rupture disk device. Rupture disks may be designed in several configurations, such as plain flat, prebulged, or reverse buckling, and can be made of either ductile or brittle material.

A rupture disk is a relatively inexpensive device that, once opened, is not subject to conditions of instability. Also, rupture disks themselves are less likely to leak prior to opening. However, they do not reclose after opening and must be replaced after each overpressure event. In some circumstances, the opening point of some types of disk may be affected by pressure pulsations.

Rupture disks are permitted by Sections I, VIII, and X. Section I allows the installation of a rupture disk only under the safety valve of an organic fluid vaporizer. The rupture disk helps prevent leakage to the environment and protects the valve from the organic fluid, which can polymerize in service and cause valve malfunction. In an unusual departure from normal Section I coverage of new construction only, PVG-12.2 requires yearly removal of safety valves on organic fluid vaporizers for inspection, repair (if necessary), and testing. When rupture disks are used in combination with pressure-relief valves, an overpressure event causes rupture of the disk followed by opening of the valve. When pressure returns to normal, the pressure-relief valve closes and prevents further loss of system fluid. The space between the rupture disk and the valve must be provided with a pressure gage, a try cock, a drain, a free vent with an excess flow check valve, or a telltale indicator, so that pressure cannot build up undetected. A buildup of pressure between the rupture disk and the valve inlet will alter the bursting pressure of the rupture disk.

Other types of nonreclosing devices are permitted, but for Section VIII use only. These devices are **nonreclosing spring-loaded pressure-relief valves, breaking pins**, and the relatively new **buckling pin device**, first allowed in the early 1990s for Section VIII, Division 1 use by Code Case 2091.

The **nonreclosing spring-loaded pressure-relief valve** is designed so that the spring-loaded portion of the device opens at the specified set pressure and remains open until manually reset. Such a device may not be used in combination with any other type of pressure-relief device.

A **breaking pin device** is a nonreclosing pressure-relief device actuated by inlet static pressure. It functions by the failure in tension of a breaking pin that supports a pressure-containing or closure member within the breaking pin device. When set pressure is reached, the breaking pin breaks, allowing the device to open and flow its rated capacity. The pin must be replaced before the device can be reclosed. Breaking pin devices may not be used alone, but only in combination with a spring-loaded pressure-relief valve. Like rupture disks, they may be used to protect the valve from potentially harmful fluids or to reduce the likelihood of valve leakage.

A **buckling pin device** is one in which a piston resisting the vessel pressure is held in place by a slender pin in compression. The length and diameter of the pin are closely controlled so that its elastic buckling load in compression can be accurately predicted. When pressure in the vessel reaches set pressure, the pin buckles, allowing the piston to move sufficiently to open a discharge port, thereby relieving the pressure. The pin must be replaced before the device can be returned to service.

The ASME Code normally does not prohibit specific designs. One exception to this rule, however, is the stipulation that a weight-loaded (as opposed to spring-loaded) valve may not be used for pressure-relief. A weight-loaded pressure-relief valve is similar to a spring-loaded valve, except that the spring is replaced by a weight. The problem with this type of valve is that any mechanical interference in the valve or tampering with the weight can compromise its pressure-relieving capability. However, such valves are not prohibited internationally and are used for low-pressure vents in the U.S. and elsewhere. A second prohibited type, in Section IV, is a bottom-guided or wing-guided disk design in which the guidance is provided on the process side of the disk. The concern in this case is possible fouling or sticking of the guide due to the process fluid.

The types of pressure-relieving devices allowed by each book section of the Code are summarized in Table 12.1.

DESIGN OF PRESSURE-RELIEF VALVES

Although the primary emphasis of the ASME Code is on the design of boilers and pressure vessels, pressure-relief devices are recognized as important appurtenances of the vessels. The Code provides guidance for the operational integrity of these devices, but with the exception of Section III, provides only limited design guidance for them. Most notably (again with the exception of Section III), design rules are not provided for determining wall thickness of pressure-relief valves. Consequently, the manufacturers must look elsewhere—for instance, to the design bases published in such standards as *ASME/ANSI B16.34, Valves—Flanged, Threaded, and Welding End*, and *ASME/ANSI B16.5, Pipe Flanges and Flanged Fittings*, or to other methods, from proof testing to design-by-analysis using finite element methods, to establish pressure and temperature ratings.

Design guidance for nonreclosing devices is even more limited in regard to what materials may be used for rupture disk holders and what sampling and testing methods must be used to verify the stamped bursting pressure at coincident temperature.

Mechanical Requirements

The mechanical design rules focus on the operability and reliability of the valve. To ensure the integrity of the spring, for example, the designer must limit the full lift spring compression to no greater than 80%

of nominal solid deflection. As an added safeguard, the spring must not exhibit a permanent set of more than 0.5% of free height after being compressed to solid height three times. These rules are intended to ensure that the spring does not relax under load, causing a variation in valve set pressure.

The operating or moving parts of a valve must be free at all times. Thus, drains are generally required below valve seats, to avoid collection of liquids or deposits that might inhibit free operation. Sections I, IV, VIII, and X also require lifting levers that allow a valve to be mechanically operated from time to time to verify that it is free to operate. Except for Section IV valves, the lever must be able to lift the disk when inlet pressure is above 75% of set pressure. Levers on Section IV valves must be able to lift the disk at least 1/16 inch with no pressure in the boiler. Levers are required for all Section I and IV valves, and for Section VIII and X valves in steam, air, and hot water (over 140°F) service. Many users are concerned about the release of undesirable fluids into the atmosphere, so levers are not generally used for fluids other than steam, air, and water. The requirement for a lifting device on a pilot-operated pressure-relief valve may be satisfied by providing a connection for applying pressure to the pilot to verify that the critical moving parts are free to move.

Other rules relate to the integrity of external adjustments. After a manufacturer, assembler, or other authorized agent adjusts a pressure-relief valve, the adjustment must be sealed, generally with a lead seal on wire. Except for Section IV valves, the seal must be marked to identify the organization making the adjustment, and the design must be such that the seal has to be broken to change the adjustment.

The structural integrity of most pressure-relief valves is further assured by a Code requirement for the manufacturer to hydrostatically test the primary side of the valves.

Materials

Sections I, IV, VIII, and X of the Code define two categories of materials acceptable for pressure-relief valve construction: one for external pressure-containing parts of the device (e.g., the body and bonnet of the valve) and a second for internal parts, such as valve nozzles and disks. In the first category, materials are limited to those normally used for vessel construction, which are listed in the governing book section. The chemical composition, physical properties, allowable stresses, and weldability of these materials are well known. However, the designer of pressure-relief valves must also consider other properties important for proper valve function, such as corrosion resistance of seats, disks, springs, and adjacent sliding surfaces; antigalling characteristics for sliding surfaces; high strength for valve springs; and erosion resistance for valve seating.

Section I, Division 1 of Section VIII, and Section X also permit the use of some eighteen other material specifications for pressure-containing valve parts, under Code Case 1750. This case has evolved as a means of permitting the use of new materials without requiring the valve manufacturer to undertake the extensive and burdensome procedures normally required to qualify material for vessel construction and normally used by the appropriate Code committee to establish allowable design stresses as a function of temperature. Case 1750 thus includes no allowable stresses, leaving the designer to choose his or her own, presumably using book section criteria.

In the second category of materials (for internal parts), the Code permits a much wider latitude, limiting materials to those listed in Section II or in ASTM specifications or those controlled by a specification that provides "control of chemical and physical properties and quality at least equivalent to ASTM standards." The purpose of this requirement is to help ensure consistent quality of these components.

ESTABLISHING AND CERTIFYING RELIEVING CAPACITIES

Many codes and standards applicable to pressure-relief valves worldwide specify performance requirements similar to those found in the ASME Code, and regulatory bodies may require capacity certification and testing of valves. However, only the ASME Code requires the pressure-relief valve manufacturer to demonstrate performance and relieving capacity of production pressure-relief valves, by a test performed at an ASME-accepted test facility, supervised by an ASME Authorized Observer, and witnessed by a qualified third party designated by the ASME. Performance and capacity testing of pressure-relief valves was first mandated (by Section I of the Code) in 1937, following a program of valve testing sponsored by the National Board of Boiler and Pressure Vessel Inspectors. These tests, conducted in 1935 and 1936, in some cases showed actual discharge capacities significantly less than those claimed by the valve manufacturer.

The procedures used for capacity testing and the formulas used to determine capacity are based on the assumption that compressible flow through the valve is critical (acoustically choked), i.e., it is essentially a function of inlet pressure. For steam flow, the capacity is assumed to follow Napier's formula, which also states that flow is proportional to inlet pressure. For dry saturated steam above 1500 psi, the Code specifies a correction factor for Napier's formula. For incompressible flow of nonflashing (subcooled) liquids, the capacity is assumed proportional to the square root of the pressure drop across the valve. Pressure-relief valve capacity certification tests conducted at the relatively low pressures available at test laboratories in the 1930s are still valid today. Extrapolation of valve capacities based on low-pressure tests of small valves to higher pressures and larger sizes continues to be acceptable Code practice.

All ASME pressure-relief valves must be marked by the manufacturer with an ASME Code symbol. Before the ASME Code symbol may be applied to a pressure-relief valve, its relieving capacity must first be certified and then confirmed by tests of randomly selected production valves. This process generally involves an initial or provisional test whereby a number of valves are tested to determine the rated relieving capacity and to demonstrate consistency of valve-relieving performance, in accordance with the requirements of the applicable book section. This is followed by periodic testing (initially, and at least every 5 years thereafter) of randomly selected production valves in the presence of a representative of the National Board, which serves as what is known as an ASME-designated organization. (Such an organization meets certain ASME criteria intended to assure appropriate operating and due process procedures.) The ongoing testing verifies that production valves continue to perform as originally certified.

The initial or provisional testing of valves that are to be used on compressible fluids is performed on dry saturated steam, air, or natural gas. For Section VIII use, if the valve is intended for use on steam, at least one of the certification valves must be tested on steam. Valves that are to be used for incompressible fluid service must be tested on water.

Before each test is performed, valve set pressure must be checked for compliance with the applicable book section (i.e., it must meet the opening pressure tolerance). The capacity test is then performed at the allowable overpressure above set pressure specified in that same book section. There is some minor variation in test procedure among the various book sections.

There are three different methods used for certification testing: the **three valve method**, the **four valve method**, and the **coefficient of discharge method**.

The Three Valve Method.   If the intent is to test a *single* valve design at a *single set pressure*, three identical valves, all set at the same pressure, are tested to determine capacity. Each of the measured capacities must fall within ±5% of the average of the three capacities. This average capacity is then multiplied by 0.9 to determine the official certified capacity to be stamped on the valve, for this valve design at the particular pressure chosen.

The Four Valve Method.   This method is used to determine the capacity of a *single valve design* over a *range of set pressures*. The four valve method is also known as the slope method, since there is a straight line relationship between capacity and absolute inlet pressure for compressible fluids and between the log of capacity and the log of pressure drop across the valve for incompressible fluids. In this method, four valves of the same size are set at four different pressures, and the capacity of each valve is determined by test. For *compressible fluids*, each measured capacity is divided by the absolute flowing pressure to determine a slope, in pounds per hour per psi. The slope derived from each individual test must fall within 5% of the average value of slope calculated from the result of all four tests. The rated, or certified, relieving capacity is then determined as 0.9 times the product of the average slope times the (absolute) set pressure plus allowable overpressure. For example, the certified capacity to be stamped on a Section I valve would be 0.9 × average slope × (1.03 × set pressure + 14.7). For *incompressible fluids*, the capacities determined from the four tests are plotted on log paper against the differential test pressure (inlet pressure minus discharge pressure), and a straight line is drawn through the four points. Individual points may not depart more than 5% from the straight line. The relieving capacity is then determined from this plotted line. The certified capacity may not exceed 0.9 times the capacity taken from the line. The results of the four-valve method may be extrapolated to cover a larger range of pressures.

The Coefficient of Discharge Method. When a *complete line of valves* consisting of *various inlet sizes, orifice sizes*, and *set pressures* is to be certified, the coefficient of discharge method is used. In this case, three valves of each of three different sizes (a total of 9 valves) are tested, with each valve of a given size set at a different pressure. From each test, a coefficient of discharge is calculated equal to the actual flow measured divided by a theoretical flow determined from formulas appropriate to the fluid being used. Each of the calculated coefficients of discharge must fall within ±5% of an average coefficient of discharge calculated from the results of the nine tests. For sections III and VIII, the average coefficient of discharge is then multiplied by 0.9 to determine the rated coefficient of discharge to be used for calculating the rated relieving capacity of that line of valves. For valves certified per the rules of Section I and IV, the average coefficient is used directly and the 0.9 factor is included in the formula used for calculating the relieving capacity. Over the years, these inconsistent but equivalent methods have been a source of some confusion, and recently (in 1997) action was taken to reconcile the two by having all Code sections adopt the version used by Sections III and VIII. Henceforth, the 0.9 factor will be incorporated into the coefficient of discharge.

All provisional tests must be performed at an ASME-accepted test facility that meets the requirements of *Pressure Relief Devices, ASME Performance Test Code PTC 25*, and must be supervised by an individual (Authorized Observer) whose qualifications must also meet the requirements of that code. The facilities, procedures, and qualifications of the Authorized Observer must be accepted by ASME on the recommendation of a representative of an ASME-designated organization. (The National Board is the ASME-designated organization when pressure-relief devices are involved). Continued acceptance of the testing facility is subject to review by a representative of an ASME-designated organization within each 5-year period.

The valves selected for testing must fall within the capacity of the accepted test facility. Valves larger than those actually tested are therefore designed with flow paths geometrically similar to those of the valves actually tested. The results of capacity testing can thus be extrapolated for the stamping of larger valves.

Once the pressure-relief valve capacity has been determined and certified, and before the Code symbol stamp can be applied, two production valves must be selected by a representative of an ASME-designated organization and tested to verify operation and stamped capacity. These tests are performed at an ASME-accepted facility supervised by an Authorized Observer and witnessed by a representative of an ASME-designated organization. Thereafter, two more production valves must be selected and tested within each 5-year period. Should any valve fail to meet specified performance requirements or the stamped capacity, additional valves must be selected and tested. Two valves are selected and tested for each valve that failed. Continued or unexplained failure may cause the revocation of the manufacturer's or assembler's authorization to use the Code symbol stamp on that particular design of valve.

The valve capacity certification requirements for each of the book sections of the Code are similar. There are other requirements in each of the book sections for pressure-relief devices other than valves (e.g., rupture disks) and for those devices in combination with valves. These requirements are similar to those described for valves.

Test data from the initial or provisional test and the subsequent production test are signed by the manufacturer and Authorized Observer and submitted to the ASME-designated organization for review and acceptance. The rated capacities at various pressures, the rated capacity slope, or coefficient of discharge, as applicable for each valve design are then listed in a book entitled *Pressure Relief Device Certifications*, published annually by the National Board of Boiler and Pressure Vessel Inspectors of Columbus, Ohio.

Over the years, the National Board has issued over 1100 valve capacity certifications to more than 80 manufacturers worldwide, covering air, gas, steam, and liquid service.

Questions are sometimes asked about the so-called derating factor applied to certification test results to determine stamped capacity. There is a perception that a pressure-relief valve could actually pass considerably more fluid than the stamped capacity would indicate. In fact, valves supplied as "non-Code" valves are sometimes stamped with capacities that don't include the derating factor. Nevertheless, there are several reasons to require the use of the 0.9 factor. Individual valves that have been tested as part of the capacity certification program may have demonstrated capacities up to 5% below the average used as the reported test value. This reduces the potential difference between the stamped capacity and the tested capacity of those valves to only 5%. Moreover, it is reasonable to expect that further variation may occur due to normal manufacturing tolerances. Considering the importance of adequate pressure-relieving capacity, the concerned committees chose to apply a conservative derating factor to the reported test value, to ensure that actual flow through the device at specified overpressure will always equal or exceed its stamped capacity. However, the designer of the valve discharge system should recognize the possibility of flow (and resulting discharge forces) larger than those based on rated capacities.

QUALIFICATION OF PRESSURE-RELIEF VALVE MANUFACTURERS

A unique aspect of the ASME Code is a requirement for the manufacturer of pressure-relief valves to demonstrate that its manufacturing, production, and test facilities, and quality control system, will ensure close agreement between the performance of production valves and the performance of the valves submitted for certification. The valve manufacturing control systems must be documented in the manufacturer's quality control manual. That manual is reviewed to ensure compliance with ASME Code requirements. At the present time, the ASME-designated organization whose representatives review the manufacturers' facilities and operations is the National Board of Boiler and Pressure Vessel Inspectors. When the ASME is satisfied with a manufacturer's qualifications, a Certificate of Authorization is issued to the manufacturer, good for 3 years, allowing the manufacturer to use a Code symbol stamp to mark pressure-relief devices it manufactures, as evidence of ASME Code compliance. The review of a manufacturer's manual and system is repeated every 3 years, before its certificate can be renewed. This periodic review of pressure-relief device manufacturers differs from the usual ASME third-party inspection system, in which an Authorized Inspector monitors boiler and pressure vessel manufacturers on an ongoing basis to assure Code compliance. However, the relief device manufacturers' operations are subject to inspections at any time by a representative of an ASME-designated organization, although unannounced inspections are rare in the absence of a specific complaint.

The manufacturer must perform a series of production tests on each pressure-relief valve. Each valve is tested to verify that set pressure, seat tightness, and in some cases (Section I and IV steam valves) blowdown, meet the requirements specified in the applicable book section. The acceptance standard for seat tightness on steam and water is generally expressed as "no visible leakage." Seat leak tests performed on air must generally meet industry accepted standards.

Production testing of nonreclosing devices in which a component is required to break cannot be used to demonstrate opening of the actual component. Instead, duplicate parts, such as disks for a rupture disk device or pins for a breaking pin device, are manufactured from the same lot of material as the part to be supplied. These duplicates are tested to failure to verify proper operation of the device to be shipped.

Since the mid 1970s, the ASME has also recognized and authorized organizations known as assemblers to assemble and Code stamp pressure-relief valves using only original, unmodified parts obtained from a valve manufacturer. The assembler must assemble, adjust, test, and seal the valves under the same ASME system of oversight and quality control as the original valve manufacturer.

INSTALLATION GUIDELINES

The safety valve rules of the ASME Code deal primarily with design and functional requirements of pressure-relief devices, while providing only limited guidance on their installation. However, a few key installation requirements are specified for pressure-relief devices in each of the book sections. For example:

1. A pressure-relief valve must be installed as close as possible to the vessel or system being protected. This is to minimize inlet pressure drop, which can lead to valve malfunction such as chatter.

2. The opening through all pipe and fittings between a pressure vessel and its pressure-relieving device must have at least the area of the inlet to the pressure-relieving device. This, too, is to minimize inlet pressure drop.

3. A pressure-relief valve must generally be mounted in a vertical position to minimize any potential reseating problems caused by a lateral gravity loading.

4. Discharge lines, and in many cases valve bodies, must be provided with drains to prevent any accumulation of liquid or sediment that could interfere with proper operation.

5. With some few exceptions, no valves may be placed between a pressure-relief device and the vessel it is protecting or on the discharge side, between the device and the atmosphere, since such valves could be inadvertently left in the closed position and thus nullify the overpressure protection.

6. Section I invokes the use of ASME B31.1, Power Piping, for the design of certain boiler piping known as boiler external piping. Power Piping contains Appendix II, Nonmandatory Rules for the Design of Safety Valve Installations. This appendix provides advice on determining loads due to valve discharge and loads on vent stacks, on sizing vents to prevent blowback, and allowable stress criteria for various combinations of loads. It also includes guidance on locating safety valves away from pipe bends, to avoid flow-induced vibrations in the safety valve nozzles. Such vibrations have been known to damage the safety valve internals and cause the valve to malfunction.

7. Both divisions of Section VIII contain nonmandatory appendices (M and A, respectively), Installation and Operations, which provide useful supplementary information and guidance regarding pressure-relief devices. (Section X has a similar but shorter appendix.) In some instances, suggested alternatives differ from the mandatory rules of Section VIII; these alternative rules may be used only with permission of the local jurisdictional authorities where the vessel is installed. Among the topics covered are the use of stop valves between a pressure-relieving device and the vessel being protected or on the discharge side of the device, a suggested limit on pressure drop between a vessel and the relief device, design considerations for discharge piping, the use of common discharge headers, advice on differentials between operating and set pressure, the effects of relief valve reaction forces, and guidance on the sizing of pressure-relief devices for fire conditions. A useful list of references is also provided. Both Divisions of Section VIII also contain identical mandatory appendices (11 and 10, respectively) that provide methods for determining the capacity of pressure-relief valves on fluids other than the one on which the valve was officially rated.

SAFETY VALVES IN SERVICE

Since the ASME Code with few exceptions applies to new construction only, once new pressure-relief devices are installed on a boiler or pressure vessel, adjusted, set, and sealed by the device manufacturer or an assembler, the Code rules no longer directly apply. However, although any subsequent work on these devices is not directly covered by the Code, the local jurisdictional authorities refer to and use the Code rules as guidelines. For example, the National Board under its VR (valve repair) program authorizes organizations of proven capability to service, repair, adjust, set, and seal pressure-relief valves that have been in service. The objective is to restore and maintain the valves so they will continue to function as when new and thus continue to meet the Code rules for newly installed safety valves on new vessels.

A word of caution is appropriate here. Pressure-relief valves are tested for operation and capacity under optimum conditions: inlet pressure drop and backpressure are minimal; all adjustments on the valve are set and sealed by the manufacturer or an assembler, who is qualified to make these adjustments. When any of these conditions is not as it should be, the relieving capacity and proper operation of the valve can be compromised. That is why the Code requires that all piping and fittings leading to and from pressure-relief valves be adequate to minimize inlet pressure drop and build-up of backpressure. Excessive inlet pressure drop can lead to valve chatter and damage; excessive backpressure may reduce capacity. Tampering with adjustments by unqualified persons can cause failure to open at set pressure, failure to achieve full lift (and certified capacity), failure to reseat properly, and the possibility of damage due to unstable operation. For all these reasons, in the mid 1980s the concerned committees revised the rules to permit valve adjustment only by the manufacturer or assembler of the valve. However, since these rules apply to new construction only, they do not ensure that only qualified individuals will be allowed to adjust these valves in service. It is important that valves in service be maintained so that they will perform as when originally manufactured.

SUMMARY

The overpressure protection rules of the ASME Code are based on an integrated system of accreditation and certification that applies to the different parties involved: manufacturers of boilers or pressure vessels, manufacturers of pressure-relief devices, testing laboratories, Authorized Observers who supervise the testing, and a representative of an ASME-designated organization (the National Board) who reviews the facilities, quality control systems, and activities of organizations involved in Code construction. Overpressure protection of boilers and pressure vessels may be provided only by use of devices allowed by the section of the Code applicable to the vessel or system being protected.

The ASME overpressure protection rules, with their integrated system of oversight and quality control, have achieved an outstanding safety record. Confirmation of this record is evidenced by the adoption of similar rules in the standards of many other countries, including draft standards currently being considered by the International Standards Organization and the European Committee for Standardization.

FEEDWATER SUPPLY AND WATER LEVEL INDICATION

IMPORTANCE OF FEEDWATER

The designer of a boiler must provide adequate cooling flow of water or steam through the various pressure parts so that they will not exceed their intended design temperatures. Water-cooled furnaces must contain high-temperature combustion gases and cool them sufficiently to pass through closely spaced rows of heating surface without overheating that surface or (for coal-fired boilers) causing slagging problems in the furnace or elsewhere. Ferritic alloys (carbon and low alloy steels) are used for water-cooled furnace tubes, because of their resistance to chloride-induced stress corrosion cracking, as discussed in Chapter 3 under Special Concerns. Boiler circulation must be sufficient to cool the furnace walls to temperatures at which ferritic steels have adequate strength and oxidation resistance. Section I has never provided rules for boiler circulation, as the responsibility for functional design has traditionally remained with the boiler manufacturer. However, Section I does have rules that implicitly recognize the serious consequences resulting from the loss of adequate water supply to heat-absorbing circuits critical to the safe operation of the boiler. Firing an inadequately cooled or partially dry boiler can lead to overheating of the pressure parts, with consequent loss of strength and potentially catastrophic failures due to internal pressure. Section I addresses this problem with rules intended to assure the availability of feedwater and the presence of water in the drum.

RULES FOR FEEDWATER SUPPLY

Feedwater supply rules are found in PG-61. For boilers firing solid fuel not in suspension with more than 500 square feet of water heating surface, a redundancy of supply is required, i.e., at least two means of feeding water must be provided. Smaller units are considered to be adequately supplied by a single source. Note that this rule doesn't apply to boilers firing solid fuel in suspension, such as pulverized coal, or to boilers firing liquid or gaseous fuels, if means are provided for shutting off the heat input before the water level reaches the lowest permissible level mentioned in PG-60.1.1. In such cases the fuel in the furnace is consumed almost immediately after the fuel supply is stopped, and the problem of significant overheating is avoided. Another concern addressed in PG-60.1.1 is that the feedwater system must have adequate supply pressure to maintain flow to the boiler under all conditions. This is accomplished by specifying that each source of feeding be capable of supplying the boiler at a pressure 3% higher than the highest setting of any safety valve on the boiler. Thus feedwater would presumably continue to flow into the boiler even when it is operating at a pressure sufficient to lift the safety valves.

Boiler damage due to overheating can also occur if a particular firing method leaves an inventory of burning fuel in the furnace when the feedwater supply is lost or interrupted. This is a concern for stoker fired boilers and fluid bed boilers, where a bed of combustible material remains when the boiler is tripped due to an electrical outage or mechanical failure and feedwater flow stops. For such boilers and for boilers whose setting or heat source can continue to supply sufficient heat to cause damage to the boiler if the feed supply is interrupted, PG-61.1 requires that one means of feeding shall not be susceptible to the same interruption as the other, and each source must be able to provide sufficient water to prevent damage to the boiler. The intent is for the designer to consider potential failure modes and provide an independent backup feedwater system to cool the pressure parts that otherwise could be overheated by the setting or by any burning fuel that remains in the furnace. The backup feedwater source is only required to provide sufficient water to prevent damage to the boiler. The determination of that quantity is left to the boiler designer, since he or she is best qualified to make that decision. In 1981 Subcommittee I elaborated its position on minimum required feedwater flow and supply pressure in Interpretations I-81-15, I-81-20, and I-81-25, noting that Section I doesn't define this minimum and placing on the designer the responsibility for providing sufficient feedwater flow to satisfy all requirements of Section I. Here are those Interpretations:

> Question: *Since requirements for the flow capacity of the feedwater source are not given in PG-61.1, is it the responsibility of the designer of the feedwater system to provide arrangements that establish the feedwater flow required to satisfy the pressure requirements of PG-61?*
> Reply: *Yes.*
> Question: *The rules of PG-61.1 require that each source of feedwater be capable of supplying water to the boiler at a pressure 3% higher than the highest setting of any safety valve on the boiler. Does the phrase ''at a pressure'' apply to the pressure at the feedwater pump outlet or to the pressure in the boiler drum?*
> Reply: *The feedwater pressure requirements in PG-61.1 apply to pressures in the boiler drum, not to the pressure at the feedwater pump outlet.*
> Question: *Is the total capacity of the feedwater source based on the maximum process steam requirements, or on the maximum boiler evaporation?*
> Reply: *Section I rules do not address the capacity requirements for the boiler feedwater source.*
> Question: *What is the minimum flow rate required to satisfy the requirements of PG-61.1?*
> Reply: *Section I does not define the minimum flow rate required to satisfy the requirements of PG-61.1. It is the responsibility of the designer to provide for sufficient feedwater flow to satisfy all requirements of Section I.*

Interpretation I-89-57 dealt with required feedwater flow capacity for a forced-flow steam generator with no fixed steam or waterline:

> Question: *When determining the required feedwater flow for a forced-flow steam generator with no fixed steam or waterline, does the ''maximum designed steaming capacity'' in PG-61.5 refer to the flow of steam through the boiler with its safety valves fully open?*
> Reply: *No.*

A number of inquiries have been received about feedwater supply for boilers firing fuel partially in suspension. Interpretation I-86-62, Boilers Firing Solid Fuel, clarified this situation:

> Question: *Does the term ''solid fuel not in suspension'' used in PG-61.1 refer to any portion of the fuel which is not burned in suspension?*
> Reply: *Yes, it applies to that portion of the fuel burned in any type of bed in the furnace.*
> Question: *Do the requirements in PG-61.1 (for boilers fired by fuel not in suspension) to have two means of feeding water, one of which is not susceptible to the same interruption as the other, apply if the solid fuel is burned partially in suspension and partially in a bed?*
> Reply: *Yes.*
> Question: *For boilers that are fired with solid fuel not in suspension, do two electrically driven feed pumps, each supplied with power from independent electrical sources, meet the requirements of PG-61.1?*
> Reply: *Yes.*

A 1986 inquiry sought further guidance regarding redundancy of the feedwater lines themselves. This was provided in Interpretation I-86-18:

> Question: *When a boiler requires two means of feeding water under PG-61, are two separate feedwater lines to the boiler required?*
> Reply: *No.*
> Question: *Are the valves in the boiler external feedwater piping required to be the same size as the boiler connection and boiler external feedwater pipe?*
> Reply: *No.*

It is clear that section I requires a backup feedwater supply in the several cases just described. In 1995 an inquirer asked whether that philosophy extended to the normal operation of the boiler. Interpretation I-95-22, Redundancy of Feedwater Supply, advised as follows:

> Question: *A boiler firing solid fuel not in suspension is provided with two independent means of feeding water not susceptible to the same interruption, as required by PG-61.1. Does Section I prohibit the continued operation of the boiler following the loss of one of the independent means of feeding water?*
> Reply: *No. PG-61.1 establishes the feedwater arrangements which must be provided. Section I as a construction code does not impose additional rules covering boiler operation.*

During committee discussion of this inquiry, it was noted that requiring shutdown of the boiler on loss of one of the two required sources of feedwater would be akin to driving no further after experiencing a flat tire and mounting the spare. It was also noted that the availability of a single backup source of feedwater would take care of most situations, such that the loss of the remaining source would be a rare occurrence, with little impact on the life of the boiler, as measured by cumulative creep damage.

There are two exceptions to the PG-61.1 rule requiring backup feedwater supply for boilers firing solid fuel not in suspension or whose setting or heat source could damage the boiler if the feed were interrupted. The first of these concerns what are called black liquor chemical recovery boilers used in the Kraft pulp and paper industry. They operate with a hot, molten smelt bed in the boiler, a bed which in theory could cause damage to the boiler on loss of feedwater. However, the real danger with these boilers comes from the possibility of water leaking into the smelt bed and causing a violent explosion. Accordingly, the industry developed and uses leak detection methods that on finding a leak stop the feedwater and trigger an emergency rapid drain procedure. In the course of using this system, it was found that the boilers were sustaining very little damage and could thus be exempt from the rule requiring a continuing supply of water. This was affirmed by Interpretation I-83-03:

> Question: *Is a black liquor chemical recovery boiler equipped with an emergency rapid drain system required to have a steam driven feedwater pump to satisfy the provisions of PG-61.1?*
> Reply: *No, since it has been demonstrated that the heat source of a black liquor chemical recovery boiler equipped with an emergency rapid drain system does not supply sufficient heat to damage the boiler if feedwater is interrupted, it is not required that one of the feedwater pumps be steam driven.* (Note: The 1983 requirement for a steam-driven pump was subsequently changed to a requirement that one means of feeding not be susceptible to the same interruption as the other.)

The other potential exception to the two sources of feedwater rule appears in Code Case 2160, which covers certain types of fluid bed boilers. The concern about fluid bed boilers involves the possibility that the bed, which contains fuel and limestone, might continue to evolve sufficient heat to damage the boiler after it was tripped on loss of feedwater. Case 2160 provides controls and safeguards intended to ensure that this won't happen. It also includes a requirement that the designer provide calculations, signed by a Professional Engineer, demonstrating that even under the most severe conditions no part of the boiler will overheat to the point that the allowable stresses are exceeded.

FORCED-FLOW STEAM GENERATORS

Once-through forced-flow type boilers with no fixed steam and waterline differ from natural circulation units in that the feedwater flow rate is directly proportional to steam flow and is not controlled by boiler drum level. Since no drum or drum level indication is available, the operator has no visible assurance that sufficient water is being supplied to the boiler. These units depend on control instrumentation and interlocking alarms and trips to protect against loss of feedwater supply. For such boilers, Section I simply requires that a source of feedwater be available to the boiler at a pressure not less than the expected maximum sustained pressure at the boiler inlet corresponding to operation at maximum designed steaming capacity with maximum allowable working pressure at the superheat outlet. The determination of this pressure is left to the boiler manufacturer.

LOW WATER LEVEL FUEL CUTOFFS

The exception to the requirement for a second feedwater supply system for boilers firing gaseous, liquid, or solid fuels in suspension applies when means are furnished for shutting off the fuel (or heat input) before the water reaches the lowest permissible level. Typical fuels included in this category are natural gas, oil, or pulverized coal. The fuel shutoff is normally referred to as a low-level cutoff, trip, or interlock. Section I does not mandate such fuel cutoffs, although they are routinely furnished on large boilers.

WATER LEVEL INDICATION

Rules governing the provision of water level indicators on a boiler are found in PG-60. This paragraph also covers water columns and pressure gages and the connections between these various appurtenances and the drum. Note that this connecting piping falls in the category of boiler external piping, and its design must follow the applicable rules of ASME B31.1 (see Chapter 5).

Low water level is often responsible for pressure-part failures, because in most boilers circulation is dependent on the pressure head of water in downcomers and low water directly reduces that head. Without adequate circulation, pressure parts are subject to overheating and premature failure. Therefore, every boiler must have a water gage glass. Because knowledge of correct water level is so important to the safe operation of the boiler, PG-60 provides for redundancy in drum water level measurement for boilers operated at pressures over 400 psi, by requiring provision of two water gage glasses. (Like many others in Section I, this is an arbitrary but not unreasonable rule.)

For many years, PG-60 stated that for power boilers with MAWP at 900 psi and above, two remote level indicators could be provided instead of one of the two required gage glasses. Then, when both remote units were in reliable operation, the remaining gage glass could be shut off, but had to be maintained in serviceable condition. In 1990 a boiler manufacturer asked that the 900 psi threshold for substituting two remote level indicators for one of the two required gage glasses be lowered to 400 psi. There was nothing magic about 900 psi, and after some consideration, Subcommittee I approved Code Case 2109 to permit use of two remote level indicators for one gage glass at 400 psi, which is at the bottom of the pressure range that normally requires a second gage glass. After several years of satisfactory experience, the case was incorporated, simply by lowering the 900 psi threshold to 400 psi. This change appeared in the 1996 Addenda.

The provision of two remote level indicators for one of the two water gage glasses normally required means that the second gage glass need not be furnished at all. This was confirmed in response to a 1982 inquiry answered by Interpretation I-83-13, Gage Glass Requirements:

> Question: *For a boiler whose design pressure is 900 psi or above, PG-60.1.1 permits the provision of two remote level indicators for one of the two gage glasses normally required.*

Does this mean that the boiler may be furnished with only a single gage glass, which may then be shut off when both remote level indicators are in reliable operation?
Reply: *Yes, but note that the single gage glass must be maintained in a serviceable condition.*

A traditional operating rule is that a boiler should never be fired when there is no water in the gage glass. Section I stipulates that the lowest visible part of the gage glass shall be at least 2 inches above the level at which overheating damage can be expected. Thus if water is visible in the glass, there is always some margin against overheating. The Code leaves it to the boiler manufacturer to determine the level at which overheating damage can be expected. This was explained in 1992, by Interpretation I-92-50, Lowest Permissible Water Level:

Question: *Does Section I provide a definition of the lowest permissible water level as referenced in PG-60.1.1?*
Reply: *No. PG-60.1.1 requires the boiler manufacturer to determine the lowest permissible water level.*

When a direct reading of gage glass water level is not readily visible to the operator in the area where immediate control actions are initiated, two dependable indirect indications must be provided, either by transmission of the gage glass image or by remote level indicators. In 1991 Addenda, after a trial period by means of Code Case 1897 showed the method to be reliable, Subcommittee I added a provision to PG-60 stating that transmission of a gage glass image to the operator (in the area where he could initiate immediate control actions) by means of fiber optic cable was permitted, provided no electrical modification of the optical signal was employed. Such transmission is considered to be direct reading of gage glass water level. The committee felt that electronic black boxes would not provide the same reliability as a directly transmitted optical signal along a fiber optic cable. That is also why the committee has so far refused to accept as the equivalent of a direct reading gage glass a number of very clever devices that base their water level readings on the large difference in electrical conductivity between water and steam. These devices may be used as remote level indicators only. The committee responded to a request to permit the use of such devices as the equivalent of a gage glass in Interpretation I-86-2, Boiler Water Level Indicators:

Question: *In the case of power boilers with all drum safety valves set at or above 900 psi, may a new technology level indicator be substituted for the remaining gage glass when two remote level indicators are used in lieu of the other gage glass?*
Reply: *No.*

Note that devices using this new technology may be used as remote level indicators, even if Subcommittee I doesn't consider them the equivalent of a gage glass for simplicity and reliability. Devices using differential conductivity of steam and water have achieved a greater measure of acceptance in Europe.

Since Section I doesn't clearly define a remote level indicator, one inquirer wanted to know whether a low water alarm and trip indication would serve the purpose. In 1987 under the heading Water Level Indicators, the committee issued Interpretation I-86-50:

Question: *Is a direct reading water level gage available to a roving operator considered to be readily visible to the operator in his working area?*
Reply: *No.*
Question: *Is a low water alarm and trip indication available to the operator in his working area considered to be a dependable indirect indication of boiler water level?*
Reply: *No.*

A more successful attempt to determine whether a device could serve as a remote water level indicator was answered in late 1994 by Interpretation I-95-04:

Question: *May a pressure- and temperature-compensated differential pressure drum level transmitter with a remote readout be considered a remote level indicator?*
Reply: *Yes.*

Other safety rules related to water level control apply to the installation of the gage glass and its appurtenances. For example, shut-off valves on the gage glass connections must have an outside-screw-and-yoke or a lever-lifting type of operating mechanism that indicates whether the valve is open or closed. The rules also require through-flow valve construction to prevent stoppage by deposit of sediment. There is also a rule essentially limiting the type of connections to piping between a water column or gage glass and the drum to those used for water level indication. This rule is to minimize the possibility that activation or failure of another type of device connected to that piping could cause a false reading of water level. Since the water level reading instruments base their readings on a differential pressure of only a few inches of water, any sudden flow or other system disturbance could disrupt them.

PG-60 also provides rules for the minimum size of piping and fittings for water columns, gage glasses, pressure gages, and related components.

With the increasing use of computers, some boiler operators would like to send signals from instrumentation directly to the computer for reading on a monitor. This has led to a concern about continuous display of water level. In 1993 in response to an inquiry, the committee issued Interpretation I-92-63, Continuous Display of Remote Water Level Indication:

> Question: *If a remote computer terminal is used as one of the two remote level indicators required by PG-60.1.1, may that remote computer terminal display drum water level on demand rather than continuously, when the single gage glass is in service and is readily visible to the operator in the area where immediate control actions are initiated?*
> Reply: *Yes, provided the second remote level indicator is displayed continuously.*

This reply is consistent with the principle that a boiler whose MAWP is 400 psi or more should have two independent and reliable means of drum water level indication readily visible to the operator in his or her working area, where the operator can take any necessary control action to correct a water level problem. The fact that one gage glass was readily and continuously visible to the operator was the key to this reply, and distinguishes it from the situation described in the following interpretation.

In 1995 in response to a similar inquiry, the committee issued Interpretation I-95-07, Remote Level Indication on Computer Monitor:

> Question: *May a power plant distributed digital control system (DCS) with cathode-ray tube (CRT) screens providing graphic drum water level displays in the control room be considered to provide an independent remote level water indication in accordance with PG-60.1.1?*
> Reply: *Yes. Note that two such independent displays are required in order to omit one of the two normally required gage glasses for boilers whose MAWP is 900 psi or greater.* (Note by authors: The 900 psi threshold was lowered to 400 psi in the 1996 Addenda, as related above.)
> Question: *May a power plant distributed control system (DCS) be considered to provide an independent remote level indication as described in PG-60.1.1 if the displays are interruptible, noncontinuous displays which may require a keystroke to become visible?*
> Reply: *No.*

Another type of water level indicating device that has been the subject of inquiries uses an internal magnetic float that provides an external indication of water level. Although such devices would appear to be simple and reliable, the committee had concerns about not being able to see the water level, as well as the possibility that the float might somehow get stuck or that another magnet could be used to cause an erroneous external level indication. However, the devices were considered acceptable as remote water level indicators. In 1993 in response to an inquiry, the committee issued Interpretation I-92-69, Magnetic Level Indicator:

> Question: *May a magnetic level indicator be used to provide an indirect reading of water level as permitted by PG-60.1.1?*
> Reply: *Yes, provided all other requirements of this Section are met.*

This last proviso, that "all other requirements of this Section are met" is a time-honored but unnecessary piece of advice.

In response to a related inquiry, in 1994 the committee issued Interpretation I-92-96, Magnetic Level Gage as Remote Level Indicator:

> Question: *Is a magnetic level gage which provides a local, indirect visual indication of the boiler drum water level considered to be a gage glass as required by PG-60.1.1?*
> Reply: *No.*
> Question: *Do PG-60.1.1 and PG-60.2.6* (Note: the latter paragraph is now PG-60.3.8) *permit the installation of a magnetic level gage on the piping between a water column and a gage glass, regardless of boiler maximum allowable working pressure?*
> Reply: *Yes.*

Sometimes when a water gage glass is at some distance away from the operator in the control room, it is possible to install a system of mirrors that will provide an image of the gage glass water level. The committee has established that it considers such means the equivalent of direct reading of the gage glass, by Interpretation I-92-15, Water Level Indicators:

> Question: *Is the viewing of a gage glass via a system of mirrors considered direct reading of gage glass water level?*
> Reply: *Yes.*

WATER COLUMNS

There is sometimes confusion as to what Section I means by the term **water column**. A water column is a vertical vessel connected to the drum by two horizontal connections located in such a way that the lower connection is in the water space and the upper connection is in the steam space of the drum. The level in the water column provides a calmer environment than the inside of the drum, but it still follows the movement of drum water level. Gage glasses can then be mounted on the water column, where they can obtain a steadier water level reading than if they were mounted directly on the drum.

There have been a few Interpretations about water columns. One of these, I-86-07, was issued under the title Materials for Water Columns, although it dealt with broader issues:

> Question: *May a water column for use on a power boiler be designed as a standard pressure part?*
> Reply: *Yes. See PG-11 and PW-42.*
> Question: *May a water column for use on a power boiler be constructed of SA-36 for the nozzles or body?*
> Reply: *No. SA-36 material may not be used, except as permitted in PG-13.*
> Question: *May a water column for use on a power boiler be constructed using standard piping components and/or individually designed parts, as permitted by the rules of Section I?*
> Reply: *Yes.*
> Question: *May a water column for use on a power boiler be designed to the rules of Section VIII?*
> Reply: *No.*

The prohibition mentioned in the reply on the use of SA-36 for other uses than those permitted in PG-13 is based on a long-standing and perhaps justified belief on the part of some members of the committee that the quality of structural materials such as SA-36 is not as good as that of materials manufactured specifically for withstanding pressure. This can be seen in the Table 1-100, Criteria for Establishing Allowable Stress Values for Tables 1A and 1B of Section II, Part D. In the setting of allowable stresses for pressure retention applications for what are called ferrous materials of structural quality, a factor of 0.92 is applied to the stress as normally derived, i.e., an arbitrary penalty of 8% is imposed. Given modern steel-making practices, that penalty may no longer be justified; however, it is not likely to be dropped any time soon, especially as more steel is imported from producers whose long-term record of quality control may not be known.

A somewhat related Interpretation issued in 1988 as I-89-06 dealt with welding requirements for gage glass bodies. The questions and replies would be equally applicable to water columns, since they are both Section I appurtenances that can be furnished on the same basis.

> Question: *Are gage glasses considered to be part of the boiler proper as defined in Fig. PG-58.3.1?*
> Reply: *No, they are part of the Boiler External Piping, subject to the rules of PG-60.*
> Question: *May welded gage glass bodies be provided under the requirements of PG-11.3 by a manufacturer who does not possess a Section I Certificate of Authorization?*
> Reply: *Yes.*
> Question: *Must the welding for gage glass bodies, complying with a manufacturer's standard, meet the requirements of PG-11.3?* (PG-11.3 calls for welders and weld procedures to be qualified in accordance with PW-26 through PW-39, essentially the full qualification normally required for Section I construction.)
> Reply: *Yes.*

CREEP AND FATIGUE DAMAGE DURING BOILER LIFE

INTRODUCTION

The Foreword to Section I states that the objective of the Section I rules is "to afford reasonable protection of life and property and to provide a margin for deterioration in service so as to give a reasonably long safe period of usefulness." Sooner or later almost all equipment wears out, and boilers are no exception. The factors that cause the gradual deterioration of a boiler are many. Among them are erosion and corrosion, which cause a gradual wastage and thinning of pressure parts. Two other significant factors that can limit the life of a boiler are creep and fatigue, processes that can cause progressive deterioration of material properties without wastage and thinning. This deterioration is characterized as creep damage or fatigue damage. For practical purposes when discussing boilers, it can be said that creep damage occurs only at elevated temperature, about 700°F and above for carbon and low alloy steels and 800°F and above for the high alloy steels typically used in superheater construction. Fatigue damage can occur over the entire range of boiler operating temperature, but as will be explained later, components in elevated-temperature service are more susceptible to fatigue damage. These two damage mechanisms are discussed separately, but their effects are additive.

CREEP

In our everyday experience, metals such as steel are firm, unyielding solids. In courses on elementary strength of materials, we learned that steel deforms elastically under load, following Hooke's law. For steels at room temperature, strain is a function of stress only. For example, if a load were placed on the end of a cantilever beam, the measured deflection would remain constant, whether measured when the beam was first loaded, or the next week, or the next year. That is, the deflection is independent of time. For certain other solids, and for many metals at elevated temperature, this is not the case. These materials are found to undergo a slow plastic deformation under stress, a process known as creep. Glass creeps at room temperature, but very slowly. This was confirmed when the thickness of 1000-year-old monastery window panels was measured; they were found to be thicker at the bottom than at the top, showing that the glass had slowly deformed, or flowed, under the influence of gravity. Experiments in the 19th century showed that lead creeps at room temperature. Steels such as those used in boiler construction are known to creep at elevated temperature.

The temperature at which creep begins defines the bottom of the creep range—about 700°F for carbon and low alloy steels and about 800°F for high alloy steels such as the austenitic stainless alloys. Increasing alloy content raises the temperature at which creep begins. As a broad rule of thumb, this temperature is about one-fourth of the melting temperature, expressed on an absolute temperature scale. Much careful experimentation in the 20th century has been used to measure and predict the elevated-temperature creep behavior of structural steels, but there are as yet few theories that can accurately describe this behavior, except for conditions of constant temperature and stress. It is very difficult to predict creep behavior under varying temperatures and varying multi-axial states of stress. As a practical matter, it is also very difficult for the designer to predict with confidence the actual temperature and stress state a component may experience during its service life.

Typically, boiler steels are tested to creep failure, or creep rupture, under constant load at constant temperature using dead weight testing machines under very careful temperature control, since creep rate is very sensitive to temperature. Stress and temperature are chosen for each test to yield useful information within a reasonable time. Both elongation and rupture time are recorded. Results are needed from less than 100-hour rupture time to several hundred thousand hours. (Boiler components in service 80% of a 30-year life might spend over 200,000 hours under creep loading conditions.) Figure 14.1 shows the results of a typical creep test at constant load and temperature, with creep strain plotted as a function of time. This classic form of creep curve is composed of three parts, illustrating what is described as primary (stage I) creep, followed by a generally longer period of secondary (stage II) creep, and finally tertiary (stage III) creep, which culminates in stress rupture. From microstructural examination of creep specimens at various stages of the creep process, increasing damage can be recognized, as illustrated on Fig. 14.2. Voids begin appearing on the grain boundaries somewhere in the second stage. These voids eventually link together, finally forming microcracks. The remaining uncracked load paths must then carry additional load, and the creep damage mechanism accelerates, with the formation of larger cracks finally leading to failure. These various damage stages are sometimes used in remaining-life assessments of components to determine fitness for continued service.

Designers need to be able to predict creep rupture behavior without waiting for the completion of tests that could take many years, and fortunately researchers have discovered that for a given stress level, there is a certain equivalence between time and temperature. Thus creep rupture tests can be shortened by conducting them at higher temperatures. Perhaps the best known creep life predictive methods are those developed by two researchers named Larson and Miller.

The results of the creep tests can be plotted in a number of useful ways, some of which are used to determine stress-to-produce rupture in 100,000 hours, one of the factors used in setting the allowable stresses described in Chapter 15. One typical plot is that shown as Fig. 14.3, a stress versus time-to-rupture plot for 2¼ Cr-1 Mo steel, a common superheater alloy. (This plot is taken from Subsection NH of Section III, which covers methods of elevated-temperature design of nuclear power plant components. Subsection NH is the most comprehensive Code reference on that subject and is recommended for those interested in further study of elevated-temperature design methods. Before their appearance as Subsection NH in the 1995 Addenda to the Code, these design rules were found in a famous nuclear Code Case, Case N-47.) Note that Fig 14.3 is a log-log plot, with temperature as a parameter. At given temperatures, the stress-versus-time relations are seen to be straight lines. A little study shows how very strong the influence of temperature is; just a 50°F increase causes rupture life to shorten dramatically. For example, Fig. 14.3 shows that at a stress level of 10,000 psi, the creep ruture life shortens by a factor of about five when the temperature increases from 1000°F to 1050°F.

One way to assess creep damage is to use a so-called **linear damage rule** (also called Robinson's rule), which can be illustrated by use of Fig. 14.3. Suppose a number of steady state operating conditions are anticipated for an elevated-temperature component, in which both stress, time of loading, and temperature are known. The creep damage accumulated during each of these time periods is assumed to be linear with time and is found in the following way. The rupture life for each specified stress and temperature condition is determined. For example, suppose for one of these conditions that the predicted life is 50,000 hours, and the condition is anticipated to last 5000 hours. The cumulative damage during this condition is then defined to be 5000 hours divided by 50,000 hours, or 10%, i.e., 10% of the creep life would be used up by this operating condition. The designer seeks to keep the sum of the cumulative damage due to all anticipated operating conditions less than 100%. If he or she doesn't, this method would predict creep failure of the

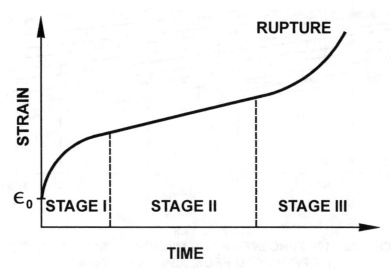

FIG. 14.1
STRAIN IN CONSTANT STRESS CREEP TEST
(Initial elastic strain ϵ_0 and three stages of creep are shown.)

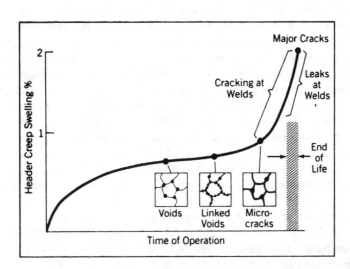

FIG. 14.2
PROGRESSION OF MICROSTRUCTURAL DAMAGE DURING COMPONENT LIFE
(REPRODUCED FROM NEUBAUER AND WEDEL (1983))

component during its service life. Unfortunately, the method is not so simple or accurate as it first might seem, because it is difficult to determine accurately past or future operating conditions (stress, time, and temperature) and also because there is considerable scatter in the data used to develop creep rupture curves. This method can be a useful tool, however, for assessing such things as damage from a short-term temperature excursion. Note also that fatigue damage is additive to creep damage, and the two damage mechanisms interact to accelerate the process of component life expenditure at elevated temperature. This phenomenon is called **creep–fatigue interaction**.

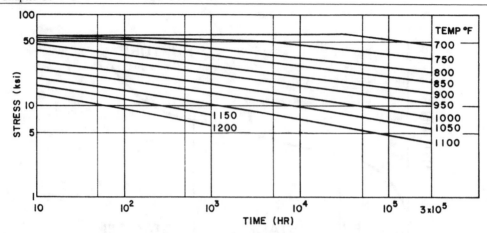

FIG. 14.3
2¼ Cr-1 Mo—100 PERCENT OF THE MINIMUM STRESS-TO-RUPTURE
(REPRODUCED FROM ASME SECTION III)

Creep considerations are important when designing boiler components such as superheater tubes, superheater outlet headers, and main steam piping, which typically operate in the creep range and are potentially subject to failure modes involving creep, such as swelling and sudden creep rupture. A creep rupture differs markedly from a tensile rupture at room temperature. In a room temperature test, a steel tensile specimen is subjected to increasing load, it yields, strain hardening takes place, and finally after the load is increased to about twice the yield strength, the specimen necks down and ruptures. Elongation at failure might be 25% or more. In this relatively fast tensile test, the deformation takes place by slip along planes in the grains of the metal and by what is described as ductile void coalescence, with transgranular failure occurring at high total strain. In a creep rupture test, a constant load is imposed and maintained (usually only a fraction of the yield load), while the temperature is kept constant at some elevated value, for example 1100°F. After a long time, and at some relatively small strain, perhaps only 1% or 2%, the specimen suddenly ruptures. In this slow creep test, failure takes place intergranularly, by the accumulation of grain boundary voids, at low total strain. (Whether a material fails transgranularly or intergranularly depends on a complex interplay among stress, strain rate, and temperature.) The specimen is no longer capable of carrying a load which it formerly carried, even though the load was well below the short-term yield strength of the metal.

All Code allowable stresses at elevated temperature are chosen with significant margins against creep failures, as explained in Chapter 15. However, it should be recognized that components that are stressed at elevated temperature have a finite life, and that this life is slowly being expended. This is particularly pertinent when life-extension studies are undertaken.

FATIGUE

Fatigue is commonly understood to mean the behavior of a material under the repeated application of stress or strain. The ASTM definition of **fatigue** is:

> *The process of progressive localized permanent structural change occurring in a material subject to conditions which produce fluctuating stresses and strains at some point or points and which may culminate in cracks or complete fracture after a sufficient number of fluctuations.*

When a body containing some zone of stress concentration is loaded, stress in a local region may exceed the yield strength, and the material can flow locally, without failure. If this process is reversed, the flow is also reversed, and if the process is repeated a sufficient number of times, the material may develop a crack due to the damage incurred by plastic strain cycling. The crack then tends to grow or propagate, under the action of the same mechanism that originated it, until the remaining portion of the section is too weak to carry the load and fractures suddenly. This is called a fatigue failure. A familiar example is the failure of a wire coat hanger when it is bent back and forth perhaps a dozen times. The region of stress concentration that initiates the fatigue cracking can be an obvious sharp surface notch or merely a metallurgical defect or discontinuity that is not apparent.

Designers use fatigue design curves, which are plots of allowable alternating stress versus number of cycles anticipated. These design curves are derived from actual fatigue tests that determine the number of cycles to cause failure for various levels of alternating stress. Appropriate safety factors are applied to the results of fatigue tests to establish fatigue design curves. As for creep, a linear damage rule, in this case sometimes called Miner's rule, is often employed, as follows: Suppose a given fatigue loading condition causes an alternating stress that the fatigue design curve shows may be permitted for 100,000 cycles. Suppose further that 5000 cycles of this loading are anticipated. This particular loading is then assumed to cause a cumulative fatigue damage of 5000 cycles divided by 100,000 cycles, or 0.05. Another way of saying this is that 5% of the fatigue life of this component will be expended during this loading. The designer tries to ensure that the cumulative fatigue damage for all anticipated load cycles adds up to less than 100%. Again, difficulties intervene to complicate this seemingly simple approach. At elevated temperature, fatigue life is shortened by the simultaneous accumulation of creep damage. The reduction in cyclic life is also quite dependent on the rate of cycling. A slow fatigue cycle provides more time for creep damage to accumulate. As noted earlier, this is the problem of creep–fatigue interaction.

At first consideration, the loads on a boiler do not seem to be of the type likely to cause fatigue failures, that is, they are not highly repetitive loads. Fatigue failures are often associated with a large number of load cycles—tens of thousands, or hundreds of thousands—although low cycle fatigue failures are not uncommon if the cycle is severe enough. One of the obvious load cycles in a boiler is the start-up and shutdown cycle. During this cycle, the pressure (and, therefore, the pressure stress) varies between zero and some maximum value. More important, and not so obvious, is the thermal stress variation that occurs as the fluid in the boiler heats up and cools down. For instance, suppose the boiler is placed in service quickly, with a fluid temperature increase of 200°F/hr. Thick-walled components such as superheater headers develop significant thermal stress due to thermal transients in the fluid passing though them. Since it takes some considerable time for the heat to flow radially through the wall of these components, a considerable radial temperature gradient develops across the wall, as the inner surface follows the fluid temperature but the outer surface lags behind. A 50°F temperature difference between the inside and outside of a thick header can cause a thermal compressive stress on the inside surface of perhaps 10,000 psi.

Another severe thermal stress cycle can occur during a so-called hot restart, in which the boiler has been shut down for a short time, but such elements as the superheater outlet header have not cooled significantly and are thus still near their normal 1000°F operating temperature. As the boiler is restarted, with minimal firing rate, the initial flow of steam reaching the still hot parts may be much colder than 1000°F, resulting in a thermal shock and very high thermal tensile stress on inner surfaces. These suddenly cooled inner surfaces would, if free, shrink to a smaller radius, but are constrained by the bulk of the surrounding hot metal. Another example of cyclic loading on a boiler is the thermal expansion and contraction of boiler piping, particularly the main steam piping, during start-up and shutdown.

Until recently, the relatively limited number of cyclic loads on boilers presented few problems for the designer. As more plants are being placed in two-shift operation and as faster starts and shutdowns are desired, fatigue becomes a more important consideration. It is also a key element in life-extension studies. At elevated temperature, fatigue damage is more severe than at room temperature, for a variety of reasons, among them the complex and not yet clearly understood interaction between creep and fatigue. There are as yet very few generally accepted methods to account for all the damage mechanisms involving the interaction of creep and fatigue under realistically varying operating conditions, although many helpful techniques are available, and considerable research effort in this area continues, as discussed next.

BOILER LIFE EXTENSION

The increase in energy costs that began with OPEC's oil embargo in 1973 has had a profound influence on the construction of new steam power and other industrial plants. Fuel costs have gone up sharply and erratically, and it is impossible to predict their future course with any certainty. Because of higher energy costs, users have changed their former wasteful practices, and new equipment is designed to use less energy. Under political pressure to hold down energy costs, public utility commissions have become much less generous than formerly in permitting utilities to recover the cost of new generating plants through rate increases. Whereas in the 1960s ever larger and more efficient plants were ordered to replace older units and meet steadily increasing electrical demands, growth in electrical usage is now much slower and less predictable than in the past. Furthermore, new units are now no longer likely to be larger or more efficient than most of those presently in service, since practical limits have been reached with respect to size and efficiency and few utilities are interested in pioneering larger units or advanced (higher temperature) cycles. As a matter of prudent business judgment, for most utilities it appears preferable to choose designs that are proven, especially in today's regulatory climate. Moreover, the additional competitive pressures on utilities and economic uncertainties caused by moves toward their deregulation make it more difficult for them to justify investing in new facilities.

As a consequence of all these related factors, owners of power plants, chemical plants, and oil refineries have a strong economic incentive to extend the normal life span of their plants. Accordingly, a great deal of recent research in the United States, Europe, and Japan has been directed at plant life extension and residual-life assessment, which are really opposite sides of the same coin. In the United States much of this research has been under the sponsorship of the Electric Power Research Institute (EPRI), the research arm of the electrical utilities. Studies based on the creep rupture strength of alloy steels typically used in elevated-temperature service show service life of this equipment can be expected to exceed 40 years, an estimate borne out by experience. This is so even though the allowable stresses at elevated temperatures are based on creep and rupture data extrapolated to 100,000 hours, a time less than 12 years. As explained in Chapter 15, this is because of the design (safety) factors used on the test data and the fact that the stresses are calculated using design temperature and pressure, both of which are significantly higher than the temperature and pressure at which most equipment actually operates during most of its service life. However, creep rupture life is quite sensitive to temperature, and individual elements such as superheater tubes may suffer creep damage and premature failure if they are subjected to prolonged service at temperatures above their design temperatures. For this reason, in large boilers, it is customary to install thermocouples at selected locations to monitor operating temperatures and provide high-temperature alarms, so that operators can take necessary corrective action.

Current plant life extension efforts typically begin with a thorough investigation of equipment condition to determine suitability for extended service. If the discussion is narrowed to boilers, the components of prime concern are the heavy-walled elements operating at the highest temperatures. As explained earlier, components operating below the creep range may also be a concern, due to wastage and thinning, but assessment of remaining life of such elements is largely one of evaluating wall thickness and corrosion or erosion rates. The material remaining is much less likely to have suffered the gradual deterioration associated with creep and fatigue damage that occurs at higher temperatures. Thus the prime candidates for investigation are superheater and reheater outlet headers, main steam and hot reheat piping, weld zones, and the tubes that operate at the highest temperatures. Although tubes have relatively thin walls and would thus appear to be less of a concern, experience has shown that they are responsible for most outages. Thus even though a failed tube can usually be replaced promptly with readily available material, frequent and recurrent tube failures indicate the need for a major tube replacement program, tube redesign, or a change in operating conditions. The principal difference between tubes and heavy-walled components such as headers, so far as thermal stress is concerned, is that the tubes are usually transferring heat, while the headers are not. In the tubes, a large radial temperature gradient causes high steady state thermal stress in the wall, which over time can add to the creep damage caused by pressure stress. On the other hand, thermal stress in walls of heavier components (which are typically outside the hot gas stream and thus not transferring heat) is more likely to be caused by temperature transients of relatively short duration, such as those occurring during start-up. In such cases the potential damage mechanism is more likely to be fatigue.

Fatigue damage leading to premature failure is sometimes found in parts of the feedwater piping where intermittent flow causes thermal cycling such as at economizer inlet header nozzles or where flow from the economizer enters the drum. As described above in the section on fatigue, thick-walled components such as superheater headers can develop significant thermal stress and possibly fatigue damage due to thermal transients in the fluid passing through them, which occur during rapid start-up or shutdown of the boiler. For components operating in the creep range (e.g., superheater outlet header, main steam piping) fatigue damage interacts with creep damage, accelerating the process of component life expenditure.

On completion of a screening process to identify components likely to limit extended service life, two general approaches are used for their study and evaluation: **history based methods** and so-called **postservice examinations**. The first of these involves trying to establish the historical record of operating temperature, pressure, and stress to which each component has been subjected. Then by simple stress calculations and conservative (minimum) estimates of material creep rupture strength, an assessment is made of the life fraction that has been expended by using the linear damage rule described above in the discussion of creep. This approach tends to be conservative, but it has some appeal in identifying components likely to require close monitoring during future inspections for signs of damage or incipient failure. Among the drawbacks of history-based methods are uncertainties in the temperature and stress history, scatter in stress rupture data, and questions as to the validity of a simple life fraction rule.

The postservice examination method is an approach in which various techniques are used to assess the remaining life of components at the time the assessment is made, i.e., at the current age of the equipment. Among the techniques used are tests of samples taken from suspect locations and tests of the material in place. For example, accelerated creep rupture tests can be conducted on small boat samples by testing at higher temperature and comparing results with those from similar tests of unused material.

Another method used to assess life fraction expenditure (and thus remaining life) is based on the slow build-up of an oxide layer on the steam side of pressure parts in elevated-temperature service. The growth of this scale has been found to correlate directly with elevated temperature service time and temperature, in a manner analogous to the process of life expenditure associated with creep damage. It is thus possible to find a relationship between time and temperature for scale thickness similar to that discovered by Larson and Miller for creep rupture life under constant stress. In practice, scale thickness is measured microscopically (or in a more recent development, by ultrasonic means). Comparing the measured scale thickness with known data relating scale thickness, time, and temperature, it is possible to estimate an effective service temperature to which the component has been subjected during its service life. Having an effective time, temperature, and stress level, it is then possible from known creep rupture data to assess life expenditure to date and thus remaining life. Fatigue damage must also be accounted for. The process can be refined further by recognizing that an increasing steamside thickness of oxide layer acts as an insulator to raise the effective temperature of the wall. Computer programs have been written to account for this effect as well as for any wall thinning. Both effects accelerate damage, and the programs handle the problem in an iterative manner. Life assessments obtained from oxide film measurements can be supplemented by comparing microstructural changes in the component with standardized samples displaying various degrees of creep damage.

Hardness-based techniques are yet another method for predicting remaining life. These techniques rely on the fact that the strength of low alloy steel changes with elevated-temperature service exposure as a function of time and temperature. Thus a measured change in strength during service (such as a change in hardness) may be used for estimating the mean temperature at which a component has been operating, and thus how much life expenditure has occurred.

Among the nondestructive methods for making microstructural evaluations is a method of **replication** that reproduces the surface microstructure of a component by means of a cellulose acetate tape applied to its surface. The basic technique consists of surface preparation by grinding, polishing, and etching, using standard metallographic procedures, replication of the prepared surface, and laboratory examination of the replica. To aid in final evaluation, the replica can be enhanced with a vapor-deposited layer of gold or carbon. Replication permits a survey of an existing component without disturbing it. The technique is relatively inexpensive and can achieve a high degree of resolution to detect creep damage from its earliest stages. A major limitation is its application to accessible surfaces only.

While replication permits determination of form and structure, it does not identify metallurgical components present on the surface. For this, a new technique is under development in which the surface to be examined is etched for a longer time than usual, so that fragments of the material are removed with the replicating tape. These fragments (carbide precipitates) can then be examined microscopically to determine which precipitates are present, in an attempt to identify what stage of creep damage has been reached. This work is still experimental, in part because there is not yet a sufficient database to relate particular metallurgical findings to the various stages of creep damage.

An alternative or adjunct to the replication technique is the use of a microscope for direct examination of the prepared surface on the component. Because of the less-than-ideal conditions in the field (e.g., vibration from nearby machinery), this technique is limited to a magnification on the order of $400\times$ to $500\times$. From a practical standpoint, this may suffice to identify a problem. If not, the replication techniques permit removal of a surface replica for further study in a laboratory, where greater resolution and magnification are available.

Still another device useful in assessing the condition of boiler components is the borescope. With the aid of a light and optics, a borescope provides a magnified image of otherwise inaccessible interior surfaces. Some advanced borescopes incorporate a miniature camera (a so-called "chip camera") at the end of a flexible probe that can send an electronic signal to a television screen, providing a large, clear image of whatever the camera sees.

As this brief overview shows, a great deal of recent progress has been made in residual-life assessment. The goals of this effort are several: enhancing safety by providing a sound basis for extending operating life, avoiding or reducing costly unscheduled outages, and eliminating unnecessary or premature equipment replacement. However, the difficulties and uncertainties inherent in the various methods should also be recognized, since they can become cumulative. Some that can be mentioned are: scatter in material property data, lack of sufficient material property data (especially elevated temperature fatigue data), uncertainties in past and future operating conditions (particularly temperature conditions), simplified assumptions utilizing uni-axial creep rupture test data and linear damage rules, and the difficulty of assessing the interaction of creep and fatigue damage under varying temperature and multi-axial stress conditions. Because of these many uncertainties, residual life assessments should always include physical evaluation of microstructural changes, to confirm theoretical predictions, to detect creep damage, and to provide a comparative basis for future inspections.

Much development is still needed in the area of residual-life assessment, which continues to be an active field of research. The Electric Power Research Institute, the research organization for the utilities, has sponsored a great deal of work in this field. Among the tools they have developed with the assistance of other organizations is a computer program called Boiler Life Evaluation and Simulation System (BLESS), which can be used to address various problems in remaining-life evaluation.

DETERMINATION OF ALLOWABLE STRESSES

GENERAL PHILOSOPHY

Each section of the Code uses design criteria and design stresses appropriate for the components that it covers. The basis for establishing allowable stresses is related to the design philosophy, the type of construction permitted, the degree of analysis required, and the amount of nondestructive examination mandated. Sections I, IV, and VIII, Division 1, use a design approach sometimes called design-by-rule. Design-by-rule is not based on detailed stress analysis. Instead, the rules generally involve the calculation of average membrane stress across the thickness of the walls of the vessels. That is not to say that these sections of the Code completely ignore other types of stress. There is a classic statement found in UG-23 of Section VIII, Division 1, which provides a rationale for design-by-rule using membrane stress alone: ''It is recognized that high localized discontinuity stresses may exist in vessels designed and fabricated in accordance with these rules. Insofar as practical, design rules for details have been written to limit such stresses to a safe level, consistent with experience.'' In the 1970s the first sentence was slightly different. At that time the term ''high localized and secondary bending stresses'' was used instead of the current description ''high localized discontinuity stresses.'' In either case, bending stress and the effects of stress concentrations are not calculated (at least not explicitly in Section I). This means, in effect, that the Code assigns allowable membrane stresses with sufficient margin (safety factor) to cover high localized stresses in normal circumstances.

The design-by-rule approach has evolved from theory, experiment, and past successful experience. Boilers and pressure vessels thus designed have proven to meet the objectives set forth in the Foreword to all sections of the Code: ''to afford reasonably certain protection of life and property and to provide a margin for deterioration in service so as to give a reasonably long safe period of usefulness.'' This generally outstanding record of safe and reliable service has been compiled under what can be described as normal conditions. Exceptional conditions, such as severe cyclic loading, high thermal stress, or operation below the temperature of transition between brittle and ductile behavior, have led to failures. Fortunately, boilers are rarely subjected to severe cyclical conditions and, except during hydrostatic testing, are operating at temperatures that preclude brittle fracture. Moreover, designers through long experience have developed means and details to cope with thermal stress.

Design stress for Section I (defined in PG-27 as S, the maximum allowable stress value) is an allowable membrane stress. It is established by using appropriate safety factors against the various potential failure modes of the pressure parts of the boiler. For example, gross yielding of a part is one possible failure mode. Except as discussed below, Section I uses a factor of safety of 1.5 on this failure mode, i.e., the allowable stress does not exceed 2/3 of the yield strength of the material. Similarly, a factor of safety (sometimes called a design margin) against bursting is provided by a factor of approximately 4 with respect to the ultimate tensile strength of the material. Thus the allowable stress may not exceed about 1/4 of the tensile strength of the material. Other factors are applied to stress rupture strength and creep rate at high temperatures where time-dependent deformation occurs.

The Section I basis for determination of allowable stress is explained in Section II, Part D, Appendix 1, where the various safety factors, or design factors, and quality factors applied to the various tensile, yield, and creep strengths are presented in Table 1-100. (For ready reference this two-page appendix is reproduced at the end of this chapter.) Appendix 1 also notes that the committee is guided by successful experience in service. Such evidence is considered equivalent to test data where operating conditions are known with reasonable certainty. This philosophy is used to make judgments in establishing stresses for variations of existing material or new materials that are similar to existing materials. Today, requests for the adoption of new materials are typically supported by a great deal of test data, much of it the result of elevated-temperature testing.

ESTABLISHING ALLOWABLE STRESSES

To establish the allowable stresses, certain information over the temperature range of intended use must be known: the tensile strength, the yield strength, the stress that causes a secondary creep rate of 0.01% per 1000 hours, and the stress to cause rupture in 100,000 hours. These data are solicited from all sources available to the Subcommittee on Materials (Subcommittee II), formerly known as the Subcommittee on Properties of Metals, the committee responsible for developing allowable stresses. The data are then plotted for the temperature ranges for which the material will be used. There is a standard procedure for normalizing the tensile and yield strength data being studied which involves the ratio of the elevated-temperature strength of the individual lots to their corresponding strength at room temperature. The resulting ratios may then be evaluated by the least-squares procedure to define an average ratio trend curve. Ratio trend curves are then plotted of the tensile strength and yield strength from room temperature to elevated temperature.

The ratio trend curves are then multiplied by the minimum room temperature yield and tensile strength of a particular specification to produce curves of expected strength anchored to those minimum specified strengths. In the high-temperature (creep) region, the creep test data are treated directly, without a normalization technique. A variety of techniques, including time–temperature parametric analysis, is employed to estimate the 100,000 hour stress rupture strength and the 0.01%/1000 hr creep rate. Curves for both the average and minimum stress to produce stress rupture in 100,000 hours are plotted. The tensile strength trend curve is then graphically reduced by a safety factor of about 4, and the yield strength trend curve is reduced by a safety factor of 1.5. The average stress to produce a creep rate of 0.01% per 1000 hours is used directly with no factor applied to it. Factors of 67% and 80%, respectively, are applied to the average and minimum stress to cause stress rupture in 100,000 hours. The tensile strength, yield strength, creep rate, and creep rupture strength curves with their respective safety factors applied are then used to generate a single smooth curve representing the lowest values of these curves over the temperature range being considered. That curve is the allowable stress curve.

PENALTIES ON ALLOWABLE STRESSES

There are some materials for which the allowable stresses are made slightly lower than those just described. For example, an additional penalty of 15% applies to longitudinally welded pipe or tube. This is apparently because historically the committee has lacked full confidence in the quality of the weld and the reliability of the NDE methods used to inspect it. Another category, subject to an 8% penalty, is structural quality ferrous material used in pressure-retaining applications. This penalty is an old one, dating from a time when structural quality may have been a lower quality than so-called pressure vessel quality. It happens that specification SA-6, which provides general requirements for structural materials, calls for somewhat less comprehensive testing than does the comparable specification SA-20, which covers general requirements for steel plates for pressure vessels. However, the manufacturing processes have improved so much over the years that this penalty may no longer be justified.

By the very nature of their manufacture, steel castings are more likely to have various imperfections than their wrought steel counterparts. PG-25 imposes what it calls a quality factor for steel castings unless all critical areas and weld end preparations pass certain nondestructive examinations outlined in that paragraph. The quality factor is 80%, i.e., a 20% penalty must be applied by the designer to the allowable stress published in Section II, Part D. If the castings pass the required NDE, no penalty need be applied to the allowable stress. Should unacceptable imperfections be discovered, they must be repaired by welding and given a postweld heat treatment. The designer may then use the full allowable stress. However, often a designer may elect to take a 20% penalty on allowable stress and avoid the time, expense, and bother entailed in the fairly extensive NDE required. This can lead to the following situation.

Sometimes a large cast part, such as a pump casing, is welded to boiler piping by a large-diameter circumferential weld. That weld requires radiography, which can reveal unacceptable flaws in the casting adjacent to the weld. Even though the pump designer had decided to accept a 20% stress penalty and forego NDE of the casting, it is generally agreed that an unacceptable flaw must be repaired. Which party (e.g., the pump manufacturer, the boiler manufacturer, or the purchaser of the boiler) has responsibility for the costs of the repair would depend on the contracts and specifications that had been agreed to, and perhaps legal issues such as merchantability, implied warranties, and fitness for service.

From the criteria used to set allowable stresses at elevated temperature, it is seen that Section I permits some high-temperature relaxation or creep. This is based on the philosophy of providing an adequate boiler design life within reasonable economic limits, since experience has shown that some small amount of permanent (creep) strain is not harmful in most circumstances.

Another example of this practical approach occurs in the temperature range in which tensile or yield strength governs the determination of stress (i.e., the temperature range in which creep phenomena do not control the choice of allowable stress). In that range, Section I permits stresses higher than 2/3 the yield strength for some nonferrous and austenitic materials if some slight permanent deformation (plastic strain) is not objectionable. The use of design stresses relatively closer to the yield strength of these materials can be justified by the fact that these alloys work harden on yielding (they grow stronger) and also that they have a very large margin between yield strength and ultimate tensile strength, i.e., they are very ductile. These alternative higher stresses are identified by a footnote in the Section II, Part D tables. These higher stresses are satisfactory for tubes or pipe, but are not recommended for the design of flanges or other applications where a small amount of creep relaxation or permanent strain might cause leaks or other problems.

DESIGN STRESSES: PAST, PRESENT, AND FUTURE

The so-called safety factors, or design margins, now used in establishing allowable stresses with respect to the various failure modes, such as yielding or creep rupture, have evolved over the life of the Code. Before World War II the factor used on tensile strength was 5. It was changed to 4 in order to save steel during the war. Starting in the late 1970s, the factor on yield strength was changed from 5/8 to 2/3, a change that was carried out over quite a long period. The factor on the 100,000-hour creep rupture strength was formerly 0.6. Around 1970, this was changed to the current factor of 0.67. These reductions in design margins, or safety factors, were adopted over time as improvements in technology permitted. These improvements included the development of newer and more reliable methods of analysis, design, and nondestructive examination. The imposition of quality control systems in 1973 and a record of long satisfactory experience also helped justify reducing some of the design conservatism.

One of the design factors the ASME uses in setting allowable stress not used by most other countries is the factor of approximately 4 on ultimate tensile strength. It happens that this design factor is a significant one, because it controls the allowable stress for many ferritic (carbon and low alloy) steels below the creep range. This has put users of the ASME Code at a disadvantage in world markets where competing designs are able to utilize higher allowable stresses based just on yield strength. This situation has caused the Code committees to reconsider the usefulness and necessity of using tensile strength as one of the criteria for setting allowable stress. In 1996 the Pressure Vessel Research Council (PVRC), a research group closely associated with the Code committees, was asked to study whether the design factor on tensile strength could safely be reduced. The PVRC prepared a report reviewing all the technological improvements in boiler and pressure vessel construction that have occurred since the early 1940s, which was when the design factor on tensile strength was last reduced, from 5 to 4. On the basis of that report's favorable recommendation, Subcommittee VIII has decided, as an initial step, to change the factor on tensile strength from about 4 to about 3.5 for pressure vessels constructed under the provisions of Section VIII, Division 1.

Subcommittee I decided to make the same change for Section I and in 1997 established a task group to investigate the potential effects of such a change and how best to implement it. That task group concluded that Section I could safely join Section VIII in increasing its allowable stresses. The actual mechanics of setting and publishing new allowable stresses took Subcommittee II some time, because it was quite a task. In order to expedite the process while Subcommittee II completed its work, new stresses for a limited group of materials were introduced by means of Code cases, one for Section I and two for Section VIII, since Code cases can be issued far more quickly than Code Addenda. The Section I case is Case 2284, Alternative Maximum Allowable Stresses for Section I Construction Based on a Factor of 3.5 on Tensile Strength. The three cases were approved for use in mid-1998, and their higher stress values are expected to be incorporated into Section II, Part D in the near future, perhaps in the 1999 Addenda. One problem with the new cases is that some jurisdictions were reluctant to accept them, or had no ready mechanism which would permit their prompt adoption. Thus widespread use of the higher stresses may have to await their incorporation into Section II, Part D.

In the future, it is possible that the design factor on tensile strength may be reduced further or eliminated altogether, depending on the results of these first steps. Note that the change in design factor applied to tensile strength from 1/4 to 1/3.5 is a 14% change. However few allowable stresses will change that much, because the higher allowable stress will probably be determined and controlled by the factor applied to yield strength, rather than tensile strength. Also note that the higher stresses will be applicable below the creep range only.

The above-described methods of setting maximum allowable (design) stress values have been used by the Code committees since the mid-1950s. During that time, new data have been obtained and analyzed to revise design stresses as appropriate, based both on new laboratory tests and reported experience from equipment in service. There have been times when the analysis of new data has resulted in a significant lowering of the allowable stresses at elevated temperature. In all but a few instances, however, the fine safety record of equipment built to the ASME Code has demonstrated the validity of the material data evaluation, design criteria, and design methods used.

APPENDIX 1 OF SECTION II, PART D
Basis for Establishing Stress Values in Tables 1A and 1B
(Reproduced From ASME Section II)

1-100

In the determination of allowable stress values for pressure parts, the Committee is guided by successful experience in service, insofar as evidence of satisfactory performance is available. Such evidence is considered equivalent to test data where operating conditions are known with reasonable certainty. In the evaluation of new materials, the Committee is guided to a certain extent by the comparison of test information with available data on successful applications of similar materials. These values are established by the Committee only.

Nomenclature

S_T = specified minimum tensile strength at room temperature, ksi

R_T = ratio of the average temperature dependent trend curve value of tensile strength to the room temperature tensile strength

S_Y = specified minimum yield strength at room temperature, ksi

R_Y = ratio of the average temperature dependent trend curve value of yield strength to the room temperature yield strength

S_{Ravg} = average stress to cause rupture at the end of 100,000 hr

S_{Rmin} = minimum stress to cause rupture at the end of 100,000 hr

S_C = average stress to produce a creep rate of 0.01%/1000 hr

NA = not applicable

The maximum allowable stress shall be the lowest value obtained from the criteria in Table 1-100. The mechanical properties considered, and the factors applied to establish the maximum allowable stresses, are as given below.

(a) At temperatures below the range where creep and stress rupture strength govern the selection of stresses, the maximum allowable stress value is the lowest of the following:

(1) one-fourth of the specified minimum tensile strength at room temperature;

(2) one-fourth of the tensile strength at temperature;

(3) two-thirds of the specified minimum yield strength at room temperature;

(4) two-thirds of the yield strength at temperature.

In the application of these criteria, the Committee considers the yield strength at temperature to be $S_Y R_Y$, and the tensile strength at temperature to be $1.1 S_T R_T$.

Two sets of allowable stress values are provided in Tables 1A and 1B for austenitic materials and specific nonferrous alloys. The higher alternative allowable stresses are identified by a footnote to the tables. These stresses exceed two-thirds but do not exceed 90% of the minimum yield strength at temperature. The higher stress values should be used only where slightly higher deformation is not in itself objectionable. These higher stresses are not recommended for the design of flanges or other strain sensitive applications.

(b) At temperatures in the range where creep and stress rupture strength govern the selection of stresses, the maximum allowable stress value for all materials is established by the Committee not to exceed the lowest of the following:

(1) 100% of the average stress to produce a creep rate of 0.01%/1000 hr;

(2) 67% of the average stress to cause rupture at the end of 100,000 hr;

(3) 80% of the minimum stress to cause rupture at the end of 100,000 hr.

Stress values for high temperatures are based, whenever possible, on representative properties of the materials under laboratory test conditions. The stress values are based on basic properties of the materials and no consideration is given for corrosive environment, for abnormal temperature and stress conditions, or for other design considerations.

(c) For structural materials used for nonpressure retention applications, such as for component and piping supports, the basis for setting stresses is the same as for all other materials. When used for pressure retention applications, such as plates used for pressure vessel shells, and bars and shapes used as shell stiffeners or staybars, the basis for setting allowable stresses is the same as for all other materials, subject to a quality factor of 0.92.

TABLE 1-100
CRITERIA FOR ESTABLISHING ALLOWABLE STRESS VALUES FOR TABLES 1A AND 1B
(REPRODUCED FROM ASME SECTION II, PART D)

Product/Material	Below Room Temperature		Room Temperature and Above						
	Tensile Strength	Yield Strength	Tensile Strength		Yield Strength		Stress Rupture		Creep Rate
Wrought or cast ferrous and nonferrous	$\dfrac{S_T}{4}$	$\tfrac{2}{3}\,S_Y$	$\dfrac{S_T}{4}$	$\dfrac{1.1}{4}\,S_T R_T$	$\tfrac{2}{3}\,S_Y$	$\tfrac{2}{3}\,S_Y R_Y$ or $0.9 S_Y R_Y$ [Note (1)]	$0.67 S_{Ravg}$	$0.8 S_{Rmin}$	$1.05 S_c$
Welded pipe or tube, ferrous and nonferrous	$\dfrac{0.85}{4}\,S_T$	$\tfrac{2}{3}\times 0.85 S_Y$	$\dfrac{0.85}{4}\,S_T$	$\dfrac{(1.1\times 0.85)}{4}\,S_T R_T$	$\tfrac{2}{3}\times 0.85 S_Y$	$\tfrac{2}{3}\times 0.85 S_Y R_Y$ or $0.9\times 0.85 S_Y R_Y$ [Note (1)]	$(0.67\times 0.85)\,S_{Ravg}$	$(0.8\times 0.85)\,S_{Rmin}$	$0.85 S_c$
Ferrous materials, structural quality for pressure retention applications [Note (2)]	$\dfrac{0.92}{4}\,S_T$	$\tfrac{2}{3}\times 0.92 S_Y$	$\dfrac{0.92}{4}\,S_T$	$\dfrac{(1.1\times 0.92)}{4}\,S_T R_T$	$\tfrac{2}{3}\times 0.92 S_Y$	$\tfrac{2}{3}\times 0.92 S_Y R_Y$	NA	NA	NA
Ferrous materials, structural quality for nonpressure retention functions	$\dfrac{S_T}{4}$	$\tfrac{2}{3}\,S_Y$	$\dfrac{S_T}{4}$	$\dfrac{1.1}{4}\,S_T$	$\tfrac{2}{3}\,S_Y$	$\tfrac{2}{3}\,S_Y R_Y$	NA	NA	NA

NOTES:
(1) Two sets of allowable stress values are provided in Table 1A for austenitic materials and in Table 1B for specific nonferrous alloys. The higher alternative allowable stresses are identified by a footnote. These stresses exceed two-thirds but do not exceed 90% of the minimum yield strength at temperature. The higher stress values should be used only where slightly higher deformation is not in itself objectionable. These higher stresses are not recommended for the design of flanges or other strain sensitive applications.
(2) The value of S_T shall be in accordance with the material specifications.

CHAPTER

16

RULES COVERING BOILERS
IN SERVICE

INTRODUCTION

As explained in Chapter 9, the signing of the completed Manufacturer's Data Report by the Authorized Inspector and the stamping of the boiler with a Code symbol stamp mark the completion of the boiler. The manufacturer has met its responsibility to supply a boiler which complies with the rules of Section I. Since Section I applies to new construction only, by its own definition in the Foreword, its direct applicability to the boiler has ended. It then becomes the responsibility of the user to maintain the boiler in accordance with requirements of the jurisdiction where the boiler is installed. As can be seen on the map in Chapter 2 showing jurisdictions in the U.S. and Canada with boiler and pressure laws, there are still a few states that have no boiler laws. Besides the U.S. and Canada, only a few industrialized nations have requirements comparable to Section I for new boiler construction. For boilers in service, requirements for inspection, maintenance, repair, and alteration vary considerably. Some jurisdictions have specific requirements or invoke those of the National Board Inspection Code (NBIC), which is described below. In some domestic and foreign locations, there are almost no such requirements. These jurisdictions manage without detailed rules by relying on the self-interest of the boiler owners and their insurance companies in inspecting and maintaining boilers in service so that the safety of the operators and the public is protected. Most of the time, this approach is successful, particularly if the boiler is constructed in accordance with Section I in the first place. Unfortunately, however, there are instances each year when lack of proper operating, maintenance, or repair procedures leads to serious accidents. The occurrence of such accidents in the 19th century prompted early legislation to promote boiler safety and eventually culminated in the development of the first edition of Section I in 1915, which comprised what were, in effect, safety rules for the construction of steam boilers.

As early as 1833, Congress passed a law establishing a steamboat inspection service, requiring steamboat boilers to be inspected every six months, and since those early days of steam, the inspector has played a key role in promoting and assuring boiler safety. Gradually, the various jurisdictions that passed laws regulating boilers established organizations to administer and enforce those laws, which typically cover both new construction and boilers in service.

Recall now the advice in the Foreword to all sections of the ASME Code that its design rules provide a margin for deterioration in service so as to give a reasonably long, safe period of usefulness. Accordingly, it is the general industry practice that the Authorized Inspection Agencies (the insurance companies) continue to accept and insure boilers in service where tube wall thicknesses have been reduced below their original design values due to wastage or corrosion, typically on the fire side. In such cases, the tubes are monitored to establish the rate at which the walls are thinning, and repair or replacement is usually accomplished before the tubes reach 50% to 60% of their original thickness. The decision on when to replace a thinning component is one normally made by the boiler owner in conjunction with his Authorized Inspection Agency or insurer. Unfortunately, it is hard to find published recommendations regarding what minimum wall thickness is still safe, because of the potential liability involved. For Kraft process black liquor chemical recovery boilers used in the pulp and paper industry, a relatively smaller amount of wall thinning may be permitted, because of the potential for an explosion caused by water leaks into the chemical recovery (smelt) bed.

ASME CODE EDITION APPLICABLE TO REPAIRS AND ALTERATIONS

One of the first questions to be decided when making repairs or alterations to an old boiler is which edition of Section I should be used—the edition which was used originally, the latest edition, or perhaps some intermediate edition. Although it is often simpler and easier to use the latest edition of Section I, repairs and alterations can, and often are, made to the rules of older editions. To provide some perspective on this issue, it should be noted that Section I is an old and settled Code, where changes now are incremental, with few significant changes from one edition to the next. Thus, in general, it makes little practical difference, for example, whether a replacement header is built to the current edition of Section I or to an edition from 20 years ago. This is not always the case, however; occasionally, allowable stress values are reduced, or calculation methods are made more conservative. (The opposite situation has also occurred: e.g., allowable stresses have been increased.)

Perhaps the next question is whether any Code or other authority forces a boiler owner, or a boiler manufacturer, to use any particular edition of Section I for repairs or alterations. The answer to this question is, with some qualification, no. The National Board Inspection Code (NBIC), in its introduction to the chapter on repairs and alterations (paragraph RC-1020) offers this general advice:

> *Where the standard governing the original construction is the ASME Code, repairs and alterations shall conform, insofar as possible, to the section and edition of the ASME Code most applicable to the work planned.*
>
> *When the standard governing the original construction is not the ASME Code, repairs and alterations shall conform, insofar as possible, to the edition of the construction standard or specification most applicable to the work. Where this is not possible or practicable, it is permissible to use other codes, standards, or specifications, including the ASME Code, provided the R Certificate holder has the concurrence of the Inspector and the jurisdiction where the pressure retaining item is installed.*

The language of the first paragraph just quoted (''shall conform insofar as possible'' and ''section and edition of the ASME Code most applicable to the work planned''), leaves considerable leeway in the choice of Code edition to be applied to repairs and alterations. Even more choice is provided when the the original construction code was not the ASME Code. Note that the concurrence of the Inspector is needed, and also that the laws of the jurisdiction might preclude the use of an obsolete design.

Although this NBIC advice may not seem definitive, it recognizes certain long-standing industry practices. One of these is what is known as ''replacement in kind,'' in which an old part is replaced with a new duplicate. The logic of this procedure is that the old part gave good, safe service for a long time; its duplicate should do the same. Consequently, building to the original Code edition is usually justified. (The old part was safe originally and gave years of good service; a new duplicate should be safe now.) By extension, the same philosophy could apply to alterations of an old boiler. That is, a case might be made for using the original Code edition for an alteration, if it were particularly advantageous to do so.

However, there are occasions when replacement in kind may not be prudent. Consider the example of a replacement for a P-11 (1¼ Cr-½ Mo) reheat outlet header designed in the early 1960s for a design temperature of 1000°F. The allowable stress at that temperature is now about 20% lower than in 1964

(6300 psi now, 7800 psi then). If a manufacturer (an S stamp holder) is asked for a replacement in kind or is asked to "build to print" (i.e., duplicate the original part in accordance with a drawing of the original), it is placed in a difficult position. On one hand, the manufacturer would simply be replacing (duplicating) a part that gave satisfactory service for almost 30 years. On the other hand, the manufacturer should know that the allowable stress is now 20% lower, and thus a duplicate part at design conditions would be 25% overstressed by today's standards. Even though most parts seldom operate at design conditions, such a situation presents the possibility that the replacement header could suffer a premature failure due to creep rupture, with potentially serious consequences to all concerned.

The jurisdiction may have rules dealing with the above-described situation. For example, Section 9(1) of the Boilers and Pressure Vessels Act of the Province of Alberta has a section dealing with "unsafe and obsolete designs," which stipulates that when the Chief Inspector determines that a design no longer meets the requirements of the regulations, he can forbid further new construction employing that obsolete design. The Alberta law doesn't explicitly address continued use of an old, obsolete design, and such use is apparently permitted, unless the provincial authorities were to rule the design unsafe.

Usually the choice of which edition of Section I to apply to a repair or alteration is uncomplicated. It is often most convenient to use the latest edition. Where design methods and stresses have changed, careful consultation among the parties (the owner/user, the parts manufacturer, the Authorized Inspection Agency, and the jurisdiction) may be needed to justify construction to an earlier edition of Section I.

Notwithstanding the availability of the comprehensive National Board system of rules covering repairs and alterations to boilers and pressure vessels, a jurisdiction may authorize organizations to engage in repairs and alterations of boilers and pressure vessels based on such (perhaps less stringent) criteria as the jurisdiction deems appropriate. Remember that the jurisdiction has the legal authority to regulate repairs and alterations. In many states and provinces, the NBIC is acceptable, but nonmandatory.

The question of what construction code to use for repairs and alterations has an interesting variation: What edition of the National Board Inspection Code may be used to cover the repair or alteration? NBIC Interpretation 95-20 addressed this:

> Question: *May the requirements of an earlier Edition and Addenda of the National Board Inspection Code be used when performing a repair or alteration?*
> Reply: *Yes.*

It happens that the National Board Inspection Code changed very slowly from the late 1970s until the early 1990s, but since then has undergone fairly rapid expansion and change, which may continue. There are some states that for one reason or other haven't modified their rules and regulations to call for the latest edition and addenda of the NBIC. They may still be calling for repairs and alterations to be done under an earlier edition, which may differ considerably from the latest version. (This fact may also explain why the National Board permits the use of earlier editions of the NBIC.) Accordingly, an organization contemplating repairs or alterations in a given jurisdiction should investigate which edition, if any, is mandatory, and plan to follow it. Otherwise, it may be possible to use any edition.

JURISDICTIONAL REQUIREMENTS

Jurisdictional requirements covering boilers in service are found in so-called "boiler laws." As explained in Chapter 2, the Uniform Boiler and Pressure Vessel Laws Society provides model legislation covering boilers and pressure vessels to jurisdictions and supports the adoption of nationally accepted Codes such as Section I and Section VIII as standards for the construction of boilers or pressure vessels.

In addition to requiring boilers to be constructed to Section I, typical boiler laws mandate periodic inspection of boilers (usually annually) in order to maintain the boiler's operating certificate. Inspections may be made by an Authorized Inspector (AI) from the jurisdiction or from an Authorized Inspection Agency authorized to operate in that jurisdiction. The jurisdiction or insurance company (AIA) may have regulations covering the Inspector's duties for boiler inspection. If not, the NBIC (see next section) provides an excellent guide covering general precautions, inspection of pressure parts, piping, and appurtenances such as safety valves, low water fuel cutoffs, pressure gages, etc., and hydrostatic testing if deemed necessary.

On completion of the inspection, the AI reports his or her findings to the owner with recommendations for correcting any deficiencies. Should the Inspector discover an unsafe condition or serious safety violation (e.g., inoperative safety valves), the AI is empowered by the jurisdiction to revoke or refuse to renew the boiler's operating certificate pending necessary corrective action.

Loss of an operating certificate is a very heavy penalty. Should an owner willfully continue to operate without a certificate, he would be violating laws of the jurisdiction and most likely the terms of his insurance coverage. If anyone were injured as a consequence of this continued operation, the owner's liability could be great.

THE NATIONAL BOARD

In 1919, the National Board of Boiler and Pressure Vessel Inspectors was formed as part of an effort to achieve widespread adoption and uniform enforcement of the then relatively new ASME Boiler Code. The National Board is composed of chief inspectors of states and municipalities in the United States and of the provinces of Canada that have adopted one or more sections of the ASME Boiler and Pressure Vessel Code. The National Board works closely with the ASME Boiler and Pressure Vessel Committee. In addition to the advisory role of the Conference Committee described in Appendix II, many members of the National Board are members of various Code committees and participate in their work.

The important role of the Authorized Inspector in assuring compliance with the rules of Section I was explained in Chapter 8. Equally important is the Inspector's role in maintaining the integrity of the boiler after it is placed in service. The National Board, as a leader in this effort, has developed a manual called the National Board Inspection Code for boiler and pressure vessel Inspectors. The stated purpose of the NBIC is to "maintain the integrity of pressure retaining items after they have been placed into service by providing rules and guidelines for inspection, repair, and alteration, thereby ensuring that these objects may continue to be safely used." A guiding principle of the NBIC is to continue to follow the rules of the ASME B&PV Code insofar as practicable. For example, the National Board requires that any welding used in repairs or alterations be done in accordance with the same strict standards Section I requires, using welders and weld procedures qualified in accordance with Section IX, except that since 1995 the NBIC has permitted the use of some ANSI/AWS standard welding procedures. The National Board Inspection Code also requires that all the safety appurtenances of a Section I boiler (e.g., safety valves, gages, etc.) be maintained in good order. Thus it could be said that the National Board Inspection Code takes up where Section I leaves off, except that unlike Section I, the National Board Inspection Code is not mandatory in many jurisdictions (although it is widely accepted). Considering that several states still don't mandate boiler construction in accordance with Section I, it is hardly surprising that many states do not consider it necessary to mandate the use of the NBIC for repairs and alterations of boilers and pressure vessels. The Uniform Boiler and Pressure Vessel Laws Society (see Chapter 2) and the National Board both continue their efforts to convince all the jurisdictions to adopt and enforce the ASME Code. Also, the National Board has had considerable success in getting the jurisdictions to mandate the use of the NBIC.

REPAIRS AND ALTERATIONS UNDER THE NBIC

The National Board Inspection Code rules for repairs and alterations are consistent with the ASME Code requirements for new construction, in mandating quality control systems and third-party inspection to assure safe construction and compliance with governing codes or regulations. The 1992 edition of the National Board Inspection Code included a number of significant changes. One of these is that the NBIC now makes a distinction between inspectors involved in new ASME code construction and those used for repairs and alterations under the NBIC. The latter are now called Inspectors (as opposed to Authorized Inspectors) and must have a Certificate of Competency to perform in-service repair and alteration inspections, must be regularly employed by an Authorized Inspection agency, and must have a commission from the National Board. As a general rule, no repair or alteration may be undertaken without the authorization of an Inspector, who must be satisfied that the welders and weld procedures are qualified, and the repair or alteration methods are acceptable. (A few exceptions to this rule are permitted for limited, routine repairs if the Inspector gives prior approval and is advised of any such repairs.)

The scope of the 1995 Edition of the National Board Inspection Code was broadened to cover the repair and alteration of boilers, pressure vessels, pressure piping, and overpressure protection devices, described collectively as **pressure-containing items** (PCI). As explained in the preceding section of this chapter, the NBIC invokes the code of original construction and applies the NBIC Repair R (repair) symbol stamp program (see below) for repair and alteration methods, inspection, and documentation. The code of original construction is no longer limited to the ASME Code; it may be any code or standard (worldwide) to which the component was originally constructed. The R stamp holder, with the concurrence of the Inspector and the jurisdiction where the pressure-containing item is installed, may select the code edition and addenda to which the work will be performed.

Repairs can vary from minor and routine weld repair or replacement of tubes to replacement of major components. A **repair** is defined by the National Board as the work necessary to restore a component or system to a safe and satisfactory operating condition, such that existing design requirements are met.

An **alteration** is defined as any change to an item described on the original Manufacturer's Data Report which affects its pressure-containing capability. A physical change such as adding or removing heating surface is an alteration. A change in maximum allowable working pressure or design temperature is also considered an alteration, even if no physical change is made. Examples of repairs and alterations are given in Appendix 6 of the NBIC.

Another key difference between a repair and an alteration is that the original design calculations are considered to remain valid for a repair, since the component is supposedly just being restored to a safe condition. For an alteration, this is not the case; something new is involved. Therefore, someone must assume design responsibility for the alteration and furnish design calculations for the altered components. There are a number of ways this used to be accomplished. For example, until the NBIC rules were changed by the publication of the 1992 Edition, it was possible for the holder of an appropriate ASME Code symbol stamp to take design responsibility for an alteration. That ASME Code symbol holder was allowed to use the National Board Form R-1, Record of Welded Repair or Alteration, to certify that any design changes associated with the alteration met the requirements of the NBIC. This is no longer the case; all repairs and alterations documented on any of the National Board forms (there are now four such forms for non-nuclear pressure-retaining items) must now be made only by an organization authorized to use the National Board R symbol stamp. Paragraph RC-3021 now stipulates that design calculations covering alterations must be completed by an organization experienced in design under the code or standard used for the alteration. These calculations must be made available for review by the Inspector accepting the design.

THE NATIONAL BOARD REPAIR STAMP PROGRAM

This program covers both repairs and alterations. The following discussion deals first with repairs and then with alterations. As noted in the preceding section, the significant difference between repairs and alterations has to do with assuring that some appropriate organization takes design responsibility for the alteration. The NBIC system for repairs and alterations is a direct parallel to the ASME system of Code construction. The ASME requires organizations engaging in Code work to have first received an ASME Certificate of Authorization to use one of the Code symbol stamps. In similar manner, the National Board R Symbol Stamp program requires organizations engaging in repair of boilers or pressure vessels to obtain a National Board Certificate of Authorization to use the R symbol stamp. Such organizations are then designated as repair organizations. Repair organizations are not limited to manufacturers and assemblers of ASME Code items; contractors and owner-users may also become repair organizations.

Repair organizations are accredited by the National Board after appropriate review of their quality control systems (which the NBIC calls Quality Systems), procedures, and facilities. This review is conducted by a National Board member jurisdiction where the facilities are located, unless that jurisdiction is the repair organization's inspection agency of record or elects not to perform the review, in which case the review is conducted by a representative of the National Board. This is also the case where there is no jurisdictional authority. Repair organizations must have in force at all times a contract with an Authorized Inspection Agency of Record to participate in reviews of the organization's quality control system for issuance or renewal of the Repair Authorization Certificate and to monitor compliance with the NBIC. Also, any changes to the QC system are subject to the acceptance of the Authorized Inspection Agency (AIA) of record.

Some aspects of the National Board R stamp program that differ from the analogous ASME program are worth noting. NBIC RA-2151 outlines the requirements for a quality system that National Board R symbol stamp holders must have. That system must include controls for repairs and alterations. The methods for performing and documenting repairs and alterations must be described. The R stamp holder must obtain the prior acceptance by the Inspector of the methods of repair or alteration intended to be used. Also, the quality control manual must describe by ASME Code section (or other original code of construction) the scope and type of repairs or alterations the organization is capable of and intends to make. Unlike the ASME Code, the NBIC since 1995 has permitted organizations making repairs and alterations to use quite a number of ANSI/AWS standard welding procedures instead of qualifying those procedures themselves. A list of these procedures is found in Appendix A of the NBIC. Finally, Part RD of the NBIC provides three welding methods that may be used as alternatives to postweld heat treatment, since PWHT may be inadvisable or impractical in some circumstances for equipment in service.

PRESSURE TESTS FOLLOWING REPAIRS AND ALTERATIONS

The QC system must also call for any pressure tests required at the completion of a repair. For repairs, these tests used to be at the discretion of the AI and, if required, would not exceed the MAWP of the boiler. This approach was modified by the 1995 Addenda to the NBIC to provide more consistency and control. The modified rules, which permit some exceptions, are found in RC-2050. Welded repairs must be subjected to pressure tests at 80% of the maximum allowable working pressure or the operating pressure, whichever is greater, using water or another liquid. If replacement parts are installed, they must be tested at their maximum allowable working pressure. In either case, if the liquid used in the pressure test might contaminate the pressure parts, or when pressure testing is not practical, two alternatives are permitted: With the concurrence of the Inspector, the owner, and the jurisdictional authority where required, a pneumatic test may be substituted for a hydrostatic test. (Suitable precautions must be taken when using pneumatic tests, which are inherently more dangerous than tests using an incompressible fluid.) Another alternative, also requiring the prior concurrence of the Inspector and sometimes the jurisdiction, is to forego a pressure test and substitute NDE methods that verify the integrity of the repair. The question of who decides whether a pressure test of a repair is practical or not was addressed in NBIC Interpretation 95-28. That interpretation advised that the R certificate holder has that responsibility, as stipulated in RC-2050(a). Presumably, the same advice would apply to the pressure testing of alterations.

The requirements for pressure tests following alterations are similar to those used following repairs. It used to be that these tests were done at 1.5 MAWP, as for new construction, but because of a concern that certain old boilers or other pressure-retaining items might no longer be able to sustain such a high pressure, the pressure test rules have been relaxed somewhat, to provide alternatives to a full hydrostatic test. RC-3030 calls for a pressure test as required by the original code of construction, which would usually be at 1.5 times the maximum allowable working pressure, but allows exceptions. With the prior acceptance of the Inspector and the jurisdiction, lower pressure tests (at MAWP) are permitted for welds attaching new replacement parts to existing parts. Or, the pressure tests can be waived altogether, provided those welds are subjected to an NDE method that verifies their integrity. As in the case of repairs, if the liquid used in the pressure test might contaminate the pressure parts, or if the R certificate holder considers it impractical to conduct a pressure test, two alternatives are permitted: Again with the concurrence of the Inspector, the owner, and the jurisdictional authority where required, a pneumatic test may be substituted for a hydrostatic test. The other alternative, also requiring the prior concurrence of the Inspector and (where required) the jurisdiction, is to forego a pressure test and substitute NDE methods that verify the integrity of the repair.

The rest of what the NBIC calls the quality system is similar to the Section I QC system. After a review shows the repair organization has met all the above-described requirements, the National Board grants a Repair Certificate of Authorization, that is, authorization to use the NB repair symbol stamp, the R stamp.

REPAIR DOCUMENTATION AND STAMPING

The repair organization must document the repair on National Board Form R-1, Report of Welded Repair, a form analogous to the ASME Manufacturer's Data Report. This form provides a summary of certain important information identifying the repair organization, the owner of the boiler, its location, original manufacturer, and identifying numbers. The NBIC edition and addenda used for the repair and the original construction code must also be identified. Space is provided to describe the repair work. As on the ASME forms, Manufacturer's Partial Data Reports or the equivalent National Board R-3 Forms covering individual replacement parts used in the repair must be listed and attached. A Certificate of Compliance box is provided for the repair organization to certify that the repair complies with the NBIC. A Certificate of Inspection box is provided for the Inspector to certify that he or she has inspected the work and believes it to have been done in accordance with the NBIC. The R-1 form must be distributed to the owner or user, the Inspector, the jurisdiction if required, and the Authorized Inspection Agency responsible for in-service inspection. Since 1995 it has been possible to file R-1 forms reporting repairs with the National Board, but this would be unusual except perhaps for major repairs.

A repair nameplate must be permanently attached near the original nameplate or stamping, marked with the name of the repair organization, its National Board R certificate number, the date of the repairs, and a stamped R symbol. In the case of multiple repairs by the same organization, this plate may be used again, by listing the date of each subsequent repair. Responsibility for the repair rests on the repair organization. Signing and distribution of the R-1 completes the repair.

ROUTINE REPAIRS

Certain repairs are considered to be of a routine nature and must be perfomed under the R certificate holder's quality system program, but need not involve the Inspector during the repair if he has given prior approval. RC-2031 of the NBIC gives examples of routine repairs, such as weld repair or replacement of tubes, or weld build-up of wasted areas not exceeding certain limits in shells and heads. There used to be many more repairs that were considered routine, but due to abuse of the concept, the NBIC now limits this category to only four types. Stamping and nameplates are usually not required for routine repairs.

ALTERATION DOCUMENTATION AND STAMPING

Until the 1992 edition of the NBIC changed the rules, an alteration had to be made and documented by an organization that held an appropriate ASME Code symbol stamp; it wasn't necessary for that organization to have an R stamp. The term ''appropriate stamp'' meant that to alter a boiler, an organization had to have an S stamp, and to alter a pressure vessel, an organization had to have a U stamp. Also, the alteration had to be within the scope of activities permitted by the stamp holder's QC system.

The National Board system is now different. Under the current NBIC rules, repairs and alterations may be made and documented only by R stamp holders, who have now assumed the role of project managers. The question of design responsibility for an alteration is addressed in RC-3020 and RC-3021. Note that it is always possible for an R stamp holder to subcontract the design function (or other aspects of the work) so long as this subcontracting is covered by his quality control system and the R stamp holder retains National Board Code responsibility for the alteration or repair. Note also that it is a relatively simple matter for an ASME Code symbol stamp holder to obtain a Certificate of Authorization for use of the R symbol stamp.

Starting in 1992, certificates of authorization to use the R stamp have had endorsements that define whether the organization is qualified to undertake design responsibility for alterations. The basic endorsements on the face of these certificates authorize either **repairs** or **repairs and/or alterations.** The participation of the Authorized Inspection Agency of Record in the work of the repair organization provides further assurance that the organization will undertake only work for which it is qualified.

Completion of Form R-2, Report of Alteration, is the responsibility of the repair organization. A copy of the original manufacturer's data report if it is available and any manufacturer's partial data reports, or their National Board equivalent Forms R-3, covering new or replacement components must be attached to Form R-2. If required by the jurisdiction, revised calculations covering the alteration must accompany the R-2 Form. The R-2 Form contains information similar to that on the R-1 Form, except that it has four certification boxes, for design, design change review, construction, and inspection. Copies of the R-2 Form with attachments must then be furnished by the R certificate holder responsible for the alteration to the Inspector, the Authorized Inspection Agency responsible for the in-service inspection of the boiler, its owner or user, the jurisdiction if required, and the National Board, if the boiler is registered with the National Board.

A nameplate or stamping adjacent to the original nameplate or stamping is used to record the alteration or rerating of a boiler or pressure vessel. This indicates compliance with the NBIC, and is done by the organization responsible for the preparation of the R-2 Report of Alteration form, but only with the knowledge and authorization of the Inspector. The nameplate or stamping shows either the word ''altered'' or ''rerated,'' the name of the R certificate holder responsible, the MAWP, and the date. Starting with the 1995 edition, it has been mandatory to include the R symbol as part of the stamping. If temperature appeared on the original stamping, it must appear on the alteration stamping. (For Section I items, temperature was required for separately fired superheaters only, until this requirement was deleted in the 1997 Addenda.)

ADVANTAGES OF FOLLOWING THE NBIC

The NBIC is written as a general document intended to encourage uniformity in the administration of rules governing boilers and pressure vessels in service. Because of the varied requirements of the many jurisdictions in the U.S. and Canada, there are occasional minor actual or implied conflicts between the NBIC and local rules, in which case the local rules take precedence. Thus, although every provision of the NBIC may not be mandatory in certain jurisdictions, this code embodies the best practices and procedures that have evolved from many years of experience in the in-service inspection, repair, and alteration of boilers and pressure vessels. It is also a living document, now updated every year on July 1; a new edition is issued every three years. It is thus important to use the current edition when contemplating repairs or alterations. However, there are some jurisdictions that invoke earlier editions of the NBIC, due to their inability to keep their regulations current. As explained earlier in this chapter, it is permissible to use earlier editions of the NBIC.

Some jurisdictions do not mandate full compliance with NBIC rules, particularly with regard to qualifying repair organizations and insistence on authorized inspection. In such cases, should the owner wish to move a repaired boiler or pressure vessel, he might be refused an operating license in a jurisdiction that enforces the NBIC.

Following the NBIC rules provides a continuity of certification and documentation over the life of the equipment, showing that it complies with national safety standards. Compliance with these standards assures long, satisfactory service. Failure to comply increases the owner's liability in the event of an accident.

POSTCONSTRUCTION ACTIVITIES OF THE ASME

Traditionally the ASME through its Boiler and Pressure Vessel Code has concentrated largely on new construction. In the mid-1990s, it became apparent that there was considerable interest in developing or augmenting existing standards dealing with equipment in service. To that end, the ASME established a committee to consider what are called postconstruction issues. That committee is called the Post Construction Committee. It reports to the Board on Pressure Technology Codes and Standards at the same level in the overall codes and standards organization as the Main Committee of the Boiler and Pressure Vessel Committee. While at first glance this committee's field of interest would appear to be similar or redundant to that of the National Board Inspection Code, this is not at all the case, as can be seen from the names of the two "book" subcommittees that the Post Construction Committee has so far established: the Subcommittee on Flaw Evaluation and the Subcommittee on Inspection Planning. Their goal is to apply available modern methods (nondestructive examination, probabalistic risk assessment, fracture mechanics, and fitness for service assessments that have been developed by the nuclear, chemical, and petroleum industries) to the in-service inspection of boilers, pressure vessels, and piping, and to the evaluation of any flaws found. The committee plans to develop standards incorporating the best available validated techniques. These standards will provide guidelines for developing and implementing inspection programs and for evaluating the findings of those inspections. The standards will be suitable for reference by existing in-service inspection codes such as the National Board Inspection Code (NBIC 23) and the American Petroleum Institute's Pressure Vessel Inspection Code (API-510).

One of the first tasks the Post Construction Committee addressed was to determine where the need exists for new postconstruction pressure technology codes and standards. They identified a number of such areas, among which are:

1. Inspection and testing of pressure-relief devices
2. Standardized procedures for incident reporting and investigation
3. Repair techniques
4. Guidelines for bolt tightening
5. NDE as an alternative to pressure testing
6. Inspection, testing, and repair of fiber-reinforced plastic vessels and other nonmetallic equipment
7. Standard creep testing procedures
8. Development of internal procedures for record keeping methods to be included in standards by standard developers

The Post Construction Main Committee also developed a reference list of existing postconstruction standards. The list is drawn from many sources, domestic and international, and includes quite a few ASME standards, e.g., Section VII, Recommended Guidelines for the Care of Power Boilers. The list also includes standards from, among others, the American Petroleum Institute, the Electric Power Research Institute, the U.S. Department of Energy, the Welding Research Council, and the National Association of Corrosion Engineers.

Based on a comparison of the lists of needed and existing standards, the committee selected the two areas mentioned above for standards development: inspection planning and flaw evaluation. The membership of the postconstruction committees includes many engineers now active on the Boiler and Pressure Vessel Committee, and it will be interesting to see the results of their work.

REFERENCES

1. Greene, Arthur M., *History of the ASME Boiler Code*, reprinted from *Mechanical Engineering* articles of 1952 and 1953, ASME, New York.

2. Cross, Wilbur, *The Code, An Authorized History of the ASME Boiler and Pressure Vessel Code*, ASME, New York, 1990.

3. Bernstein, Martin D., "Design Criteria for Boilers and Pressure Vessels in the U.S.A.," *J. Pressure Vessel Technology*, Vol. 110, No. 4, Nov. 1988, pp. 430–443.

4. Moen, Richard A., *Practical Guide to ASME Section II—1996 Materials Index*, Casti Publishing Inc., Edmonton, Alberta, Canada, 1996.

5. *Metals and Alloys in the Unified Numbering System*, 6th Ed., Society of Automotive Engineers, Warrendale, PA, 1993.

6. Timoshenko, S., and Woinowsky-Krieger, S., *Theory of Plates and Shells*, McGraw-Hill, 2nd Ed., New York, 1959.

7. *Pressure Relief Devices*, ASME Performance Test Code PTC 25, 1994.

8. Bernstein, Martin D., and Friend, Ronald G., "ASME Code Safety Valve Rules—A Review and Discussion," *J. Pressure Vessel Technology*, Vol. 117, No. 2, May 1995, pp. 104–114.

9. Viswanathan, R., *Damage Mechanisms and Life Assessment of High Temperature Components*, ASM International, Metals Park, OH, 1989.

10. Narayanan, T.V., *Criteria for Approving Equipment for Continued Operation, Welding Research Council Bulletin 380, Recommendations to ASME for Code Guidelines and Criteria for Continued Operation of Equipment*, Welding Research Council, New York, April 1993, pp. 9–25.

11. Neubauer, B., and Wedel, U., "Restlife Estimation of Creeping Components by Means of Replicas," in *Advances in Life Prediction Methods*, D. A. Woodford and J. R. Whitehead, Eds., pp. 307–314, American Society of Engineers, New York, 1983.

12. Rodabaugh, E., *Report on Evaluation of Figure PG-32 of ASME Boiler Code (1974)*, Batelle Columbus Laboratories, Columbus, OH, July 1975.

13. Langer, Bernard F., Mershon, James, and Cooper, W. E., *Tentative Structural Design Basis for Reactor Pressure Vessels and Directly Associated Components*, U.S. Navy Document, 1958.

14. Harth, G., and Sherlock, T., "Monitoring the Service Induced Damage in Utility Boiler Pressure Vessels and Piping Systems," *Proceedings 1985 PVP Conference*, Vol. 98-1, American Society of Mechanical Engineers, New York, 1985.

15. ASME, *1998 Boiler and Pressure Vessel Code, Section I, Power Boilers*, American Society of Mechanical Engineers, New York, 1998.

16. ASME, *1998 Boiler and Pressure Vessel Code, Section II, Materials, Part D—Properties*, American Society of Mechanical Engineers, New York, 1998.

17. ASME, *1998 Boiler and Pressure Vessel Code, Section III, Rules for Construction of Nuclear Power Plant Components*, American Society of Mechanical Engineers, New York, 1998.

18.   ASME, *1998 Boiler and Pressure Vessel Code, Code Cases Boilers and Pressure Vessels*, American Society of Mechanical Engineers, New York, 1998.

19.   ASME, B36.10M, *Welded and Seamless Wrought Steel Pipe*, American Society of Mechanical Engineers, New York, 1996.

20.   ASME, B16.34, *Valves—Flanged, Threaded, and Welding End*, American Society of Mechanical Engineers, New York, 1996.

APPENDIX I

EVOLUTION OF THE ASME BOILER AND PRESSURE VESSEL CODE

A major goal of this book is to provide an understanding of Section I and its application. That understanding can be enhanced by a review of the history of steam power in the 19th century and the events that led to the development of the first boiler code by the ASME in 1915.

The early 1800s saw the first widespread application of steam in waterborne transport and in factories, as an alternative to sail and water power. With steam engines, ships no longer had to depend on the wind, and mills and factories no longer had to be located next to a source of water power. In 1800, typical steam boiler pressure was about 5 psi. This gradually increased to 30 psi by 1850 and to over 200 psi by 1900. Unfortunately, the development and application of steam power was accompanied by an increasing number of boiler explosions. In the early 19th century, the cause of these explosions was not clearly understood. Reports and articles on boiler explosions included many strange and unproven theories on the cause of the failures. One of these was the formation of hydrogen by the dissolution of water into hydrogen and oxygen, followed by an explosion of the hydrogen. Another was the mistaken belief that water could somehow flash into steam at a pressure higher than the saturation pressure corresponding to the water temperature. In 1830, due to increasing public concern, the Franklin Institute of Philadelphia established a committee to investigate the causes of boiler explosions, to determine the best means of preventing or mitigating them, and of applying and enforcing these remedies. (That committee long ago recognized that a code alone was insufficient; it had to be enforceable. The concept of an enforceable code eventually became accepted and is the basis for the successful system we have today.)

With funding from the Congress of the United States (the first government support for private research), the committee conducted a large number of experiments on several small boilers, two of which were tested to destruction. One very small boiler ($7\frac{1}{2}$ inches in diameter, $14\frac{1}{2}$ inches long) was made entirely of glass and was exposed to the fire over its whole length. This seemingly dangerous device permitted the direct observation of what was happening inside. Another boiler, used for most of these experiments, was a horizontal cylinder of rolled iron, $\frac{1}{4}$-inch thick, with an inside diameter of 12 inches, a length of 2 ft. $10\frac{1}{4}$ inches, and flat riveted heads. Each head had a $2\frac{1}{2}$-inch by $1\frac{3}{8}$-inch glass window for observation of the water and steam inside. The setting was of brick. A charcoal-burning grate provided heat to the entire lower surface of the cylinder. There were three try cocks, a glass water gage, a mercury steam gage, a safety valve, thermometers extending to various depths into the shell, a fusible plate, and means to feed water into the steam space and also into the water near the bottom of the shell, by using a hand pump.

While these boiler experiments were being conducted, another Franklin Institute committee undertook a comprehensive study of the strength of materials used in the construction of steam boilers. This work took 5 years and culminated in over a 100 tables covering the properties of a variety of materials at different temperatures.

At the conclusion of the investigations, the committee reported that the experiments "have shown not only what are some of the causes of explosions, but, which is quite important, what are certainly not causes." Five major causes of explosions were then discussed:

1. Excessive pressure that built up gradually
2. Overheated metal
3. Defective construction of the boiler or its appurtenances
4. Carelessness or ignorance of the operators
5. Collapse of boiler or flue due to vacuum within

The committee made a number of recommendations for preventing boiler explosions. Among the recommendations were these: All boilers should be provided with a proper pressure gage; a glass water gage; two adequately sized safety valves, one of which was inaccessible for tampering; and a fusible metal device not in contact with the steam. The committee also stressed the danger from a buildup of boiler scale, the need for frequent internal inspections, and the need for low-water alarms.

The work and conclusions of the Franklin Institute were well received and gave impetus to attempts to legislate rules and regulations for boiler safety. Earlier efforts had been largely unsuccessful, due to lack of general agreement on what should be done and because the proposed legislation was applicable only to cities or states rather than the whole country. In 1833 Congress passed a law establishing a steamboat inspection service. Boilers were required to be inspected every six months, and the inspectors had to certify their soundness. More comprehensive laws were passed in 1838 and 1843, culminating in the Steamboat Act of 1852, which provided the first rules and regulations for the design and construction of boilers. In the latter half of the 19th century, further amendments continued to incorporate improvements in boiler design and safety.

As rules for the design and inspection of boilers evolved, there also developed organizations involved in the inspection and insurance of boilers. As will be seen later, the inspector came to play a key role in the successful functioning of the code. In Europe in the 19th century, similar developments were taking place, with the participation of various governments and marine underwriters.

The rapid growth and industrialization of the United States during the 19th century would not have been possible without the ever widening application of steam power. Unfortunately, there was a darker side to this great progress—a harrowing record of lost lives and damaged property from boiler explosions throughout the United States. One of the worst on record occurred on the steamboat *Sultana* on the Mississippi River on April 27, 1865, two weeks after the end of the Civil War. This large ship had been built in 1860 and had a crew of 200. At Vicksburg it picked up 2000 Federal soldiers who had just been released from prison camps in Alabama and Georgia. Conditions in all prison camps were harsh in those days, but those in the South were notorious. It is thus doubly tragic that most of the 2000 soldiers who had managed to survive the prison camps, and were at last on their way home, were killed when one of the boilers exploded at 3:00 A.M., engulfing the engine room in flames from the wood-fired boilers. The wooden boat burned to the water level in 20 minutes, with the loss of 1500 lives. Considering the country's population then and now, an equivalent accident today would take almost 7000 lives.

In 1889, the American Boiler Manufacturers Association was formed for the purpose of raising the standards of boiler manufacture and preventing the manufacture and sale of unsafe boilers. The ABMA developed a standard specification for boilers, based on what was considered the best current American and European practice. This specification covered much of the subject matter subsequently included (25 years later) in the first ASME Code. Unfortunately, however, the manufacturers treated the standard specification principally as a device to permit uniformity in bidding, and attempts to promote uniform adoption by the states were thwarted by some members of the Association who could not look beyond the interests of their own companies.

By 1890 typical steam pressure used in prime movers had reached 80 psi. By 1900 pressure in the most advanced designs had reached 250 psi. At this time, despite a general understanding of the principle of safe boiler design, laws mandating inspection of steam boilers and licensing of boiler operators, the existence of government and underwriter rules for marine boiler construction and the standard specification for stationary boilers developed by the American Boiler Manufacturers Association, no state yet had a legal code covering safe stationary boilers. Boiler explosions were occurring at the rate of one a day in the United States.

This situation continued until two major shoe factory boiler explosions occurred in Massachusetts in 1905 and 1906, killing 59 persons and injuring over 100 people. The calamity finally prompted the chief inspector of boilers for Massachusetts to ask for authority to develop a state boiler law governing the construction of most boilers manufactured in or coming into the state. With the support of the governor and the legislature, the first such law was passed in 1907. It was only three pages long, but further rules were added, so that by 1909 the Massachusetts law was fairly comprehensive. It had been formulated by a committee representing a variety of interests: manufacturers, users, inspectors, and a professor from the Massachusetts Institute of Technology.

In 1911, Ohio adopted the Massachusetts rules with a few changes and additions. It had taken 80 years since the Franklin Institute's pioneering and public-spirited investigation of boiler explosions to establish the basis on which the ASME was to develop the ASME Boiler Code over the next four years. The Massachusetts and Ohio laws on construction and operation of steam boilers were a major milestone. However, other state and municipal laws dealing primarily with the inspection of boilers differed widely, resulting in a non-uniform, cumbersome system. For example, a boiler built for a given state had to be inspected by an inspector licensed by that state. That same boiler could not be installed in another state that did not recognize the competency of the first state's licensed inspector or that required different construction details. States with no laws became the dumping ground for boilers that couldn't pass muster elsewhere. What was clearly needed was a uniform set of rules for the construction of steam boilers that could be adopted by all the states as the basis for their boiler laws. What was also needed was a uniform standard of qualification for inspectors that would be accepted by all the states.

The president of the ASME in 1911 was a farsighted individual named Col. E. D. Meier, who was also president of a boiler company and a member of the American Boiler Manufacturer's Association. He had long advocated a uniform set of rules for the construction of steam boilers and believed that a commercially disinterested organization like the ASME, because of its reputation and broad scientific interests, could produce a set of rules that might be adopted by all the states. Col. Meier convinced the Council of the ASME to appoint a committee "to formulate a standard specification for the construction of steam boilers and other pressure vessels and for the care of the same in service." The committee was chosen to represent different interests—consulting engineers, boiler users, boiler manufacturers, steel manufacturers, and the insurance companies.

The committee studied the Massachusetts and Ohio rules and, with few changes, adopted them as the basis of the committee's first draft recommendations. More than 2000 copies of this draft were sent to engineers, engineering educators, engineering societies, boiler manufacturers, material manufacturers, boiler users, and boiler operators all over the world. Recipients were asked for their criticism and suggestions. Many replies were received, and several hearings were held at which all who wished were given an opportunity to speak. As a result of all the comments, many changes were made in the earlier draft, eventually culminating in a final draft that was approved by the Council on March 12, 1915. Although not finally approved and published until 1915, it became what was called the 1914 edition. The title of this final draft was "Rules for the Construction of Stationary Boilers and for Allowable Working Pressures." It was 105 pages long, with a 28-page index. It was made up of Part I for new installations (73 pages on power boilers and 7 pages on heating boilers) and Part II for existing installations (5 pages). An appendix of 20 pages contained methods of computing the efficiency of riveted joints, formulas for braced and stayed surfaces, safety valve capacities, dimensions of standard flanged fittings, and fusible plug requirements. This was the historic first edition of what is now called Power Boilers, Section I of the ASME Boiler and Pressure Vessel Code. Many provisions of that first edition remain in Section I today. Subcommittee I (SC I), the committee that administers, interprets and revises Section I, is a subcommittee of the ASME Boiler and Pressure Vessel Committee. It is a direct descendant of the original committee that developed the first edition of Section I.

The authors of this book have been priviledged to serve on Subcommittee I since the 1970s. Like soldiers who serve in a famous regiment, they feel great pride in this service and do their best to live up to its traditions.

APPENDIX II

ORGANIZATION AND OPERATION OF THE BOILER AND PRESSURE VESSEL COMMITTEE

ORGANIZATION AND RESPONSIBILITY

In 1911 the ASME set up a committee for the purpose of formulating standard rules for the construction of steam boilers and other pressure vessels. The committee is now known as the Boiler and Pressure Vessel Committee. From one small group in 1911, the Boiler and Pressure Vessel Committee has grown to a 1997 membership of 807 individuals on the overall committee structure. This consists of the Main Committee (Note: In September 1998 the Main Committee was renamed the Standards Committee), the Executive Committee, the Conference Committee, the Marine Conference Group, 14 subcommittees, and various lower tier committees called subgroups, working groups, and special committees.

Recent figures show a membership breakdown as follows: there are 30 members of the Main Committee, about 300 on subcommittees, and over 1200 on related lower tier committees. (The total number of committee positions is larger than the membership of 807 because many individuals serve on more than one committee.)

At the bottom of the committee structure are the subgroups and working groups. Typically, these groups are responsible for a specific technical field or a specific part of a section of the Code, for example, the Subgroup on Radiography (a Section V subgroup) or the Subgroup on Design (a Section I subgroup). At the subcommittee level, the responsibilities broaden to include a complete section of the Code, such as Section I, or a complete technical field, such as Section V, Nondestructive Examination. At the highest level is the Main Committee, consisting of a maximum of 30 members. This committee is responsible for every technical action taken by the Boiler and Pressure Vessel Committee. It deals with all sections of the Code, Code Cases, and Accreditation. It also hears appeals arising from technical or accreditation activities when these matters can't be resolved at the subcommittee level.

There are two other committees that act in an advisory capacity to the Main Committee. These are called the Conference Committee and the Marine Conference Group. These two committees represent legal jurisdictions or other authorities who have made the Code a legal requirement. Each state in the U.S., each province in Canada, and certain large cities that have adopted one or more sections of the ASME Code and maintain a department that enforces the Code are invited to appoint a representative to act on the Conference Committee. There are about 50 such representatives on the committee. An analogous committee is the Marine Conference Group, composed of representatives of marine interests who promulgate and enforce regulations based on the ASME Code. The five members of the Marine Conference Group represent the American Bureau of Shipping, the U.S. Coast Guard, the U.S. Department of the Navy, Lloyds Register of Shipping, and the Canadian Coast Guard.

Both Conference committees have direct access to the Main Committee, and can bring to it any problems with respect to implementation of Code requirements. Conference Committee members are entitled to participate in discussion and voting at the Main Committee, and in letter ballot on the minutes for items that are receiving first consideration (explained below under Voting at the Main Committee). On items receiving reconsideration, Conference Committee members' participation is limited to discussion, without vote. This participation by the regulatory authorities fosters their willingness to accept Code rules in their jurisdictions and assists in uniform administration of the Code.

Many members of ASME may not have a clear picture of its organizational structure and just how and where the Boiler and Pressure Vessel Committee fits in. The top ASME level of authority is the Board of Governors. Reporting to the Board of Governors are 5 Councils that supervise the next level of the organization, 37 Boards and Technical Groups. These in turn supervise Committees and Technical Divisions (the Boiler and Pressure Vessel Committee is at this level). This general organizational structure is shown in Fig. A.II.1. The particular portion of the structure encompassing the Subcommittee on Power Boilers is shown in Fig. A.II.2. The chain of command leading down to Subcommittee I, the Subcommittee on Power Boilers, is as follows: the Board of Governors, the Council on Codes and Standards, the Board on Pressure Technology Codes and Standards, the Main Committee (of the Boiler and Pressure Vessel Committee), and finally the Subcommittee on Power Boilers, Subcommittee I. Subcommittee I in turn has five subgroups: Piping, Design, Materials, General Requirements, and Fabrication and Examination. Until 1996 there was another subgroup, the Subgroup on Care of Power Boilers, which was traditionally active only for major revisions of Section VII, Recommended Guidelines for the Care of Power Boilers. Thus Section VII is a creature of Section I and has remained virtually unchanged except for major revisions, which occur infrequently, at 10 or 15 year intervals. In 1996, the Subgroup on Care of Power Boilers was disbanded, and its duties were turned over to the Subgroup on General Requirements, where a task group was assigned to review Section VII for any needed updating.

The personnel of these committees are listed in the front of all book sections. The majority of the subgroup members are also members of Subcommittee I. The Main Committee is made up of a few members from each of the subcommittees, usually the chairman, vice chairman, and possibly some subgroup chairmen or other senior members of the subcommittee. This arrangement of overlapping membership facilitates the work of the Main Committee since certain members of the Main Committee are quite familiar with items originating in their respective subcommittees, and can thus explain and answer questions about the items when they are considered by the Main Committee.

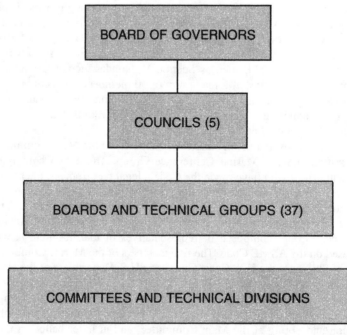

FIG. A.II.1
ORGANIZATION OF ASME (CONDENSED)

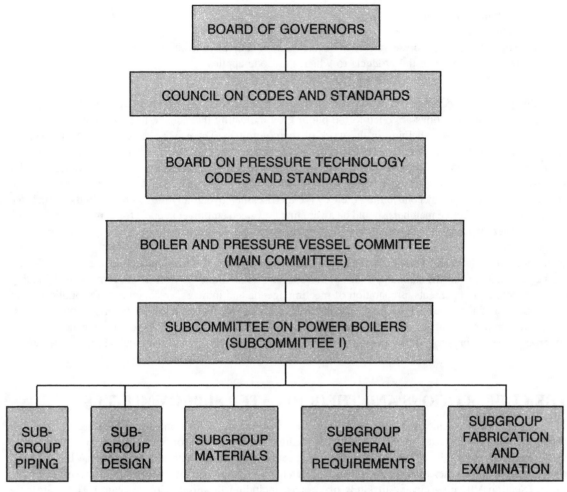

FIG. A.II.2
PORTION OF ASME ORGANIZATION ENCOMPASSING SUBCOMMITTEE ON POWER BOILERS (SUBCOMMITTEE I)

A BALANCE OF INTERESTS

Since its inception in 1911 when the Committee was established, it has been ASME policy that the members should represent a balance of interests, to avoid domination by any one interest group. This is one of the ways by which the ASME tries to ensure that actions of the Committee represent a valid technical consensus, fair to all and free of any commercial bias. Above all, the goal of the Committee is to promote the welfare and safety of the public. In furtherance of this goal, each committee member must sign an agreement to adhere to the ASME policy on avoidance of conflict of interest and to conform to the ASME Canon of Ethics. The ASME has also established procedures to provide for due process in Committee operation (e.g., hearings and appeals), thus safeguarding the members and the ASME against any charges of unfairness.

Members of the Committee are categorized according to the interests they represent. ASME has designated 18 categories of interest involved in codes and standards activities. Seven of these categories are represented on Subcommittee I:

1. Designer/Constructor
2. General Interest, such as consulting engineers and educators

3. Insurance/Inspection
4. Manufacturer
5. Regulatory, e.g., representatives of local, state, or federal jurisdictions
6. User, i.e., a user of the products to which the Code applies
7. Utility

Individuals typically become members of the Boiler and Pressure Vessel Committee by attending committee meetings as guests (meetings are open to the public), by indicating their desire to join, by participating in discussions, and otherwise helping in the work of the committee. There is a practical limit to the size of various committees, and as openings arise, the chairman chooses members to maintain a balance of interests on the committees and, occasionally, seeks out individuals with particular expertise. New members usually start by joining a subgroup or working group, and as they gain experience in committee operations and demonstrate their ability by contributing their own expertise, they eventually move up within the committee organization. Prospective members should be aware that they need employer support for committee participation, to cover expenses incidental to meeting attendance and to provide time to carry out committee assignments.

In addition to the many volunteer members of the committee, who are supported in these activities by their companies, the ASME maintains a staff of secretaries who facilitate the work of the committees by taking care of meeting arrangements, preparation of meeting agenda and minutes, arrangements for publication of the Code, scheduling, record keeping, correspondence, and telephone inquiries from the public. Staff secretaries prepare the agenda and take minutes at the Main Committee and subcommittee level. At the subgroup and working group level, one of the volunteer members of the committee usually serves as secretary.

THE CODE SECTIONS AND THEIR RELATED SUBCOMMITTEES

The formulation of "standard rules for the construction of steam boilers and other pressure vessels" on which the committee started in 1911 eventually became the first edition of Section I of the ASME Boiler and Pressure Vessel Code, in 1915. That first edition actually dealt only with boilers. Section VIII, covering pressure vessels for other than steam, was added later, in 1925, as part of the expanding coverage of the Code. (Section VIII now covers all kinds of vessels, including those containing steam.) There are now eleven sections of the Code, designated by Roman numerals I through XI. In 1997 a new subcommittee, Subcommittee XII, was appointed to develop a twelfth Code section, on Transport Tanks. When completed, this section will cover the design, construction, and continued operation of tanks used to carry dangerous materials by all means of transport. The various sections of the ASME Code (sometimes called the book sections) and the committees directly responsible for each are shown in Table A.II.1.

THE SERVICE COMMITTEES

In addition to the ten subcommittees governing the various book sections, there are two subcommittees under the Main committee, called service committees because they serve the book sections.

The Subcommittee on Safety Valve Requirements (SC-SVR) deals with the design, construction, testing, and certification of pressure relief devices. There is no separate book section on safety valves; the book sections (I, III, IV, VIII, and X) provide appropriate rules for these devices. Inquiries that pertain to safety valves are usually referred by these book committees to the Subcommittee on Safety Valve Requirements. Replies approved by that committee are returned to the book committees for further approval and action.

Until 1989, a service committee known as the Subcommittee on Properties of Metals (SC-P) established the allowable stress for all the materials used throughout the Code. In 1989, this committee was merged with the Subcommittee on Material Specifications (SC II) into a new committee called the Subcommittee on Materials (SC II), which carries out all the duties formerly handled by the two separate committees.

TABLE A.II.1
THE BOOK SECTIONS OF THE ASME B&PV CODE

CODE SECTION	GOVERNING COMMITTEE
Section I (1915)* Rules for the Construction of Power Boilers	Subcommittee on Power Boilers (Subcommittee I)
Section II (1924) Materials	Subcommittee on Materials (Subcommittee II)
Section III (1963) Nuclear Power Plant Components	Subcommittee on Nuclear Power (Subcommittee III)
Section IV (1923) Heating Boilers	Subcommittee on Heating Boilers (Subcommittee IV)
Section V (1971) Nondestructive Examination	Subcommittee on Nondestructive Examination (Subcommittee V)
Section VI (1971) Recommended Rules for Care and Operation of Heating Boilers	Subgroup on Care and Operation of Heating Boilers (A subgroup of Subcommittee IV)
Section VII (1926) Recommended Guidelines for the Care of Power Boilers	Subgroup on General Requirements (A subgroup of Subcommittee I)
Section VIII (1925) Rules for the Construction of Pressure Vessels	Subcommittee on Pressure Vessels (Subcommittee VIII)
Section IX (1941) Welding and Brazing Qualifications	Subcommittee on Welding (Subcommittee IX)
Section X (1961) Fiberglass-Reinforced Plastic Pressure Vessels	Subcommittee on Reinforced Plastic Pressure Vessels (Subcommittee X)
Section XI (1970) Rules for Inservice Inspection of Nuclear Power Plant Components	Subcommittee on Nuclear Inservice Inspection (Subcommittee XI)
Section XII, Transport Tanks (In preparation in 1998)	Subcommittee on Transport Tanks (Subcommittee XII)

*Year shown is first publication as a separate Code Section.

The Subcommittee on Design (SC-D) deals with special design problems and helps the other subcommittees in formulating design rules. Among the many subjects covered by this subcommittee are the design of openings, design for external pressure, elevated-temperature design, creep, fatigue, and the interaction of creep and fatigue.

THE ACCREDITATION COMMITTEES

As explained in the discussion of the various Code symbol stamps, no organization may do Code work without first receiving from the ASME a Certificate of Authorization to use one of the Code symbol stamps. The accreditation committees issue these certificates to applicants found to be qualified by ASME review teams. The Subcommittee on Boiler & Pressure Vessel Accreditation (SC-BPVA) handles this work for boiler and pressure vessel activities. The Subcommittee on Nuclear Accreditation (SC-NA) does the same for nuclear activities. Any disagreements as to the qualifications of applicants and any allegations of Code violations are dealt with by one or the other of these two accreditation committees, in deliberations that are not open to the general public. An ASME Certificate of Authorization can be revoked for cause, following hearing and appeal procedures.

COMMITTEE OPERATIONS

Since 1986, the Boiler and Pressure Vessel Committee has had four major meetings a year, during four weeks known as Codeweeks. The Committee used to meet six times a year, but decided to reduce the number of meetings as an economy measure. One of these meetings is known as the out-of-town Codeweek, because all the rest used to be held at ASME headquarters in the United Engineering Center in New York City. By 1996, for various reasons (expense and snowstorms being high on the list), only the September meeting was still held in New York City at the United Engineering Center. (Unfortunately, that building was sold in 1997, and the ASME had to find other quarters. The possibility that any future Codeweek meetings will be held in New York City is uncertain.) What used to be the only out-of-town meeting is traditionally held during the first week of May at various locations around the United States and Canada. The out-of-town meeting is also held jointly with the annual meeting of the National Board of Boiler and Pressure Vessel Inspectors. The chief inspectors of the various states and provinces who comprise the membership of the National Board are the top officials who enforce those sections of the Code that are adopted into the laws of their jurisdictions. The out-of-town meeting provides an opportunity for them to observe and participate as guests at the various Code committee meetings. The Main Committee always meets on Friday; most subcommittees (including Subcommittee I) meet on Thursdays. Subgroups and working groups usually meet earlier in the week. This arrangement facilitates an orderly and timely flow of information from the lower committees upward to the Main Committee.

HOW THE COMMITTEE DOES ITS WORK

The Boiler and Pressure Vessel Committee administers the Code. The major technical work of the Committee falls into three categories: providing interpretations of the Code in response to inquiries, revising the Code, and adding new provisions to it. This work usually starts at the lower levels of the committee structure, the subgroups and working groups. Proposals developed there are then considered at the subcommittee level. Many items (Code changes for instance) require consideration by the Main Committee. Actions of the Main Committee are subject to approval by one or the other of the two Boards above the Main Committee (one for nuclear and the other for non-nuclear items). Technical items are then subject to public review through publication in *Mechanical Engineering* magazine and the *ANSI Reporter*. Thus these items have received very careful technical consideration within the Committee and are also open to review by the public to avoid any inequity, hardship, or other problem that might result from a Committee action. Any comments received during public review delay an item until the originating subcommittee considers those comments. The several categories of the committee work are now described.

CODE INQUIRIES AND INTERPRETATIONS

Anyone who has used the Code knows the aptness of the second paragraph of the preamble to Section I: "The Code does not contain rules to cover all details of design and construction." What it contains rather are many rules for what might be called standard construction at the time the Code was written in the early years of this century. The Code has undergone gradual change since then, as modern steam power plants have evolved, presenting new situations, new arrangements, and new equipment. It is thus not surprising that so many inquiries are received by the Committee, asking for guidance in the application of specific provisions of the Code.

The ASME has established procedures and controls on responding to inquiries and publishes the questions and replies for the guidance of all users of the Code. These procedures are intended to protect the committee members and the ASME from any inference that a specific industry or company has an undue influence in the formulation of the questions or replies, or may benefit to the detriment of others. Sometimes inquirers ask questions that the Committee can't answer, for various reasons. The Committee is not in the business of consulting engineering. It does not have the resources to study plans and details sent in by inquirers and pass judgment on those designs. It also is in no position to undertake the potential liability for making such judgments. Accordingly, the Boiler and Pressure Vessel Committee Operating and Administrative Procedures provide four form letters for responding to the most common types of questions considered inappropriate: Indefinite questions that don't address some particular Code requirement; semicommercial questions; questions that would involve review or approval of a specific design; and questions that ask for the basis or rationale of Code rules. These form letters explain that the Committee cannot or does not answer such questions and advises the inquirer to pose only general questions addressing particular Code requirements, or to make specific recommendations for any proposed Code changes with supporting technical reasons or data.

In 1983, to reduce the work involved in replying to inquiries, mandatory appendices that give instructions on how to prepare technical inquiries were added to the various book sections. (See, for example, Appendix I of Section I). Inquiries are supposed to be sent to the Secretary of the Committee, but in actual practice they are often sent to an Assistant Secretary, who is the secretary of the subcommittee involved. That secretary gives the inquiry an item number and reviews the files of previous inquiries to see if the same question or a similar one has previously been answered. If such a reply is found, it is sent to the new inquirer. If not, there are three ways to handle the question. The usual method, used by the secretary of Subcommittee I, is to forward the inquiry to the Chairman of the Section I Subgroup responsible for the subject matter of the inquiry, with a request that it be considered at the next meeting of the Subgroup. Sometimes the secretary will formulate and attach a proposed reply for consideration by the Subgroup. Copies of this package are sent to all members of the cognizant subgroup, all members of Subcommittee I (at this time, for information only), and the chairman and vice chairman of the Main Committee.

An alternative method for dealing with an inquiry, not used by Subcommittee I, is to send it to a special committee of at least five members, who must agree unanimously on a proposed reply for it to be sent. A third possibility is to bring the technical inquiry to the Main Committee. This is quite a rare occurrence.

Once an inquiry is mailed to a Subcommittee subgroup, the chairman or any member may propose a reply for consideration at the next meeting. Otherwise, at the next meeting, the chairman assigns the item to a task group, usually consisting of three members, one of whom is designated as chairman. Task group members are chosen to maintain a balance of interests. The task group is expected to have a proposed reply ready for presentation at the next meeting, at which time it may be discussed and voted upon. Often there are disagreements about both the question and the proposed reply, particularly if the inquirer hasn't made her- or himself clear. Revisions to both may be offered, in the interest of clarity, and to make both question and reply as generally applicable as possible. Sometimes no agreement is reached, or the task group chairman is absent, and the item is carried over to the next meeting.

The goal of all this work at the subgroup level is, if possible, to prepare a clear and precise proposal to send on to Subcommittee I for consideration at its next meeting. It is nominally a Subcommittee I requirement that proposals passed by subgroups during one Codeweek are not considered by Subcommittee I until the next Codeweek. If this procedure is followed, a complete information package (the letter of inquiry, any written comments, the proposed reply, pertinent previous replies, etc.) can be included in the Subcommittee I meeting agenda, so that all Subcommittee members can study the information before the meeting. However, items that are approved on Wednesday of Codeweek by a subgroup can often be added to the subcommittee agenda on the following day, especially if the inquirer expresses urgency or if the question and reply appear to be simple and straightforward. It is now the practice that new inquiries are included on the Subcommittee I meeting agenda, and answers to most of them are formulated and approved at subgroup meetings earlier in the week. The subcommittee is thus able to approve the majority of simple inquiries on their first appearance on the agenda. In a good number of cases, however, the inquiry is not so simple, and a subgroup task group is appointed to try to develop an answer upon which the committee can agree.

Inquirers occasionally attend both subgroup and subcommittee meetings, and are always welcome to participate in the discussion of their items.

At the subcommittee level, a 75% plurality is required to approve a reply. If a majority less than 75% is in favor of the reply, it may be sent to the Main Committee for approval (also by a 75% plurality), or it may be returned to the subgroup for further work. The latter is the current practice at Subcommittee I. Negative voters are requested to provide written explanations of their negative votes to guide the subgroup in its next attempt to formulate a reply.

PUBLICATION OF INTERPRETATIONS OF THE CODE

Before 1977, there was no formal mechanism for the publication and dissemination of replies by the ASME Boiler and Pressure Vessel Committee to inquiries on the interpretation of the Code. Starting in January 1977 and continuing until June 1982, the ASME twice a year published a paperback volume containing all the written replies by the ASME staff on behalf of the Committee to inquiries on the interpretation of the Code. For the convenience of the reader, these were presented chronologically in separate groups identified by the Code Sections to which they apply.

To receive the paperback volumes of Code Interpretations, it was necessary to subscribe and pay a fee. Most purchasers of the various Code books did not subscribe, defeating the major goal of their publication, namely, making this information widely available to users of the Code. Thus, starting with June 1982, the ASME began issuing the interpretations twice a year in loose-leaf form to all purchasers of Code books. (Each section is issued only interpretations pertaining to that Section.) A very useful feature of this service is a cumulative index by paragraph number and also by subject, although the subject may not accurately describe the essence of the interpretation. The Foreword to the interpretations contains a disclaimer to the effect that the interpretations are not a part of the Addenda or the Code. In 1995, a Main Committee member raised concerns as to the official status of Interpretations. This prompted a reconsideration of this disclaimer. It was explained by the ASME staff that the disclaimer was originally intended to mean only that buying the Code books didn't entitle a purchaser to receive the Interpretations without paying extra. In 1996, the Main Committee approved the following Interpretation, identified as item number BC96-134:

> Question: *The cover page of Interpretations Volumes and page iii in each Section of the Code state, "Interpretations are not part of the Code or the Addenda." Do interpretations establish new Code rules?*
>
> Reply: *No. The quoted statement refers to the publication and distribution service for interpretations.*
>
> *Interpretations explain or clarify rules in specific Editions or Addenda of the Code.*

Thus although the Interpretations may not literally be a part of the Code, they serve a number of useful purposes. The published replies often contain detailed guidance and explanation about some specific topic for which Section I (or another section) provides only general rules, or none at all. As explained earlier, Section I is neither a textbook nor a design manual; it is a system of rules with very limited explanation of intent. Thus it is particularly helpful to have what is, in effect, a current consensus of Subcommittee I on the meaning or interpretation of some Section I rule, or on how to treat some situation not explicitly covered by Section I. The interpretations also provide a good indication of what the reply to a similar question might be. Authorized Inspectors are usually guided by Interpretations in deciding whether to accept a particular form of construction.

Finally, most inquiries to the Committee pertain to matters that are not clear, or are not covered specifically, or concern some new development in boiler design. The published replies are, therefore, likely to be of interest and benefit to many others besides the inquirer.

One caveat: An interpretation applies either to the Edition and Addenda in effect on the date of issuance of the interpretation or the Edition and Addenda stated in the interpretation. Subsequent revisions to the Code may supersede the interpretation.

ADDITIONS AND REVISIONS TO THE CODE

The Code is subject to continuous change—some provisions are revised, others deleted, still others added. Although some changes originate high in the committee structure (e.g., the mandatory appendices in each book section on preparation of technical inquiries), most start at the subgroup level, in response to an inquirer's request for a change or a request by members of the subcommittee to clarify, update, or expand existing Code provisions.

The development of a Code change follows a path similar to that of a technical inquiry. The cognizant subgroup chairman assigns a task group to do the work. Often, more than one subgroup may be involved, and the task group may include members from more than one subgroup. In appointing the task group, the chairman tries to maintain a balance of interests while making sure to include members with the specific expertise appropriate for the task. If and when the subgroup approves the change proposed by the task group, the proposal is forwarded to Subcommittee I for consideration, with documentation giving the background of the proposed change.

During deliberation on the item by Subcommittee I, the item may be approved, or further changes may be suggested, or strenuous objections may be raised, in which case the chairman may accept the presentation as a so-called "progress report" and return the item to the subgroup for further work. The subgroup then attempts to modify the proposal to overcome the objection raised at the subcommittee level. Usually, this process of refinement leads to a proposal that is eventually approved by Subcommittee I, and the secretary forwards the item to the Main Committee for inclusion on the agenda of the next meeting, normally the next Codeweek.

The secretary or one of the Subcommittee members also writes a paragraph of background explanation that appears with the item on the Main Committee agenda, in what is called "an action box" for the item. If there were negative votes on the item at the subcommittee, those voters have to provide their objections in writing, and the secretary summarizes these objections as part of the explanation sent to the Main Committee. This explanation is very helpful since the first time a Main Committee member sees an item that hasn't come from his own subcommittee is when it appears on the Main Committee agenda. Voting on Main Committee items is explained later in this appendix, after the next topic.

CODE CASES

One of the many definitions of the word **case** is "a special situation," which is a good description of the circumstances covered by what the Committee calls a Case. In the early application of the first edition of the Code, users sought Committee guidance for circumstances not specifically covered by the Code and when the intent of the Code rules was not clear. The Committee considered a number of these special situations and issued formal guidance in the form of what it called a Case. This is the origin of the practice of issuing numbered Code Cases. In those days, no distinction was yet made between a Case and what we now call an Interpretation.

Today, Code Cases are used for several purposes:

1. To provide for early implementation of new or revised Code rules, since a Code Case can be approved and published more quickly than the text of the Code can be revised. A text change can take well over a year after approval by the Main Committee before it is published, while a Code Case can be approved in a single Codeweek, after which further approval is needed only by the Board of Pressure Technology Codes and Standards. Thus a Code Case may be usable two or three months after approval by the Main Committee.

2. To permit the use of new materials or new forms of construction not covered or not otherwise permitted by existing Code rules, when the need is urgent.

3. To gain experience over a period of time with new materials, new forms of construction, or new design rules before changing the Code to include them. This is particularly useful for rapidly evolving technology.

Code Cases have a limited life, usually three years. They automatically expire then unless the Committee reaffirms them for another three years. Code Cases may also be annulled at any time by the Committee. Usually Code Cases are annulled six months after their contents have been incorporated and published in Addenda to the Code.

Some other special aspects of Code Cases are these:

- Code Cases are nonmandatory; they are permissive only. In effect, they are an extension of the Code.
- If a Code Case permits some deviation from normal Code rules, it usually stipulates some compensating extra requirements. An example of such a case was the provision (now included in Section I, PW-54.3) permitting some nonpressure parts to be welded to pressure parts after the hydrostatic test if five conditions were met for welding, inspection, and documentation. Usually, when a Code Case is used, the Code Case number must be listed on the Data Report.
- All Code Cases must clearly indicate their limits of applicability.
- Code Cases have an inquiry and reply format. The inquiry describes the circumstances and asks for guidance. The reply always starts with the words "It is the opinion of the Committee that . . . ." This opinion phraseology is used because presumably there are no explicit rules in the Code covering the particular circumstances, and thus the committee can only give its opinion.
- Code Cases may not contain references to other Code Cases, since Code Case life is usually short and the reference might soon be annulled.
- Although there used to be pressure to incorporate all Code Cases into the Code as soon as adequate experience had been gained with their use, it is now recognized that some cases do not lend themselves to incorporation and will be left as Code Cases indefinitely. An example is Case 1918, Attachment of Nozzles or Couplings to a Boiler Vessel after Postweld Heat Treatment. The Main Committee refused to incorporate this case because of concerns that it could be routinely used to avoid postweld heat treatment of nozzles on boiler vessels.

Code Cases are approved in the same way as revisions to the Code. Often some urgency is involved, and the interested parties do much of the work, such as preparing the initial draft of the case. Sometimes more than one subcommittee is involved. Many cases involve use of new materials, or use of materials at higher temperatures than before. In such instances, the Subcommittee on Materials, Subcommittee II, establishes the allowable stresses to be listed in the case.

The ASME publishes Code Cases in two Code Case books: Boilers and Pressure Vessels, and Nuclear Components. Supplements with the latest cases and annulments are now sent quarterly to purchasers of the Code Case books, until the publication of the next edition of the Code, when a new Code Case book must be purchased. Issuing the new cases quarterly theoretically enables the publication of new Code Cases as soon after each Main Committee meeting as possible. Unfortunately, publication is often delayed for several months. However, an interested inquirer can call the secretary of the Main Committee and find out whether the Board on Pressure Technology Codes and Standards has officially approved a particular Code Case. If the Board has approved it, it may be used immediately. A 1998 list of Section I Code Cases can be found at the end of this appendix.

VOTING AT THE MAIN COMMITTEE

Items coming before the Main Committee are considered within two categories: first consideration and reconsideration, usually called second consideration. A new item appearing for the first time is given "first consideration'' by the Committee. A single negative vote (a voice vote at the meeting, confirmed in writing as a negative ballot within a certain time after the meeting) is sufficient to stop the item and return it to the originating subcommittee for reconsideration. A negative voice vote must be supported by a written negative ballot explaining the reasons for the negative ballot or else it is considered void.

After each Main Committee meeting, the Secretary collects and mails copies of all negative letter ballots to the members. The negative ballots are also sent to the subcommittees from which the items came, where the objections of the negative voters can be considered. Often, the items originated at a lower committee level, such as a subgroup, and it is there that a response to a negative ballot is developed. Sometimes, the objector offers some constructive suggestions that will satisfy his concerns, and if these are acceptable to the subcommittee originating the item, the changes are made, approved by the subcommittee, and the item is returned to the Main Committee. At other times, the subcommittee may consider the objections invalid and refuse to make any change. In that case, the subcommittee votes to reaffirm the item, often with an accompanying rebuttal to the objections of the negative voter.

When the item is returned to the Main Committee, its status may be either first or second consideration, depending on whether its technical substance has been changed significantly or whether the change has been merely editorial. If the item is returned with some technical change, its status remains first consideration (in the changed form). It is then subject to another vote in which one negative ballot can again return it for reconsideration. If the item is returned to the Main Committee technically unchanged, its status becomes ''second consideration.'' During second consideration, four negative ballots are required to stop and, in effect, kill the item. If the originating subcommittee wants to pursue the matter further, it must start all over, usually by making sufficient revision to satisfy the objections raised. The new appearance of the item would be a new first consideration. On the other hand, if on second consideration an item receives less than four negative votes, it is considered approved by the Main Committee, and it proceeds to the next two approval levels, Board review and public review. Occasionally, a Board member supports the position of a negative voter at the Main Committee, and a similar reconsideration process begins. It takes six negative votes to stop an item on second consideration at the Board level. This happens only rarely.

Most of the items considered by the Main Committee are proposed changes in the various book sections of the Code. Fairly regularly, some items fail to pass because of strong objections by other Main Committee members who perceive the change as having negative consequences to safety or representing an unworkable situation when applied to other comparable circumstances. This is part of the give-and-take of committee actions, which are intended to achieve a technical consensus of the membership, but with concern for safety always being paramount.

It should be evident that the committee work required to answer inquiries, to approve Code Cases, and to make revisions to the Code can sometimes be a very complex and time-consuming process, especially when, as is often the case, many different committees are involved. In 1996, the ASME established a task force to study the process by which codes and standards are approved with a view to redesigning and streamlining it. The task force identified over 85 aspects of the system where improvement might be achieved. Among the task force proposals is one to maximize the simultaneous consideration of items when several committees are involved. Another is to appoint a person who would be responsible for facilitating the efficient flow of individual work items, so that the information necessary for their consideration by the various committees could be available in a timely fashion. At this writing (1998), the task force has started pilot improvement programs involving several committees, to test its recommendations. It is likely that the codes and standards development process will be improved as recommendations of the task force are adopted.

DUE PROCESS

Persons who consider themselves injured by an action of the Committee regarding a technical revision, response to an inquiry, or the refusal to issue a certificate of authorization, can request a hearing to present their side of the story. Such hearings start at the subcommittee that originated the item. Appeals that can't be resolved at the subcommittee level may be referred to the Main Committee. If the Main Committee can't reach a mutually acceptable solution, the appeal may be submitted to the appropriate supervisory board and, if necessary, to the Board on Hearings and Appeals of the Council on Codes and Standards. This careful attention to due process is the result of an unfortunate event that happened in 1971, the infamous Hydrolevel Corporation case. Here is the essence of that case.

Section IV stipulates that boilers must have an automatic low-water fuel cutoff that stops the fuel supply when the surface of the water falls to the lowest visible part of the water gage glass. Hydrolevel had developed a new probe-type low-water fuel cutoff that relied on an electrode on the probe. Water covering the electrode completed a circuit that maintained fuel flow. When the water level fell below the electrode and uncovered it, the circuit was broken and the fuel was stopped.

At that time, another manufacturer dominated the low-water fuel cutoff market with a float-operated device. That rival manufacturer happened to have a representative serving as vice-chairman of the Section IV committee. Court records subsequently showed that three officers of the rival manufacturer, including that vice-chairman, met with the chairman of the committee to draft an inquiry to the committee. The inquiry asked whether a low-water cutoff with a time-delay feature met the Code. The Subcommittee chairman at that time had the authority to respond to the inquiry on the ASME's behalf without the endorsement of the full committee. His letter of response implied that the device did not meet Section IV requirements and would not provide adequate safety. Hydrolevel subsequently alleged that the inquiry was deliberately intended to put the probe-type of device in a bad light and that copies of the ASME response were used by the rival manufacturer's sales force to discredit Hydrolevel's device. When a former Hydrolevel customer reported this to Hydrolevel in 1972, Hydrolevel complained to the ASME and asked for a clarification of the ruling. This time the ruling was put before the entire Section IV subcommittee (the vice president of the rival manufacturer had by this time become chairman of the committee), where it was reconfirmed, perhaps because of the subcommittee's belief that the Code required the fuel to be cut off as soon as the water level was no longer visible in the water gage glass (and not after a time delay). However, the Main Committee reversed the ruling and issued an official communication to Hydrolevel saying that the Section IV paragraph in question did not prohibit the use of a low-water cutoff with a time delay.

In 1975 Hydrolevel sued the parties, including the ASME, alleging conspiracy in restraint of trade. The other parties settled, but the ASME contested the charge, in the understandable belief that it had done no wrong. A district court judge awarded Hydrolevel $7.5 million in damages. The ASME appealed, lost that appeal, and then appealed to the U.S. Supreme Court, which affirmed the appellate court's decision. The essence of the court's finding was that the ASME had put certain committee members in positions where they appeared to represent the ASME and had thereby conferred on those agents the ASME's so-called apparent authority. Even though the ASME is a nonprofit professional organization, it was found liable for the willful, anticompetitive, wrongful conduct of its agents. With interest on the triple damages called for by the antitrust act, ASME had to pay almost 10 million dollars (in addition, of course, to legal fees). This was a heavy price for an educational nonprofit organization that gets much of its financial support from the dues of its members. In an ironic twist of fate, the principal owner of Hydrolevel died of a heart attack shortly after hearing the news of the Supreme Court decision. Following that decision, the ASME developed improved procedures in an attempt to ensure the fairness of interpretations and to provide for hearings and appeals for anyone who considers himself injured by an action of the Code committee, such as an Interpretation or a proposed Code change. These procedures should prevent any further cases like the Hydrolevel case.

1998 List of Section I Code Cases
(Reproduced From ASME Code Cases Boilers and Pressure Vessels)

Appendix III

Overview of Code Sections II–XII and Piping Codes B31.1 and B31.3

Users of Section I soon find that they need to know something about the other Code sections and piping codes involved in boiler and pressure vessel construction. Although some of these codes are described in detail elsewhere in the text, as needed, a brief overview of all of them is provided here, for ready reference.

When Section I was first published, in 1915, it was the entire ASME boiler code, and all the boiler construction rules were contained in a single volume. That volume eventually expanded to the extent that it became logical and convenient to establish separate sections covering components other than boilers and various aspects of Code construction, such as welding and nondestructive examination. Consequently, we now have the following sections, first published in the years indicated:

Section I	Power Boilers (1915)
Section II	Materials (1924)
Section III	Nuclear Power Plant Components (1963). (The designation Section III had earlier been used for locomotive boilers.)
Section IV	Heating Boilers (1923)
Section V	Nondestructive Examination (1971)
Section VI	Recommended Rules for Care and Operation of Heating Boilers (1971)
Section VII	Recommended Guidelines for the Care of Power Boilers (1926)
Section VIII	Pressure Vessels (1925)
Section IX	Welding and Brazing Qualifications (1941)
Section X	Fiberglass-Reinforced Plastic Pressure Vessels (1961)
Section XI	Rules for Inservice Inspection of Nuclear Power Plant Components (1970)
Section XII	Transport Tanks (in preparation in 1998)

There are two piping codes most commonly used to design the piping associated with boilers and pressure vessels. These are ASME B31.1, Power Piping, and ASME B31.3, Process Piping, which until 1996 was known as Chemical Plant and Petroleum Refinery Piping.

Section I has been covered in detail in this book. Here is a brief overview of the other book sections and of the two piping Codes, starting with Section II, Materials.

Section II, Materials

This is a four-part compendium of materials data, almost a foot thick in its entirety. The four volumes of Section II are as follows:

Part A—Ferrous Material Specifications
Part B—Nonferrous Material Specifications
Part C—Specification for Welding Rods, Electrodes, and Filler Metals
Part D—Properties

Parts A, B, and C are, as their names indicate, a compendium of ASME material specifications for all the pressure vessel and welding materials permitted for use in Code construction by Sections I, III, IV, and VIII. These specifications are first adopted by ASTM (the American Society for Testing Materials) and are subsequently adopted, sometimes with minor modification, by ASME. It is important to remember that not every material listed in Section II can be used in Section I (boiler) construction. (This is explained in Chapter 3.) In general, only materials listed in the material paragraphs of Section I—PG-5 through PG-13—may be used for Section I pressure-retaining parts. PG-11 also permits use of most materials listed in the ANSI standards recognized by Section I in PG-42.

The last part of Section II, Part D, was first issued in 1992. It lists material properties for all materials accepted by the book sections I, III, and VIII. Most important, it lists allowable stresses as a function of temperature. Until 1992, these stresses had been listed separately in those individual book sections. The ostensible purpose of Part D was to consolidate all this information in a single volume, organized into subparts covering stress tables, physical property tables, and external pressure charts. Also included are several appendices. One describes the basis for establishing the allowable stress and design stress intensity values (the latter are used in the design of so-called Class 1 nuclear power plant components and Section VIII, Division 2 vessels). Another describes how charts for external pressure design are established and what criteria are used to determine allowable compressive stresses when designing for external pressure. Unfortunately, Part D lacks a useful index, and although efforts are underway to make it easier to use, finding any needed information in the various tables, such as a particular material and grade, is difficult. (A brief description of how the various materials are sorted and listed is provided in Chapter 3: Materials.)

One useful feature of Section II, Part D is its physical property tables, which provide such information as coefficients of thermal expansion, thermal conductivity, thermal diffusivity, moduli of elasticity, Poisson's ratio, modulus of rigidity, density, melting point range, and specific heat for a fairly large variety of materials. Much of this information is provided over a large temperature range. Also included are tables of yield and ultimate tensile strength as a function of temperature, up to 1000°F.

For various historic and technical reasons, the several Code book sections that deal with the design of metal pressure vessels (Section I, III, IV, and VIII) do not all use the same allowable stresses. Section I; Section VIII, Division 1; and B31.1, Power Piping, establish allowable stresses based on the same criteria. Section IV, Heating Boilers, uses allowable stresses that are (for most components) lower than those permitted by Sections I and VIII, Division I. Section III has many different classes of construction, and the allowable stress basis varies. For Class 1 components, which are designed using a method called design-by-analysis (see below, under Section III), allowable stresses are somewhat higher than those permitted by Section I and Section VIII, Division 1. These same higher stresses are permitted by Section VIII, Division 2, which also requires design-by-analysis. The higher stresses come at the price of enhanced care in manufacture. All these different allowable stress criteria require many different stress tables in Section II, Part D. See Chapter 15 and the next topic in this appendix for further discussion of allowable stress.

Section III, Nuclear Power Plant Components

This section of the Code was first published in 1963, as a single slender volume. It was based in large part on a 1958 U.S. Navy document entitled *Tentative Structural Design Basis for Reactor Pressure Vessels and Directly Associated Components* by three distinguished pioneers in the development of nuclear power: Bernard F. Langer, then with Westinghouse Bettis Laboratories; James Mershon, then with the U.S. Navy Bureau of Ships; and W.E. Cooper, then with GE's Knolls Atomic Power Laboratory.

This seminal document first set out in systematic fashion the methods and principles of what has come to be known as the concept of design-by-analysis, because of the important role stress analysis plays in this design process. This concept, based on fundamental considerations of analysis and material behavior, was a truly great step in the evolution of pressure vessel design. It is now the basis for the structural evaluation requirements of Section III and Section VIII, Division 2, and has been widely adopted in other countries. Design-by-analysis is a unified concept of design that accounts for most possible failure modes and provides rational margins of safety against each type of failure.

This concept permits the use of higher allowable stresses with no reduction in safety, by requiring a rigorous analysis and classification of all types of stresses and loading conditions, in addition to better control of manufacturing and inspection procedures. The sections of the Code that use design-by-analysis require a detailed design specification, defining among other things all anticipated loading conditions, and a complete design report; both items must be certified by registered Professional Engineers experienced in pressure vessel design.

Section III expanded greatly during the heyday of nuclear power. It now comprises ten volumes in two divisions, one for steel structures and the other for concrete reactor vessels and containments.

Most of the design rules and allowable stresses of Section III extend only to temperatures at which creep effects are insignificant (~800°F), i.e., these stresses are based on time-independent tensile or yield strength properties. Thus these rules do not extend to elevated-temperature service. However, such service was covered by a famous nuclear Code Case, Case N47, Class I Components in Elevated Temperature Service, a major design document of well over 100 pages. After about 30 years of development and use, that Code Case was incorporated into Section III as Subsection NH, with the 1995 Addenda. Subsection NH provides complex rules for overcoming the difficulty of designing in the creep range. The difficulty arises from the inherent complexity of the creep process, which cannot be characterized precisely, particularly under varying temperatures and multiaxial states of stress. Creep behavior complicates the detailed stress analysis because the distribution of stress varies with time as well as with the applied loads. Even with the NH rules, elevated-temperature design is difficult at best. See Chapter 14 for a further discussion of these difficulties.

Section IV, Heating Boilers

This section of the Code provides rules for the design, construction, installation, instrumentation, and certification of a somewhat diverse group of relatively low-pressure steam boilers and water heaters. The preamble to Section IV describes these as steam heating boilers, hot water heating boilers, hot water supply boilers, and potable water heaters.

The design pressures of these boilers varies from 30 psi to 160 psi and design temperatures do not exceed 250°F. (Above these limits a boiler would, by definition, be a power boiler, falling within the scope of Section I.) Section IV steam boilers have an **operating** limit of 15 psi, so the safety valves must be set at 15 psi. Potable water heaters are designed for pressures up to 160 psi (with a minimum of 100 psi) and for temperatures not to exceed 210°F. The purpose of the 210°F limit is to prevent any water discharged from relief valves or other connections from flashing to steam.

Aside from the lower pressures and additional coverage of installation and instrumentation, the most significant difference between Sections I and IV is that Section IV does not require any nondestructive examination. Accordingly, Section IV, in general, has a lower allowable stress (a margin of 5:1 versus 4:1 based on ultimate tensile strength).

Section V, Nondestructive Examination

This Code book starts with a two-page Article 1 on General Requirements. The first paragraph (T-110 Scope) of Article 1 gives a good description of what it covers:

T-110 Scope
a) This Section of the Code contains requirements and methods for nondestructive examination which are Code requirements when and to the extent they are specifically referenced and

required by other Code Sections. Methods include radiographic examination, ultrasonic exami-
nation, liquid penetrant examination, magnetic particle examination, eddy current examination,
visual examination, leak testing, and acoustic emission.
b) Subsection A describes the nondestructive examination methods to be used.
c) Subsection B contains standards covering nondestructive examination methods which have
been accepted as standards. These standards are intended to be informative only and are
nonmandatory unless specifically referenced in whole or in part in Subsection A or as indicated
in other Code Sections.
d) The acceptance standards for these examinations shall be as stated in the referencing Code
Section (see also T-170).

Article 1 goes on to discuss the manufacturer's responsibility to establish suitable examination procedures and to have the examination performed by qualified personnel. The duties of the Authorized Inspector are then explained. Further advice is given regarding examination procedures and the qualification of personnel who carry them out. The rest of Section V comprises Subsection A (12 articles describing in great detail nondestructive methods of examination), Subsection B (7 articles consisting of standards for different types of nondestructive testing, e.g., SE-94 Standard Guide for Radiographic Testing), and an appendix that provides standard definitions of terms.

One point to remember is that the NDE methods of Section V are mandatory only when and to the extent they are invoked in one of the other book sections. Also, the acceptance standards are provided in the referencing book section. Note also that Section I calls for only some of the many NDE methods listed in Section V. These are radiographic, ultrasonic, magnetic particle, dye penetrant, and visual examinations. Section I makes no mention of acoustic emission examination of metallic vessels during pressure testing. Acoustic emission testing is relatively new, and its application to Section I construction does not appear needed or potentially useful. It is sometimes applied for nuclear in-service inspection and for fiber reinforced plastic pressure vessels, where it can presumably detect crack growth.

Section VI, Recommended Rules for the Care and Operation of Heating Boilers

Notice first that this Code section is nonmandatory, i.e., these are only recommended rules for the care and operation of heating boilers. Nevertheless, the rules represent a consensus on what is considered good practice, and if good practice is not followed, an owner/user's potential liability following an accident would increase. The rules of Section VI apply only to Section IV boilers, which are limited to operation in these ranges:

   a. Steam boilers for operation at a pressure not exceeding 15 psi, and
   b. Hot water heating and hot water supply boilers for operation at pressures not exceeding 160 psi and/or temperatures not exceeding 250°F.

Section VI provides a great deal of detailed information and advice on various boiler types and their accessories such as safety valves, circulating pumps, low-water fuel cutoffs, valves, gages, etc. It also covers fuels; fuel burning equipment and burner controls; water treatment; and boiler operation, maintenance, and repair.

Section VII, Recommended Guidelines for the Care of Power Boilers

This volume of so-called recommended guidelines is also a nonmandatory compendium of what is considered good practice in the industry. Section VII is, in a sense, a child of Section I, since a subgroup of Subcommittee I, the Subgroup on Care of Power Boilers, used to write Section VII for review and eventual approval by its parent committee. The Subgroup on Care of Power Boilers revised Section VII infrequently, perhaps every 15 years or so, because there was no need for more frequent updating; the methods used by the industry for the care and operation of power boilers change rather slowly. The last major rewrite was in the early 1980s; no significant changes have been made to the text since then. Since the Subgroup had nothing to do between major revisions, in 1996 it was disbanded and its duties were turned over to the Subgroup on General Requirements, where a task group was assigned to review Section VII for updating, as appropriate.

Section VII has a preamble that describes its purpose and contents quite well. The stated purpose of the guidelines is to promote safety in the use of power boilers. The guidelines are intended for use by those directly responsible for operating, maintaining, and inspecting power boilers. The emphasis of the guidelines is on industrial-type boilers because of their widespread use. Section VII is divided into nine subsections, as follows:

C1.   Fundamentals
C2.   Boiler Operation
C3.   Boiler Auxiliaries
C4.   Appurtenances
C5.   Instrumentation, Controls, and Interlocks
C6.   Inspection
C7.   Repairs, Alterations, and Maintenance
C8.   Control of Internal Chemical Conditions
C9.   Preventing Boiler Failures

Few engineers have the opportunity to gain first-hand experience in boiler operation, maintenance, and repair. Section VII provides a good alternative to such experience. It is recommended to those interested in expanding their knowledge about boiler operation, maintenance, and repair.

Section VIII, Pressure Vessels

This is a well-known and important ASME Code book. Years ago the title of Section VIII was Unfired Pressure Vessels (that's the origin of the U symbol used to stamp Section VIII vessels), but it is now called just Pressure Vessels, since it covers fired as well as unfired vessels. Section VIII now comes in three volumes, Division 1 (the regular rules) and Division 2, Alternative Rules. A third volume, Division 3, Alternative Rules for Construction of High-Pressure Vessels was published in 1997. The Division 2 Alternative Rules comprise a less extensive version of the design-by-analysis rules of Section III's Subsection NB. Division 2, with its higher allowable stresses, is not often used, because the cost of additional stress analysis, tighter control of manufacturing, and extra inspection usually outweighs any saving due to thinner vessel walls. However, its design-by-analysis methods are often useful in justifying some form of Section I or Section VIII design and construction that for one reason or other isn't included in the regular rules.

Section VIII, Division 1 is a large volume and covers so many different types of vessels and forms of construction that only the briefest of overviews can be offered here.

The Scope paragraph (U-1) explains a number of things. It notes that pressure vessels are containers for either external or internal pressure and that the pressure can come from an external source or be caused by the application of heat, or it can come from some combination of the two. The range of design pressure covered by Section VIII is from 15 psi to no upper limit. Above 3000 psi additional requirements are invoked.

Division 1 is divided into three Subsections, Mandatory Appendices, and Nonmandatory Appendices. Subsection A covers general requirements applicable to all pressure vessels. Subsection B covers specific requirements applicable to the various permitted fabrication methods, such as welding, forging, or brazing. Subsection C covers specific requirements applicable to the different classes of materials used in pressure vessel construction (such as carbon and low alloy steel, nonferrous metals, high alloy steels, cast iron, etc.). The 29 mandatory appendices cover a great variety of subjects from design formulas to NDE methods, to QC system requirements, to the design of expansion joints, among many others. The 20 nonmandatory appendices cover another large group of interesting subjects, such as allowable loads on tube-to-tubesheet joints, illustrative examples of the application of code formulas and rules, preheating before welding, and design considerations for bolted flanges. One appendix, Appendix M, deals with the installation and operation of pressure vessels. Normally, installation and operation activities are the prerogative and responsibility of the local law enforcement authorities (sometimes called the jurisdiction) where the vessel is installed.

Appendix M treats a number of subjects: corrosion, access to vessel marking, pressure-relief valve installation and other aspects of overpressure relief. What is unusual, but not necessarily unsafe, is that Appendix M suggests what are, in effect, waivers of certain rules in the mandatory portion of the text of Section VIII. Note (paragraph M-1) that the waiver is conditioned on special permission for it being granted by the legal authority having jurisdiction over the installation of pressure vessels. An example of such a waiver is this: Paragraph UG-135(e) stipulates that (in general) there shall be no intervening stop valves between a vessel and its pressure-relieving device. However, appendix paragraph M-5(a) permits such an installation ''for inspection and repair purposes only. When such a stop valve is provided, it shall be so arranged that it can be locked or sealed open, and it shall not be closed except by an authorized person who shall remain stationed there during that period of the vessel's operation within which the valve remains closed, and who shall again lock or seal the stop valve in the open position before leaving the station.''

Appendix M also contains useful information on safety valve set pressures and the pressure-relieving capacities needed under fire conditions.

Section VIII, Division 1 covers a great variety of vessel types, and provides many more design rules and details than does Section I. Although the general design approach and system of Code construction under third-party inspection employed by both these sections are quite similar, some differences are worth noting:

1. Section VIII has far more materials available.
2. Section VIII permits the use of austenitic stainless steel in water-wetted service. Section I limits such use, due to a concern for the likelihood of IGSCC (intergranular stress corrosion cracking) caused by the concentration of even minute amounts of dissolved chlorides where boiling takes place.
3. Section I permits the design of vessels for only one maximum allowable level of stress and requires that the long seam of a vessel be radiographed. For ordinary service (as opposed to what is known as lethal service), Section VIII varies the allowable stress level (by means of a so-called joint efficiency) in accordance with the amount of radiographic examination provided and the type of weld joint used. For example, a fully radiographed double-welded butt joint would have the same basic allowable stress in both Sections I and VIII. However, Section VIII provides the option of spot radiography or no radiography, in which cases the basic allowable stress is penalized by 15% and 30%, i.e., joint efficiencies of 0.85 and 0.70 are imposed in the design formula used to establish vessel thickness.
4. The Scope paragraph of Section VIII, paragraph U-1, contains a number of interesting exemptions. Vessels with internal or external operating pressures not exceeding 15 psi fall outside the scope of Section VIII. So do ''vessels having an inside diameter, width, height, or cross section diagonal not exceeding 6 in., with no limitation on length of vessel or pressure.'' This latter provision can be used to exempt such components as small-diameter heat exchange coils from Code coverage. There is a converse to the above situation: any exempt vessel can be built and stamped to the Code if it meets the Code rules.
5. Differing from Section I (whose components are usually warm when under design pressure), Section VIII in the mid-1980s imposed a requirement that a minimum design metal temperature (MDMT) be established and marked on every vessel. This is the lowest temperature at which it is safe to apply a specified design pressure without fear of brittle fracture due to the metal temperature being too close to the temperature at which its behavior changes from ductile to brittle. The designer must choose the MDMT based on the circumstances anticipated. If the choice of material and MDMT fails to satisfy certain exemption curves in paragraph UCS-66, impact tests are required to demonstrate that the material has adequate fracture toughness at the MDMT. Also differing from Section I practice, Section VIII recommends

a minimum hydrostatic test temperature 30°F above the minimum design metal temperature, where there is little danger of brittle fracture.

6. Postweld heat treatment rules in Section VIII are more lenient that those in Section I for P No. 1 materials (plain carbon steel).

7. On one question Sections I and VIII agree: Vessels known as unfired steam boilers can be built to either section. See discussion in Chapter 1.

Section IX, Welding and Brazing Qualifications

This section also has a subtitle: Qualification Standard for Welding and Brazing Procedures, Welders, Brazers, and Welding and Brazing Operators. It is a challenge to provide a simple, clear, and concise description of Section IX. Some general prefatory remarks may help. Ever since welding began to replace riveting as a means of joining together the plates and heads comprising pressure vessels, it has been recognized that the welds were potential weak points. Thus, the goal in welding has always been the achievement of sound welds, welds that are as strong as the base metal being joined. Today, the ASME approaches this goal in a two-step process. It first requires a manufacturer to develop and qualify a so-called Weld Procedure Specification (WPS), and thus prove that the particular procedure produces satisfactory welds. It then requires the individual welders to prove that they have the ability to produce sound welds when using a qualified weld procedure. Thus the welding must be done using qualified welders and qualified weld procedures. Moreover, a welder must have proven his or her ability (been qualified) for each process he or she uses. With this explanation in mind, it is possible to use the following excerpts from the Introduction to Section IX to explain its purpose and organization:

> Section IX of the ASME Boiler and Pressure Code relates to the qualification of welders, welding operators, brazers, and brazing operators, and the procedures employed in welding or brazing in accordance with the ASME Boiler and Pressure Vessel Code and the ASME B31 Code for Pressure Piping. Section IX is a document referenced for qualification by various construction codes such as Section I, III, IV, VIII, etc. These particular construction codes apply to specific types of fabrication and may impose additional welding requirements or exemptions from Section IX qualifications. Qualification in accordance with Section IX is not a guarantee that procedures and performance qualifications will be acceptable to a particular construction code.
>
> The purpose of the Welding Procedure Specification (WPS) and Procedure Qualification Record (PQR) is to determine that the weldment proposed for construction is capable of having the required properties for its intended application. It is presupposed that the welder or welding operator performing the welding procedure qualification test is a skilled workman. The procedure qualification test is to establish the properties of the weldment or brazement and not the skill of the personnel performing the welding or brazing. In addition, special consideration is given when notch toughness is required by other Sections of the Code. The notch-toughness variables do not apply unless referenced by the construction codes.
>
> In Welder Performance Qualification, the basic objective is to determine the ability of the welder to deposit sound weld metal.

Section IX describes each welding process in terms called variables that affect the welding. Examples of variables include the type of welding, thickness of metal being joined, P-number (a means of categorizing material), and postweld heat treatment.

Section IX is divided into two parts: welding and brazing. Each part is then divided into four articles, as follows:

a. General requirements (Article I Welding and Article XI Brazing)

b. Procedure qualifications (Article II Welding and Article XII Brazing)

c. Performance qualifications (Article III Welding and Article XIII Brazing)

d. Data (Article IV Welding and Article XIV Brazing)

These articles contain general references and guides that apply to procedure and performance qualifications. They explain the manufacturer's responsibilities, the type and purpose of the various mechanical tests used in procedure and performance qualification, and acceptance criteria for those tests. Also listed are the welding variables used in procedure and performance qualification.

The subject of welding and further explanation of Section IX can be found in Chapter 6.

Section X, Fiber-Reinforced Plastic Vessels

This section of the Code provides construction rules for plastic vessels, which have advantages over metal vessels in special circumstances. These fiber-reinforced plastic pressure vessels have excellent corrosion resistance, they are light in weight, and are advantageous in certain processes where great purity and cleanliness is required. They can, with relative ease, be made in very odd or unusual shapes by using inexpensive wooden forms to shape them.

The fibers used for reinforcing are made of glass, Kevlar, or carbon. The plastic portion of the matrix is a thermoset resin.

The use of fiber-reinforced plastics for the manufacture of pressure vessels presents unique material considerations in design, fabrication, and testing. Metallic vessels, made from materials that are normally isotropic and ductile, are designed using well-established allowable stresses based on measured ductility and tensile properties. In contrast, fiber-reinforced plastics are usually anisotropic, and the physical properties are dependent upon the fabrication process, the placement and orientation of the reinforcement, and the resin matrix.

It is not possible to fabricate a reinforced plastic pressure vessel of a single basic material for which there is an ASTM specification. The vessel parts are made up of various basic materials, such as fiber reinforcement and resin, which are joined in the presence of a catalyst to create a composite material that is formed into a vessel or vessel part by a specified process. The composite material will often have directional properties that must be considered in design.

Reinforced plastic vessels are designed by one of two basic methods:

a. Class I Design—Qualification of a vessel design by pressure testing one or more prototype vessels. This testing is done in two stages. First, the pressure is cycled from atmospheric pressure to design pressure 100,000 times. No leakage may occur during this stage. Second, following a successful pressure cycling test, the vessel is subjected to another pressure test, really a proof test, to determine its so-called qualification pressure. This pressure must be at least 6 times the vessel design pressure. Thus the prototype has proven the design is acceptable if it can sustain 100,000 full design pressure cycles and still contain a pressure 6 times the design pressure. This particular prototype vessel may not be stamped and put into service, since it can be considered used, and well used at that. Subsequent duplicate vessels must be built in strict accordance with the procedure specification used in fabricating the prototype, to assure that they will be as good as the prototype. These Class 1 vessels are often mass-produced.

b. Class II Design—Code-specified mandatory design rules followed by nondestructive examination and acceptance testing using acoustic emission testing while the vessel is subjected to programmed increasing stress levels up to the stress caused by 1.1 times design pressure. No leakage is permissible during the acceptance testing.

Occasionally, some part of one of these plastic vessels is made of metal, for example, to facilitate a bolted attachment. In such cases, the metal parts must be constructed in accordance with Section VIII. Although not commonly used, fiber-reinforced plastic vessels serve well in special applications.

Section XI, Inservice Inspection of Nuclear Plant Components

Normally the ASME Code covers new construction only. Section XI is the major exception; it provides rules for the examination, testing, and inspection of components and systems in operating (or shut down) nuclear power plants. The application of Section XI rules begins when the construction of the plant is completed, that is, they begin where Section III rules leave off. Unlike Section VI and Section VII, which are only suggested rules for the care and maintenance of heating and power boilers, the Section XI rules are mandatory requirements for maintaining a nuclear power plant and for safely and quickly returning it to service following outages. The rules provide for a program of periodic examination, testing, and inspection, to furnish ongoing evidence of safety. The rules also stipulate duties of the Authorized Nuclear Inservice Inspector in order to verify the successful completion of the programmed inspection and testing and permit the plant to return to service.

Section XI has three major divisions, dealing with the three major types of nuclear power plants: light-water cooled plants, gas-cooled plants, and liquid-metal cooled plants. For light-water cooled plants, some of the major headings covered are these:

- General Requirements
- Requirements for Class 1 Components
- Requirements for Class MC Components (the metal containment vessels or liners)
- In-service Testing of Pumps
- In-service Testing of Valves

Just within the category of General Requirements are these major headings:

- Scope and Responsibility
- Examination and Inspection
- Standards for Examination Evaluation
- Repair and Replacement
- System Pressure Tests
- Records and Reports

Section XI uses techniques such as flaw growth analysis to determine acceptability of components for continued service. Section XI also provides detailed repair procedures and permits temper bead welding to obviate postweld heat treatment. The inspections and tests are scheduled on a 10-year cycle. Some inspections or tests may be required only once in 10 years; others are more frequent.

An unusual aspect of Section IX is that the edition in effect when the plant was completed may continue to be used, or the owner may choose a later or the latest edition. However, the Nuclear Regulatory Commission (NRC) requires owners to update their inspection and testing program to the latest edition of Section XI endorsed by the NRC (in 1998 it was the 1989 edition). The update has to be completed by the beginning of the next 10-year inspection cycle.

Section XII, Transport Tanks

The large tanks used to transport hazardous materials have long been governed by U.S. Department of Transportation regulations. In 1997, the ASME formed a new committee, Subcommittee XII, to develop a new Code section, Section XII, Transport Tanks. This section will provide rules for the design, construction, and continued operation of tanks used to carry dangerous goods by all means of transport (road, rail, air, and sea). The committee's charge is to develop codes and standards suitable for reference by regulatory authorities and safety organizations worldwide. As is usual in book sections covering pressure vessels, the term **construction** is an all inclusive one intended to cover materials, design, fabrication, examination, inspection, testing, certification, and pressure relief. Since Subcommittee XII has just started its work, it may take several years before Section XII is approved and published.

ASME Code for Pressure Piping, B31

This is a large code, covering many types of piping. Individual code sections are the following:

B31.1   Power Piping
B31.3   Process Piping
B31.4   Liquid Transportation Systems for Hydrocarbons
B31.5   Refrigeration Piping
B31.8   Gas Transmission and Distribution Piping Systems
B31.9   Building Services Piping
B31.11 Slurry Transportation Piping Systems

A related manual is *Manual for Determining the Remaining Strength of Corroded Pipelines: A Supplement to B31, Code for Pressure Piping.*

The two sections of the piping code most used in association with power boilers are B31.1, Power Piping, and B31.3, Process Piping. These are briefly described below.

B31.1, Power Piping

The Power Piping Code is intended for use in conjunction with Power Boilers, Section I of the ASME Code. The stated philosophy of B31.1 is to parallel the provisions of Section I as they can be applied to power piping systems. The allowable stress basis for both codes is the same. Some engineers feel that B31.1 is a more conservative piping code than some other piping codes, and its foreword states that this reflects the need for long service life and maximum reliability in power plant installations. However, it is arguable that other codes (B31.3, for example) achieve comparable conservatism through other means. The foreword to B31.1 also notes that the Code often warns a piping designer, fabricator, or erector against possible pitfalls, but that it is not a handbook and cannot substitute for education, experience, and sound engineering judgment. (Similar advice was recently added to the forewords of Sections I through XI of the ASME Code.) The foreword to B31.1 goes on to explain that "a designer who is capable of a more rigorous analysis than is specified in the Code may justify a less conservative design, and still satisfy the basic intent of the Code." This provision can sometimes be very valuable, provided the designer is able to demonstrate the validity of his or her approach and obtain its acceptance by oversight or regulatory personnel.

Sometimes it is not clear which section of the B31 piping code covers a specific piping system. In the early 1990s, the introduction to the Piping Code was revised to note that in such cases the owner or user may choose any section determined to be generally applicable, with the caveat that additional requirements may be necessary to provide a safe system. The Introduction to B31.1 says of such choices: "Technical limitations of the various sections, legal requirements, and possible applications of other codes or standards are some of the factors to be considered by the user in determining the applicability of any section of this Code."

As explained in its introduction, the major topics covered by B31.1, Power Piping are these:

1. Requirements for design of components and assemblies, including pipe supports
2. Requirements and data for evaluation and limitation of stresses, reactions, and movements associated with pressure, temperature changes, and other forces
3. Guidance and limitations on the selection and application of materials, components, and joining methods
4. Requirements for the fabrication, assembly, and erection of piping
5. Requirements for examination, inspection, and testing of piping
6. References to acceptable material specifications and component standards, including dimensional requirements and pressure–temperature ratings

The design rules of B31.1 are little changed from those of the 1955 edition of the Code for Pressure Piping, in which Chapter 1 set forth rules for Power Piping. Some consider them to be conservative, but others defend this conservatism by noting that there is no assurance that the power piping systems covered by B31.1 will receive adequate long-term inspection, maintenance, and repair. Moreover, personnel working near these exposed piping systems are at risk from any failures. This contrasts with the possibly less hazardous situation of a tube failure within the boiler enclosure. The basic design approach used in the B31.1 code is to size piping components for internal pressure, using formulas and stresses very similar to those in Section I. The allowable stresses appear, conveniently, in the appendix.

B31.1 sizes components for internal pressure based on an allowable circumferential stress (that stress is a membrane stress). Longitudinal stress due to the combined effects of pressure, weight and other sustained loads is limited to the basic allowable stress in tension at design temperature. A separate calculation, with a higher allowable stress, is used to limit the stresses due to bending moments caused by cyclic loads (most often due to thermal expansion of the piping system). A simplified fatigue analysis applies if more than 7000 cycles of thermal expansion stress are anticipated, and reduces the allowable stress range in proportion to the number of stress cycles above 7000. At 100,000 cycles or more, the so-called stress range reduction factor reaches a minimum value of 0.5. However, the value of 0.5 should not be construed as establishing an endurance limit, suitable for evaluation of high cycle vibration stress.

B31.1 has a category known as Occasional Loads (paragraph 104.8.2). When occasional loads are combined with sustained loads, the allowable stress is increased by as much as 20%, depending on the duration of the occasional load. Thus stresses up to 20% higher than normally allowed are permitted for short-term loadings, such as those produced by seismic events, in combination with other, sustained loads.

B31.1 contains 13 mandatory and nonmandatory appendices covering allowable stresses, materials data (coefficients of thermal expansion and elastic moduli), flexibility and stress intensification factors for various piping components and configurations, and reference standards. There is a nonmandatory appendix, Appendix II, that deals with safety valve installations, including the calculation of safety valve discharge forces. Other nonmandatory appendices deal with nonmetallic piping, corrosion control, and design of restrained underground piping. Also included is a useful 12-page section, Recommended Practice for Operation, Maintenance, and Modification of Power Piping Systems. Like the book sections of the Boiler and Pressure Vessel Code, B31.1 issues Code Cases and Interpretations, and these are issued with the B31.1 Code as a supplement.

B31.1 has the technical responsibility for the rules covering materials, design, fabrication, installation and testing of certain boiler piping known as boiler external piping. This situation is explained in Chapter 4.

B31.3, Process Piping (Until 1996 known as Chemical Plant and Petroleum Refinery Piping)

This piping code dates from 1959, when the refinery chapter (Chapter 3 of the Code for Pressure Piping) emerged as a stand-alone document, Petroleum Plant Piping. In 1976, a draft Code section for Chemical Plant Piping was merged with the existing Code Section for Petroleum Plant Piping and renamed B31.3, Chemical Plant and Petroleum Refinery Piping. In 1980, B31.3 assimilated a draft Code section for Cryogenic Piping, and in 1996, the name of the document was simplified to Process Piping. As can be seen from its introduction, B31.3 is in many respects similar to B31.1, but there are notable differences. Different criteria are used in setting allowable stresses, which are essentially the same as those used in Section VIII, Division 2 for temperatures below the creep range and the same as Division 1 stresses for temperatures within the creep range. (See 302.3.2 of B31.3 for further particulars.) In general, B31.3 takes a broader approach to piping design than B31.1 since it deals with a greater variety of piping applications. It covers a variety of fluid services, with different requirements for each service, depending on how dangerous the fluid is considered to be. Categories of fluid service (e.g., Category M fluid service) are established to describe the combination of fluid properties (e.g., reduced hazard, flammable, toxic), operating conditions (low or high pressure and/or temperature), and other factors that establish different bases for the design of a piping system.

B31.3 in its paragraph 300(b) establishes the responsibilities of six entities involved in the construction of a process piping system. These are the owner; the designer; the manufacturer, fabricator, and the erector of the system; and the owner's inspector. In particular, the owner has overall responsibility for compliance with the B31.3 Code, for establishing all requirements necessary for compliance, and for designating the piping into appropriate categories of fluid service.

In its scope paragraph 300.1.1, B31.3 with some exceptions asserts coverage for all fluids, including:

1. Raw, intermediate, and finished chemicals
2. Petroleum products
3. Gas, steam, air, and water
4. Fluidized solids
5. Refrigerants

One difference between B31.3 and B31.1 is that the former establishes hardness limits for welds (as a means of assuring that the postweld heat treatment was carried out properly and to prevent certain types of stress corrosion cracking such as in what is known as sour oil service). These limits require testing to determine hardness and also present the problem of what to do when the weld is too hard, as can happen with P-91, a P-No.5 material. P-91 is a very strong (and therefore very hard) material, not even invented when the hardness limits for P-No. 5 materials were established. If additional postweld heat treatment is used to soften the welds, the long-term creep strength may be adversely affected. This problem has yet to be resolved.

APPENDIX IV

DESIGN EXERCISES

DESIGN EXERCISE NO. 1, DESIGN OF A HEADER

This first problem illustrates the use of Section I's method of design-by-rule, using the rules of PG-27 to design a header. The reader is advised to review Chapter 4 and PG-27 before attempting to solve the problem or looking at the solution. When reviewing PG-27, note that slightly different formulas are used for designing pipe and tubes.

It has been decided to make a certain cylindrical header from pipe, using 10-inch nominal pipe size (NPS) of appropriate wall thickness for a maximum allowable working pressure of 1020 psi. Applying Section I formula PG-27.2.2 with a 10.75-inch outside diameter and an allowable stress of 10,800 psi (at an 800°F design temperature) for SA-106 Grade B pipe material, determine the minimum wall thickness required. Then select the lightest pipe adequate for this design pressure from a table showing standard pipe sizes and wall thicknesses varying according to schedule number. Assume the constant C (for threading and structural stability) in formula PG-27.2.2 is zero, and determine the temperature coefficient y from the table in Note 6 of PG-27.4. Assume also that no corrosion allowance is needed, that the hydrostatic head which normally must be included in the design pressure when designing pipe is negligible in this case, and that there are no openings or connections which would require compensation or ligament efficiency calculations. For ready reference, PG-27 and a table of pipe sizes have been included at the end of this appendix.

SOLUTION TO DESIGN EXERCISE NO. 1

After using Note 1 of PG-27.4 to determine the efficiency, E, for seamless pipe to be 1.00, Note 3 of PG-27.4 to verify that the constant C is zero, and the table in Note 6 of PG-27.4 to determine the temperature coefficient y to be 0.4, we have all factors needed to solve formula PG-27.2.2. We can then find from that formula that the minimum thickness required is 0.49 inches, as follows:

$$t = \frac{PD}{2SE + 2yP} + C = \frac{(1020)(10.75)}{2(10,800) + 2(0.4)(1020)} = 0.49 \text{ inches}$$

We now go to a standard table of pipe sizes and thickness that derives from ANSI Standard B36.10, Welded and Seamless Wrought Steel Pipe. Note that for any diameter, the wall thickness varies in accordance with the so-called schedule number of the pipe. The table shows that schedule 60 (formerly called extra-strong) NPS 10 appears to have just about the wall thickness we need: 0.50 inches. However, this is *not* the case, because the thickness listed in the table is nominal and is subject to an undertolerance, as explained in Note 7 of PG-27.4. This 12.5% undertolerance on thickness is given in ASME/ASTM Specification SA-530 covering general requirements for carbon and alloy steel pipe. It is also found in the individual pipe specifications listed in PG-9, for example, in paragraph 19.3 of the SA-106 pipe specification. Accordingly, we must go to the next commonly available wall thickness among the specifications approved by ASME: Schedule 80, with a nominal wall thickness of 0.594 inches. Even with an undertolerance of 12.5%, the minimum thickness at any point in this pipe would be 0.875 × 0.594 inches = 0.52 inches, which satisfies our required minimum of 0.49 inches. Our final choice would therefore be an NPS 10 schedule 80 SA-106 Grade B pipe of whatever length is required.

DESIGN EXERCISE NO. 2, DESIGN OF A HEAD

Design Exercise No. 1 consisted of the design of a cylindrical header, which turned out to be an NPS 10 schedule 80 SA-106 Grade B pipe (with a 10.75-in. OD and a nominal wall thickness of 0.594 in.). In this exercise, various heads for that header will be designed, starting with a so-called dished head, using the rules of PG-29.1. (Dished heads are usually standard catalog items furnished as standard pressure parts under the rules of PG-11.) Again, the reader is advised to review the portion of Chapter 4 dealing with dished heads, and PG-29, before proceeding. For ready reference, PG-29 is also included at the end of this appendix.

(A) For the same maximum allowable working pressure of 1020 psi and design temperature of 800°F, determine the required thickness of a blank unstayed dished head with a 10-inch radius L on the concave side of the head. Assume the head is made of SA-516 Grade 70 material. The maximum allowable working stress, S, from Section II, Part D is determined to be 12,000 psi. Verify this value if you have a copy of Section II, Part D available. Notice that the allowable stress for this material is higher than that of the pipe material.

(B) PG-29.7 provides a rule for the design of a semiellipsoidal shaped head. What minimum thickness would be needed if this type of head were used?

(C) PG-29.11 provides the rules for designing a full hemispherical shaped head. What minimum thickness would be required if this type of head were used?

SOLUTION TO DESIGN EXERCISE NO. 2 (A)

$$t = \frac{5PL}{4.8S} = \frac{5(1020)(10)}{4.8(12,000)} = 0.89 \text{ inches}$$

SOLUTION TO DESIGN EXERCISE NO. 2 (B)

$$t = 0.49 \text{ inches}$$

Explanation: PG-29.7 says that a semiellipsoidal head shall be *at least as thick as the required thickness of a seamless shell of the same diameter*. Thus, the answer derived in Design Exercise No. 1 applies.

SOLUTION TO DESIGN EXERCISE NO. 2 (C)

$$t = \frac{PL}{1.6S} = \frac{1020(10.75 - 2 \times 0.594)}{2(1.6)(12,000)} = 0.25 \text{ inches}$$

Note that L is the inside radius of the head. In this case, it was chosen to match the nominal inside radius of the piece of pipe used for the shell in Design Exercise No. 1.

From the results of these three head thickness calculations, it is clear that the hemispherical head is the most efficient shape, requiring a thickness only a fraction of that required for a dished or an ellipsoidal head, and only about half that of the header to which it is attached. This is because a hemispherical head carries pressure loads by developing predominantly membrane stresses, while the ellipsoidal and especially the dished heads develop significant bending stresses in addition to the membrane stresses.

Note also that PG-16.3 establishes the minimum thickness of boiler plate under pressure as ¼ inch in most cases.

DESIGN EXERCISE NO. 3, CHOICE OF FEEDWATER STOP VALVE

In Chapter 11, under the heading Valves and Valve Ratings, it was explained that most boiler valves are furnished in compliance with the ASME/ANSI product standard for valves, ASME/ANSI B16.34, Steel Valves—Flanged, Threaded, and Welding End. This standard provides a series of tables, for various material groups, giving pressure–temperature ratings according to what are called pressure classes. The classes typically vary from the 150 class (the weakest) to the 4500 class. For each class, the allowable pressure is tabulated as a function of temperature. The allowable pressure at any temperature is the pressure rating of the valve at that temperature. Temperatures range from the lowest zone, $-20°F$ to $100°F$, to $850°F$ and higher, depending on the material group. The allowable pressure for the valve falls with increasing temperature, since it is based on the allowable stress for the material.

Also as explained in Chapter 11, the designer of a valve for Section I service must select a valve whose pressure rating at design temperature is adequate for the maximum allowable pressure (MAWP) of the boiler. However, feedwater and blowoff piping and valves are designed for a pressure higher than the MAWP of the boiler, because they are in a more severe type of service called shock service. This complicates the choice of valve somewhat, and to clarify the choice, the committee issued Interpretation I-83-91, which is discussed in Chapter 11. The following example illustrates the choice of a feedwater stop valve following the procedure outlined in the fourth reply of that interpretation.

The problem here is to select a feedwater stop valve for a boiler whose maximum allowable working pressure (MAWP) is 1500 psi. Note that the design rules for a valve that is part of the boiler external piping are found in paragraph 122 of B31.1, Power Piping.

SOLUTION TO DESIGN EXERCISE NO. 3

The first step in this problem is to determine the design pressure and temperature required for this valve. By paragraph 122.1.3 of B31.1, the design pressure of a feedwater valve must exceed the MAWP of the boiler by 25% or 225 psi, whichever is the lesser:

$$1500 \text{ psi} + 225 \text{ psi} = 1725 \text{ psi}$$
$$1500 \text{ psi} \times 1.25 = 1875 \text{ psi}$$

The design pressure P_F is taken as the lesser value, 1725 psi.

The design temperature is found from paragraph 122.1.3(B) of B31.1 to be the saturation temperature of the boiler. That temperature for a MAWP of 1500 psi is found from any steam table to be 596°F. Thus we must now find a valve designed for 1725 psi at 596°F.

Next turn to the tables of valve materials and ratings from the governing valve standard ASME/ANSI B16.34, Valves—Flanged, Threaded, and Welding End (for ready reference, these pages are included at the end of this appendix). Assume the valve material is a forging, SA-105, which is a carbon steel suitable for the relatively low temperature of feedwater service. Notice that the standard uses the ASTM designation A-105. It is seen from the Table 1 list of material specifications that A-105 falls in what is known as material group 1.1.

The task now is to find the lowest valve class adequate for the design conditions we have established. Turning to Table 2-1.1, the ratings for group 1.1 materials, we enter with the design temperature, rounded up to 600°F, and look for a valve class adequate for 1725 psi. The first such class is a 1500 (standard class) valve. However, an alternative choice is available from the ratings table for special class valves. From that table, it is seen that a 900 special class valve would also be adequate. (Remember that the difference between the ratings of these two classes is based on the amount of NDE used in their manufacture.) The final choice of valve then becomes one of cost and delivery time.

DESIGN EXERCISE NO. 4, DESIGN OF A TUBE FOR LUG LOADING

The PW-43 method for determining the allowable load on a tube lug is explained in Chapter 4, under the heading Loading on Structural Attachments. When PW-43 was revised in 1992, the sample problems in the Section I Appendix paragraphs A-71 to A-74 illustrating the determination of allowable loading on tube lugs were revised to show the new method. Somehow, the explanation of the examples as printed was not as clear as the subcommittee had intended, and the examples are a little hard to follow. Accordingly, the first of the sample problems, that shown in A-71, is explained here to serve as a guide to the method. The problem is described as follows.

A tube is suspended by a welded attachment with a 1500-lb. design load and the dimensions shown in Fig. A-71. This is a condition of direct radial loading on the tube. The allowable lug loading is calculated for the following conditions:

Tube material is SA-213 T-22
MAWP = 2258 psi
T = 800°F
Tube diameter D = 4.0 in.
Tube wall thickness t = 0.30 in.
Lug thickness = ¼ in.
Lug attachment angle (the angle subtended by the lug width, see Table PW-43.1) = 7°
Design stress for tube, S_a = 15,000 psi

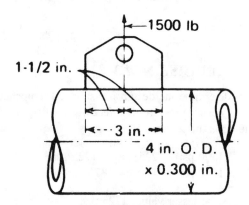

FIGURE A-71
LUG LOAD ON TUBE
(REPRODUCED FROM ASME SECTION I)

SOLUTION TO DESIGN EXERCISE NO. 4

The first step in the solution to this problem is to determine K, the tube attachment angle design factor, by interpolation in Table PW-43.1. K is found to be 1.07. Also needed is the value of the variable X, which is defined in PW-43.2. In this example,

$$X = D/t^2 = (4.0/0.3^2) = 44.4$$

The load factor L_f is now determined, either by reading it from the plot in Fig. PW-43.1 or by using the appropriate load factor equation in PW-43.2.1 or PW-43.2.2, depending on whether the lug is applying tension or compression loading to the tube.

The load factor equations are as follows:

For compression loading,

$$L_f = 1.618X^{[-1.020-0.014(\log X)+0.005(\log X)^2]}$$

For tension loading,

$$L_f = 49.937X^{[-2.978+0.898(\log X)-0.139(\log X)^2]}$$

In either case, the load factor is a function of X, which has already been determined to be 44.4. It is hard to read the plot with an accuracy greater than two significant figures (which ought to be good enough for the design of a tube lug). However, if the equations are used, greater accuracy can be obtained. Note that the log terms in the equations are logarithms to the base 10. Substituting $X = 44.4$ in those equations gives a tension load factor $L_f = 0.0405$ and a compression load factor $L_f = 0.0326$. The problem as stated concerns a tension load on the lug, but the values of both load factors are derived in order to illustrate that there can be a significant difference in allowable load, depending on the direction of the load.

The next step is to determine S_t, the amount of the allowable stress that is available for the lug loading. S_t is found from the equation in PW-43.2.3:

$$S_t = 2.0\,S_a - S$$

S_t is thus seen to be twice the allowable tube stress less the membrane pressure stress S at MAWP determined from equation PG-27.2.1. (Notice that if there were no stress due to pressure, equation PW-43.2.3 would give a value of S_t equal to twice the allowable membrane pressure stress. This is not unreasonable because the bending stress in the wall of the tube caused by lug loading is considered a secondary bending stress.) In the present example, S is found to be 15,000 psi. Thus

$$S_t = 2(15,000) - 15,000 = 15,000$$

L_a, the maximum allowable unit load on the attachment, can now be determined from equation PW-43.2.4:

$$\text{Tension } L_a = K(L_f)S_t = (1.07)(0.0405)(15,000) = 650 \text{ lb/in.}$$

The comparable unit load in compression is:

$$\text{Compression } L_a = K(L_f)S_t = (1.07)(0.0326)(15,000) = 523 \text{ lb/in.}$$

Since the design load is 1500 lb in tension, applied at the center of a 3-inch-long lug, the actual unit load is only 500 lb/in., which is less than the maximum permitted. The design is thus seen to be satisfactory.

PG-27 CYLINDRICAL COMPONENTS UNDER INTERNAL PRESSURE (Reproduced From ASME Section I)

PG-27.1 General. The formulas under this paragraph shall be used to determine the minimum required thickness or the maximum allowable working pressure of piping, tubes, drums, and headers in accordance with the appropriate dimensional categories as given in PG-27.2.1, PG-27.2.2, and PG-27.2.3 for temperatures not exceeding those given for the various materials listed in Tables 1A and 1B of Section II, Part D.

The calculated and ordered thickness of material must include the requirements of PG-16.2, PG-16.3, and PG-16.4. Stress calculations must include the loadings as defined in PG-22 unless the formula is noted otherwise.

When required by the provisions of this Code, allowance must be provided in material thickness for threading and minimum structural stability. (See PWT-9.2 and PG-27.4, Notes 3 and 5.)

PG-27.2 Formulas for Calculation
PG-27.2.1 Tubing—Up to and Including 5 in. (127 mm) Outside Diameter

$$t = \frac{PD}{2S + P} + 0.005D + e$$

$$P = S\left[\frac{2t - 0.01D - 2e}{D - (t - 0.005D - e)}\right]$$

See PG-27.4, Notes 2, 4, 8, and 10.

PG-27.2.1.1 For tubes of the materials listed in its title, Table PWT-10 may be used in lieu of the formula for determining the minimum wall thickness of tubes where expanded into drums or headers, provided the maximum mean wall temperature does not exceed 700°F (371°C).

PG-27.2.1.2 The wall thickness of the ends of tubes strength-welded to headers or drums need not be made greater than the run of the tube as determined by this formula.

PG-27.2.1.3 The wall thickness of the ends of tubes permitted to be attached by threading under the limitations of PWT-9.2 shall be not less than t as determined by this formula, plus $0.8/n$, where n equals the number of threads per inch.

PG-27.2.1.4 A tube in which a fusible plug is to be installed shall be not less than 0.22 in. (5.6 mm) in thickness at the plug in order to secure four full threads for the plug (see also A-20).

PG-27.2.1.5 Bimetallic tubes meeting the requirements of PG-9.4 shall use as an outside diameter D in the equation in PG-27.2.1 no less than the calculated outside diameter of the core material. The outside diameter of the core material shall be determined by subtracting the minimum thickness of the cladding from the outside diameter of the bimetallic tube, including the maximum plus tolerance. The minimum required thickness t shall apply only to the core material.

PG-27.2.2 Piping, Drums, and Headers. (Based on strength of weakest course)

$$t = \frac{PD}{2SE + 2yP} + C \quad \text{or} \quad \frac{PR}{SE - (1 - y)P} + C$$

$$P = \frac{2SE(t - C)}{D - 2y(t - C)} \quad \text{or} \quad \frac{SE(t - C)}{R + (1 - y)(t - C)}$$

See PG 27.4, Notes 1, 3, and 5 through 9.

PG-27.2.3 Thickness Greater Than One-Half the Inside Radius of the Component. The maximum allowable working pressure for parts of boilers of cylindrical cross section, designed for temperatures up

249

to that of saturated steam at critical pressure [705.4°F (374.1°C)], shall be determined by the formulas in A-125.

PG-27.3 Symbols. Symbols used in the preceding formulas are defined as follows:

t = minimum required thickness, in. (see PG-27.4, Note 7)
P = maximum allowable working pressure, psi (see PG-21)
D = outside diameter of cylinder, in.
R = inside radius of cylinder, in.
E = efficiency (see PG-27.4, Note 1)
S = maximum allowable stress value at the design temperature of the metal, as listed in the tables specified in PG-23, psi (see PG-27.4, Note 2)
C = minimum allowance for threading and structural stability, in. (see PG-27.4, Note 3)
e = thickness factor for expanded tube ends (see PG-27.4, Note 4)
y = temperature coefficient (see PG-27.4, Note 6)

PG-27.4 Notes. Notes referenced in the preceding formulas are as follows:

Note 1:

E = 1.00 for seamless or welded cylinders
   = the efficiency from PG-52 or PG-53 for ligaments between openings

Note 2:

For tubes the temperature of the metal for selecting the S value shall be not less than the maximum expected mean wall temperature (sum of outside and inside surface temperature divided by 2) of the tube wall, which in no case shall be taken as less than 700°F for tubes absorbing heat. For tubes which do not absorb heat, the metal temperature may be taken as the temperature of the fluid within the tube but not less than the saturation temperature.

Note 3:

Any additive thickness represented by the general term C may be considered to be applied on the outside, the inside, or both. It is the responsibility of the designer using these formulas to make the appropriate selection of diameter or radius to correspond to the intended location and magnitude of this added thickness. The pressure- or stress-related terms in the formula should be evaluated using the diameter (or radius) and the remaining thickness which would exist if the "additive" thickness had not been applied or is imagined to have been entirely removed.

The values of C below do not include any allowance for corrosion and/or erosion, and additional thickness should be provided where they are expected. Likewise, this allowance for threading and minimum structural stability is not intended to provide for conditions of misapplied external loads or for mechanical abuse.

Type of Pipe	Value of C^b, in.
Threaded steel, or nonferrous pipe[a]	
¾ in. nominal and smaller	0.065
1 in. nominal and larger	Depth of thread h^c
Plain end[d] steel, or nonferrous pipe	
3½ in., nominal and smaller	0.065
4 in., nominal and larger	0

(*a*) Steel or nonferrous pipe lighter than Schedule 40 of ASME B36.10M, Welded and Seamless Wrought Steel Pipe, shall not be threaded.

(*b*) The values of C stipulated above are such that the actual stress due to internal pressure in the wall of the pipe is no greater than the values of S given in Table 1A of Section II, Part D, as applicable in the formulas.

(c) The depth of thread h in in. may be determined from the formula $h = 0.8/n$, where n is the number of threads per inch or from the following:

n	h
8	0.100
11½	0.0696

(d) Plain-end pipe includes pipe jointed by flared compression couplings, lap (Van Stone) joints, and by welding, i.e., by any method which does not reduce the wall thickness of pipe at the joint.

Note 4:

    $e = 0.04$ over a length at least equal to the length of the seat plus 1 in. for tubes expanded into tube seats, except

    $= 0$ for tubes expanded into tube seats provided the thickness of the tube ends over a length of the seat plus 1 in. is not less than the following:

        0.095 in. for tubes 1¼ in. O.D. and smaller

        0.105 in. for tubes above 1¼ in. O.D. and up to 2 in. O.D., incl.

        0.120 in. for tubes above 2 in. O.D. and up to 3 in. O.D., incl.

        0.135 in. for tubes above 3 in. O.D. and up to 4 in. O.D., incl.

        0.150 in. for tubes above 4 in. O.D. and up to 5 in. O.D., incl.

    $= 0$ for tubes strength-welded to headers and drums

Note 5:

While the thickness given by the formula is theoretically ample to take care of both bursting pressure and material removed in threading, when steel pipe is threaded and used for steam pressures of 250 psi and over, it shall be seamless and of a weight at least equal to Schedule 80 in order to furnish added mechanical strength.

Note 6:

    $y =$ a coefficient having values as follows:

	Temperature, °F							
	900 and below	950	1000	1050	1100	1150	1200	1250 and above
Ferritic	0.4	0.5	0.7	0.7	0.7	0.7	0.7	0.7
Austenitic	0.4	0.4	0.4	0.4	0.5	0.7	0.7	0.7
Alloy 800	0.4	0.4	0.4	0.4	0.4	0.4	0.5	0.7
800H	0.4	0.4	0.4	0.4	0.4	0.4	0.5	0.7
825	0.4	0.4	0.4	...	...	...	...	...

Values of y between temperatures listed may be determined by interpolation. For nonferrous materials, $y = 0.4$.

Note 7:

If pipe is ordered by its nominal wall thickness, as is customary in trade practice, the manufacturing tolerance on wall thickness must be taken into account. After the minimum pipe wall thickness t is determined by the formula, this minimum thickness shall be increased by an amount sufficient to provide the manufacturing tolerance allowed in the applicable pipe specification. The next heavier commercial wall thickness may then be selected from Standard thickness schedules as contained in ASME B36.10M. The manufacturing tolerances are given in the several pipe specifications listed in PG-9.

Note 8:

When computing the allowable pressure for a pipe of a definite minimum wall thickness, the value obtained by the formulas may be rounded out to the next higher unit of 10.

Note 9:

Inside backing strips, when used at longitudinal welded joints, shall be removed and the weld surface prepared for radiographic examination as required. Inside backing rings may remain at circumferential welded seams of cylinders provided such construction complies with requirements of PW-41.

Note 10:

The maximum allowable working pressure P need not include the hydrostatic head loading, PG-22, when used in this formula.

PG-29 DISHED HEADS (Reproduced From ASME Section I)

PG-29.1 The thickness of a blank unstayed dished head with the pressure on the concave side, when it is a segment of a sphere, shall be calculated by the following formula:

$$t = 5PL/4.8S$$

where

t = minimum thickness of head, in.
P = maximum allowable working pressure, psi (hydrostatic head loading need not be included)
L = radius to which the head is dished, measured on the concave side of the head, in.
S = maximum allowable working stress, psi, using values given in Table 1A of Section II, Part D.

PG-29.2 The radius to which a head is dished shall be not greater than the outside diameter of flanged portion of the head. Where two radii are used the longer shall be taken as the value of L in the formula.

PG-29.3 When a head dished to a segment of a sphere has a flanged-in manhole or access opening that exceeds 6 in. (152 mm) in any dimension, the thickness shall be increased by not less than 15% of the required thickness for a blank head computed by the above formula, but in no case less than ⅛ in. (3.2 mm) additional thickness over a blank head. Where such a dished head has a flanged opening supported by an attached flue, an increase in thickness over that for a blank head is not required. If more than one manhole is inserted in a head, the thickness of which is calculated by this rule, the minimum distance between the openings shall be not less than one-fourth of the outside diameter of the head.

PG-29.4 Except as otherwise provided for in PG-29.3, PG-29.7, and PG-29.12, all openings which require reinforcement, placed in a head dished to a segment of a sphere, or in an ellipsoidal head, or in a full-hemispherical head, including all types of manholes except those of the integral flanged-in type, shall be reinforced in accordance with the rules in PW-15 and PW-16.

When so reinforced, the thickness of such a head may be the same as for a blank unstayed head.

PG-29.5 Where the radius L to which the head is dished is less than 80% of the diameter of the shell, the thickness of a head with a flanged-in manhole opening shall be at least that found by making L equal to 80% of the diameter of the shell and with the added thickness for the manhole. This thickness shall be the minimum thickness of a head with a flanged-in manhole opening for any form of head and the maximum allowable working stress shall not exceed the values given in Table 1A of Section II, Part D.

PG-29.6 No head, except a full-hemispherical head, shall be of a lesser thickness than that required for a seamless shell of the same diameter.

PG-29.7 A blank head of a semiellipsoidal form in which half the minor axis or the depth of the head is at least equal to one-quarter of the inside diameter of the head shall be made at least as thick as the required thickness of a seamless shell of the same diameter as provided in PG-27.2.2. If a flanged-in manhole which meets the Code requirements is placed in an ellipsoidal head, the thickness of the head shall be the same as for a head dished to a segment of a sphere (see PG-29.1 and PG-29.5) with a dish radius equal to eight-tenths the diameter of the shell and with the added thickness for the manhole as specified in PG-29.3.

PG-29.8 When heads are made to an approximate ellipsoidal shape, the inner surface of such heads must lie outside and not inside of a true ellipse drawn with the major axis equal to the inside diameter of the head and one-half the minor axis equal to the depth of the head. The maximum variation from this true ellipse shall not exceed 0.0125 times the inside diameter of the head.

PG-29.9 Unstayed dished heads with the pressure on the convex side shall have a maximum allowable working pressure equal to 60% of that for heads of the same dimensions with the pressure on the concave side.

Head thicknesses obtained by using the formulas in PG-29.11 for hemispherical heads and PG-29.7 for blank semielliipsoidal heads do not apply to heads with pressure on the convex side.

PG-29.10 When a flange of an unstayed dished head is machined to make a close and accurate fit into or onto the shell, the thickness shall not be reduced to less than 90% of that required for a blank head.

PG-29.11 The thickness of a blank unstayed full-hemispherical head with the pressure on the concave side shall be calculated by the following formulas:

$$t = \frac{PL}{1.6S} \tag{1}$$

$$t = \frac{PL}{2S - 0.2P} \tag{2}$$

where

  t = minimum thickness of head, in.
  P = maximum allowable working pressure, psi
  L = radius to which the head was formed, measured on the concave side of the head, in.
  S = maximum allowable working stress, psi, using values given in Table 1A of Section II, Part D

Use Formula 1; however, Formula 2 may be used for heads exceeding ½ in. (13 mm) in thickness that are to be used with shells or headers designed under the provisions of PG-27.2.2 and that are integrally formed on seamless drums or are attached by fusion welding, and do not require staying. Where Formula 2 is used, the other requirements of both of these paragraphs shall also apply.

The above formulas shall not be used when the required thickness of the head given by these formulas exceeds 35.6% of the inside radius, and instead, the following formula shall be used:

$$t = L(Y^{1/3} - 1)$$

where

$$Y = \frac{2(S + P)}{2S - P}$$

Joints in full-hemispherical heads including the joint to the shell shall be governed by and meet all the requirements for longitudinal joints in cylindrical shells, except that in a buttwelded joint attaching a head to a shell the middle lines of the plate thicknesses need not be in alignment.

PG-29.12 If a flanged-in manhole which meets the Code requirements is placed in a full-hemispherical head, the thickness of the head shall be the same as for a head dished to a segment of a sphere (see PG-29.1 and PG-29.5), with a dish radius equal to eight-tenths the diameter of the shell and with the added thickness for the manhole as specified in PG-29.3.

PG-29.13 The corner radius of an unstayed dished head measured on the concave side of the head shall be not less than three times the thickness of the material in the head; but in no case less than 6% of the diameter of the shell. In no case shall the thinning-down due to the process of forming, of the knuckle portion of any dished head consisting of a segment of a sphere encircled by a part of a torus constituting the knuckle portion (torispherical), exceed 10% of the thickness required by the formula in PG-29.1. Other types of heads shall have a thickness after forming of not less than that required by the applicable formula.

ASME B36.10M-1996   WELDED AND SEAMLESS WROUGHT STEEL PIPE

TABLE 2   DIMENSIONS AND WEIGHTS OF WELDED AND SEAMLESS WROUGHT STEEL PIPE
(EXCERPT—REPRODUCED FROM ASME B36.10M)

Customary Units			Identification		SI Units		
NPS [Note (1)]	Outside Diameter, in.	Wall Thickness, in.	Plain End Weight, lb/ft	Standard (STD) Extra-Strong (XS) Double Extra-Strong (XXS)	Schedule No.	DN [Note (2)]	Outside Diameter, mm	Wall Thickness, mm	Plain End Mass, kg/m
8	8.625	0.750	63.08	. . .	. . .	200	219.1	19.05	93.98
8	8.625	0.812	67.76	. . .	140	200	219.1	20.62	100.92
8	8.625	0.875	72.42	XXS	. . .	200	219.1	22.23	107.92
8	8.625	0.906	74.69	. . .	160	200	219.1	23.01	111.27
8	8.625	1.000	81.44	. . .	. . .	200	219.1	25.40	121.33
10	10.750	0.134	15.19	. . .	5	250	273.0	3.40	22.63
10	10.750	0.156	17.65	. . .	. . .	250	273.0	3.96	26.28
10	10.750	0.165	18.65	. . .	10	250	273.0	4.19	27.78
10	10.750	0.188	21.21	. . .	. . .	250	273.0	4.78	31.63
10	10.750	0.203	22.87	. . .	. . .	250	273.0	5.16	34.09
10	10.750	0.219	24.63	. . .	. . .	250	273.0	5.56	36.68
10	10.750	0.250	28.04	. . .	20	250	273.0	6.35	41.77
10	10.750	0.279	31.20	. . .	. . .	250	273.0	7.09	46.51
10	10.750	0.307	34.24	. . .	30	250	273.0	7.80	51.03
10	10.750	0.344	38.23	. . .	. . .	250	273.0	8.74	56.98
10	10.750	0.365	40.48	STD	40	250	273.0	9.27	60.31
10	10.750	0.438	48.24	. . .	. . .	250	273.0	11.13	71.90
10	10.750	0.500	54.74	XS	60	250	273.0	12.70	81.55
10	10.750	0.562	61.15	. . .	. . .	250	273.0	14.27	91.08
10	10.750	0.594	64.43	. . .	80	250	273.0	15.09	96.01
10	10.750	0.625	67.58	. . .	. . .	250	273.0	15.88	100.73
10	10.750	0.719	77.03	. . .	100	250	273.0	18.26	114.75
10	10.750	0.812	86.18	. . .	. . .	250	273.0	20.62	128.38
10	10.750	0.844	89.29	. . .	120	250	273.0	21.44	133.06
10	10.750	0.875	92.28	. . .	. . .	250	273.0	22.23	137.52
10	10.750	0.938	98.30	. . .	. . .	250	273.0	23.83	146.48
10	10.750	1.000	104.13	XXS	140	250	273.0	25.40	155.15
10	10.750	1.125	115.64	. . .	160	250	273.0	28.58	172.33
10	10.750	1.250	126.83	. . .	. . .	250	273.0	31.75	188.97
12	12.750	0.156	20.98	. . .	5	300	323.8	3.96	31.25
12	12.750	0.172	23.11	. . .	. . .	300	323.8	4.37	34.43
12	12.750	0.180	24.17	. . .	10	300	323.8	4.57	36.00
12	12.750	0.188	25.22	. . .	. . .	300	323.8	4.78	37.62
12	12.750	0.203	27.20	. . .	. . .	300	323.8	5.16	40.56
12	12.750	0.219	29.31	. . .	. . .	300	323.8	5.56	43.65
12	12.750	0.250	33.38	. . .	20	300	323.8	6.35	49.73
12	12.750	0.281	37.42	. . .	. . .	300	323.8	7.14	55.77
12	12.750	0.312	41.45	. . .	. . .	300	323.8	7.92	61.71
12	12.750	0.330	43.77	. . .	30	300	323.8	8.38	65.20
12	12.750	0.344	45.58	. . .	. . .	300	323.8	8.74	67.93
12	12.750	0.375	49.56	STD	. . .	300	323.8	9.53	73.88
12	12.750	0.406	53.52	. . .	40	300	323.8	10.31	79.73
12	12.750	0.438	57.59	. . .	. . .	300	323.8	11.13	85.84
12	12.750	0.500	65.42	XS	. . .	300	323.8	12.70	97.46
12	12.750	0.562	73.15	. . .	60	300	323.8	14.27	108.96

VALVES — FLANGED
THREADED, AND WELDING END

ASME B16.34-1996

TABLE 1   MATERIAL SPECIFICATION LIST (REPRODUCED FROM ASME B16.34)
Applicable ASTM Specification

GROUP 1 MATERIALS

Material		Product Form									
		Forgings		Castings		Plates		Bars		Tubular	
Group No.	Nominal Designation	Spec. No.	Grade	Spec. No.	Grade	Spec. No.	Grade	Spec. No.	Grade	Spec. No.	Grade
1.1	C C–Si C–Mn–Si	A 105 A 350	LF2	A 216	WCB	A 515 A 516 A 537	70 70 Cl. 1	A 675 A 105 A 350 A 696	70 LF2 C	A 672 A 672	B 70 C 70
1.2	C–Si 2½Ni 3½Ni C–Mn–Si	A 350	LF3	A 352 A 352 A 216 A 352	LC2 LC3 WCC LCC	A 203 A 203	B E	A 350	LF3	A 106	C
1.3	C C–Si 2½Ni 3½Ni C–Mn–Si			A 352	LCB	A 515 A 203 A 203 A 516	65 A D 65	A 675	65	A 672 A 672	B 65 C 65
1.4	C C–Si C–Mn–Si					A 515 A 516	60 60	A 675 A 350 A 696	60 LF1 B	A 106 A 672 A 672	B B 60 C 60
1.5	C–½Mo	A 182	F1	A 217 A 352	WC1 LC1	A 204 A 204	A B	A 182	F1	A 691	CM-70
1.6	C–½Mo ½Cr–½Mo 1Cr–½Mo					A 387 A 387 A 387	2 Cl. 1 2 Cl. 2 12 Cl. 1			A 335 A 369 A 691	P1 FP1 ½CR
1.7	C–½Mo ½Cr–½Mo Ni–½Cr–½Mo ¾Ni–Mo–¾Cr	A 182	F2	A 217 A 217	WC4 WC5	A 204	C	A 182	F2	A 691	CM-75

[Notes follow at end of table]

[Table 1 continues on the next page]

VALVES — FLANGED
THREADED, AND WELDING END

TABLE 2-1.1   RATINGS FOR GROUP 1.1 MATERIALS
(REPRODUCED FROM ASME B16.34)

A 105 (1)(6)	A 515 Gr. 70 (1)	A 675 Gr. 70 (1)(4)(5)	A 672 Gr. B70 (1)
A 216 Gr. WCB (1)	A 516 Gr. 70 (1)(2)	A 696 Gr. C	A 672 Gr. C70 (1)
A 350 Gr. LF2 (1)	A 537 Cl. 1 (3)		

NOTES:
(1) Upon prolonged exposure to temperatures above 800°F, the carbide phase of steel may be converted to graphite. Permissible, but not recommended for prolonged use above 800°F.
(2) Not to be used over 850°F.
(3) Not to be used over 700°F.
(4) Leaded grades shall not be used where welded or in any application above 500°F.
(5) For service temperatures above 850°F, it is recommended that killed steels containing not less than 0.10% residual silicon be used.
(6) Only killed steel shall be used above 850°F.

TABLE 2-1.1A   STANDARD CLASS (REPRODUCED FROM ASME B16.34)

Temperature, °F	Working Pressures by Classes, psig							
	150	300	400	600	900	1500	2500	4500
−20 to 100	285	740	990	1,480	2,220	3,705	6,170	11,110
200	260	675	900	1,350	2,025	3,375	5,625	10,120
300	230	655	875	1,315	1,970	3,280	5,470	9,845
400	200	635	845	1,270	1,900	3,170	5,280	9,505
500	170	600	800	1,200	1,795	2,995	4,990	8,980
600	140	550	730	1,095	1,640	2,735	4,560	8,210
650	125	535	715	1,075	1,610	2,685	4,475	8,055
700	110	535	710	1,065	1,600	2,665	4,440	7,990
750	95	505	670	1,010	1,510	2,520	4,200	7,560
800	80	410	550	825	1,235	2,060	3,430	6,170
850	65	270	355	535	805	1,340	2,230	4,010
900	50	170	230	345	515	860	1,430	2,570
950	35	105	140	205	310	515	860	1,545
1000	20	50	70	105	155	260	430	770

TABLE 2-1.1B   SPECIAL CLASS (REPRODUCED FROM ASME B16.34)

Temperature, °F	Working Pressures by Classes, psig							
	150	300	400	600	900	1500	2500	4500
−20 to 100	290	750	1,000	1,500	2,250	3,750	6,250	11,250
200	290	750	1,000	1,500	2,250	3,750	6,250	11,250
300	290	750	1,000	1,500	2,250	3,750	6,250	11,250
400	290	750	1,000	1,500	2,250	3,750	6,250	11,250
500	290	750	1,000	1,500	2,250	3,750	6,250	11,250
600	275	715	950	1,425	2,140	3,565	5,940	10,690
650	270	700	935	1,400	2,100	3,495	5,825	10,485
700	265	695	925	1,390	2,080	3,470	5,780	10,405
750	240	630	840	1,260	1,890	3,150	5,250	9,450
800	200	515	685	1,030	1,545	2,570	4,285	7,715
850	130	335	445	670	1,005	1,670	2,785	5,015
900	85	215	285	430	645	1,070	1,785	3,215
950	50	130	170	260	385	645	1,070	1,930
1000	25	65	85	130	195	320	535	965

APPENDIX V

ILLUSTRATIONS OF VARIOUS TYPES OF BOILERS

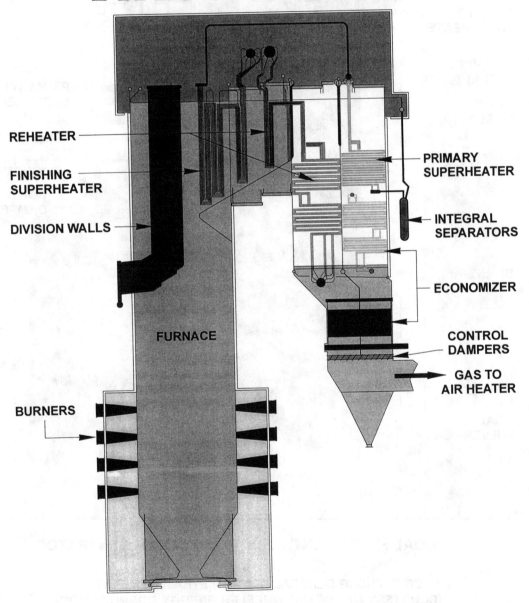

REHEATER

FINISHING
SUPERHEATER

DIVISION WALLS

FURNACE

BURNERS

PRIMARY
SUPERHEATER

INTEGRAL
SEPARATORS

ECONOMIZER

CONTROL
DAMPERS

GAS TO
AIR HEATER

FIG. A.V.1
SUPERCRITICAL ONCE-THROUGH COAL FIRING STEAM GENERATOR
(COURTESY OF FOSTER WHEELER ENERGY CORPORATION)

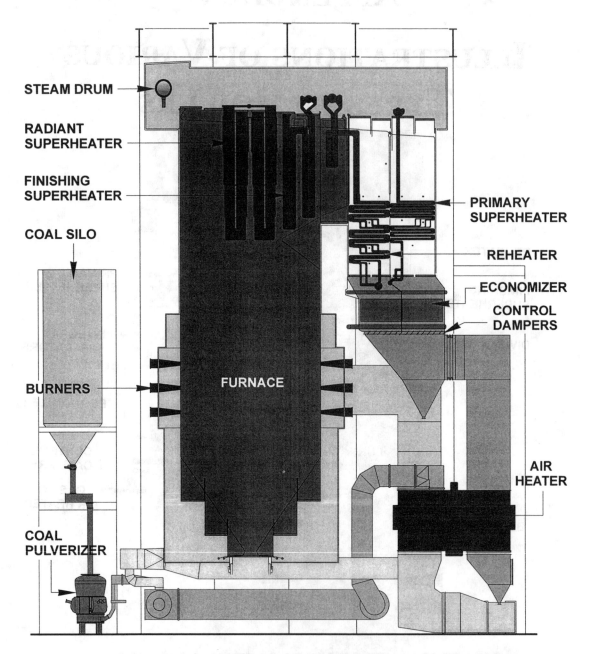

COAL FIRING CENTRAL STATION STEAM GENERATOR

FIG. A.V.2
COAL FIRING CENTRAL STATION STEAM GENERATOR
(COURTESY OF FOSTER WHEELER ENERGY CORPORATION)

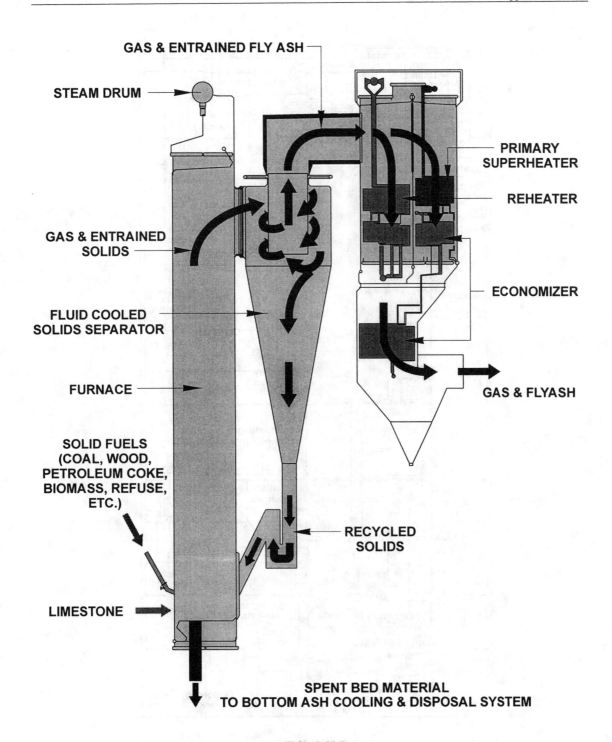

GAS & ENTRAINED FLY ASH

STEAM DRUM

PRIMARY
SUPERHEATER

REHEATER

GAS & ENTRAINED
SOLIDS

FLUID COOLED
SOLIDS SEPARATOR

ECONOMIZER

FURNACE

GAS & FLYASH

SOLID FUELS
(COAL, WOOD,
PETROLEUM COKE,
BIOMASS, REFUSE,
ETC.)

RECYCLED
SOLIDS

LIMESTONE

SPENT BED MATERIAL
TO BOTTOM ASH COOLING & DISPOSAL SYSTEM

**FIG. A.V.3
CIRCULATING FLUIDIZED BED BOILER
(COURTESY OF FOSTER WHEELER ENERGY CORPORATION)**

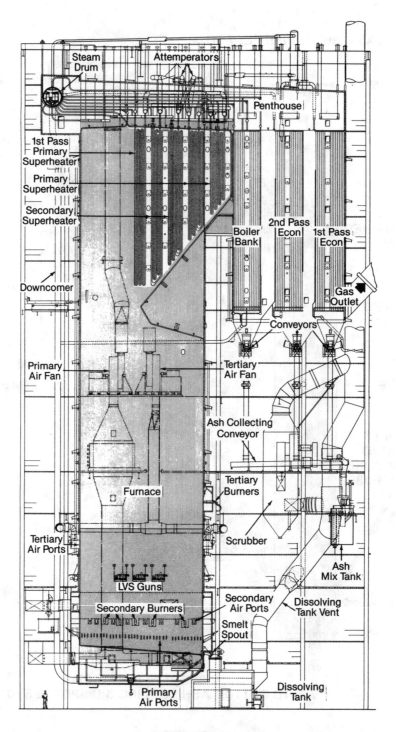

FIG. A.V.4
KRAFT PROCESS BLACK LIQUOR RECOVERY BOILER
(COURTESY OF THE BABCOCK & WILCOX COMPANY)

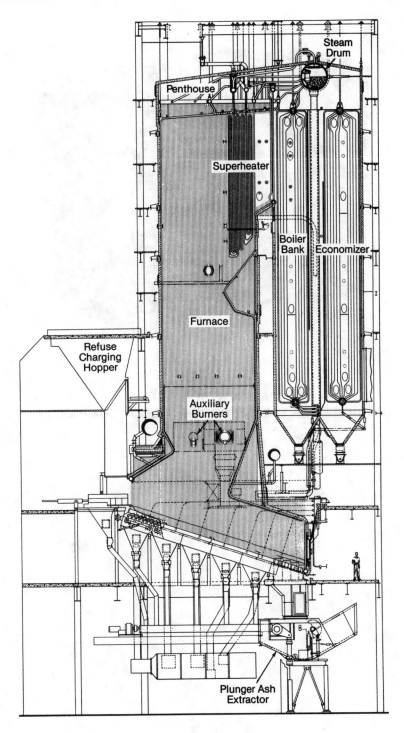

FIG. A.V.5
REFUSE FIRED BOILER
(COURTESY OF THE BABCOCK & WILCOX COMPANY)

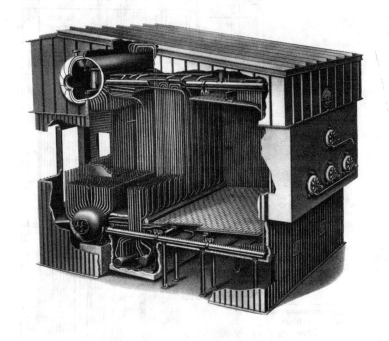

FIG. A.V.6
PACKAGE BOILER
(COURTESY OF THE BABCOCK & WILCOX COMPANY)

INDEX OF INTERPRETATIONS REFERENCED IN TEXT, THEIR MAJOR SUBJECTS AND ABSTRACTS

Interpretation No.	Major Subjects and Abstracts	Page No.

Note: Refer to full text of Interpretations for further details.

Note: Refer to full text of Interpretations for further details.

Note: Refer to full text of Interpretations for further details.

INDEX